Spectral Properties of Lipids

Chemistry and Technology of Oils and Fats

Series Editor: R.J. Hamilton

A series which presents the current state of the art in chosen areas of oil and fat chemistry, including its relevance to food technology. Written at professional and reference level, it is directed at chemists and technologists working in oil and fat processing, the food industry, the oleochemicals industry and the pharmaceutical industry, at analytical chemists and quality assurance personnel, and at lipid chemists in academic research laboratories. Each volume in the series provides an accessible source of information on the science and technology of a particular area.

Titles in the Series:

Lipid Synthesis and Manufacture
Edited by F.D. Gunstone

Spectral Properties of Lipids
Edited by R.J. Hamilton and J. Cast

Spectral Properties of Lipids

Edited by
RICHARD J. HAMILTON and JOHN CAST
School of Pharmacy and Chemistry
Liverpool John Moores University

CRC Press

First published 1999
Copyright © 1999 Sheffield Academic Press

Published by
Sheffield Academic Press Ltd
Mansion House, 19 Kingfield Road
Sheffield S11 9AS, England

ISBN 1-85075-926-X

Published in the U.S.A. and Canada (only) by
CRC Press LLC
2000 Corporate Blvd., N.W.
Boca Raton, FL 33431, U.S.A.
Orders from the U.S.A. and Canada (only) to CRC Press LLC

U.S.A. and Canada only:
ISBN 0-8493-9736-7

Printed on acid-free paper in Great Britain by
Bookcraft Ltd, Midsomer Norton, Bath

British Library Cataloguing-in-Publication Data:
A catalogue record for this book is available from the British Library

Library of Congress Cataloging-in-Publication Data:
Spectral properties of lipids / edited by Richard J. Hamilton and John Cast
     p.   cm. -- (The chemistry and technology of oils and fats)
  Includes bibliographical references and index.
  ISBN 0-8493-9736-7 (alk. paper)
  1. Lipids --Spectra.   2. Lipids -- Analysis.   I. Hamilton, R. J.
(Richard John)   II. Cast, John.   III. Series.
QP751.S68 1998
572'.5736--dc21   98-4747
                                          CIP

Preface

The last two decades have seen some excellent volumes published in the area of oils and fats. While analytical methods have been included in some of these, the present volume is unique in providing coverage of the wide range of spectroscopic methods now available, which provide the basis for quality assurance.

This book, written by authors from respected laboratories around the world, is directed at practising chemists, chemical engineers and food scientists in the oils and fats industry, the food industry and the cosmetics industry. The inclusion of some essential background theory for the techniques makes it also suitable for lecturers and advanced students in the educational sector.

The stability of oils is often affected adversely by metal ions, and it is therefore appropriate that atomic spectroscopy is the first technique considered, in chapter 2.

The applications of chemiluminescence have been attracting renewed interest, and the theory and instrumentation, together with the applications of chemiluminescence to hydroperoxides and antioxidants, are well described in chapter 3.

The combination of NMR, UV and mass spectrometry provides a powerful combination for the identification of individual marine lipids. Chapter 4 also illustrates that no single technique can solve all structural elucidation problems.

One of the major changes in the food industry has been the determination of solid fat content by pulse-NMR. Chapter 5 outlines the theory and considers the variables which need to be controlled in commodities as diverse as chocolate and whole seeds.

Chapters 6 and 7 detail the linking of mass spectrometry to GC and HPLC. Mass spectral breakdown patterns for derivatives of fatty acids and derivatives of triacylglycerols in Salatrim and milk provide an invaluable source of data.

The ability to handle samples which may be solid or opaque has encouraged the wide use of near infra red spectroscopy (NIR) by companies dealing in grains and oilseeds. Chapter 8 covers NIR, FTIR and handling techniques. Many of the official analytical methods, such as the determination of saponification value, iodine value, *trans* content, peroxide value and free fatty acids, are compared to the results obtained using infra red spectroscopy.

Chapters 9 and 11 cover electron spin resonance and X-ray diffraction,

because it is evident that these analytical techniques are gaining in importance. Indeed, on-line process control X-ray equipment is being tested at present.

Among the earliest spectroscopic techniques applied to lipids were ultra
violet spectroscopy and colorimetry. The colour of a crude oil is one of the
properties included in the specification for oil purchase. One of the classes of
compound providing colour is the carotenoid group of compounds, whose
determination has been made easier with the advent of HPLC and UV detection, applications which are considered in chapter 10. The use of colorimetry
is well illustrated in chapter 12.

R.J. Hamilton
J. Cast

Contributors

Mr Alex A. Belbin The Tintometer Ltd, Waterloo Road, Salisbury, Wiltshire SP1 2JY, UK

Dr Debra L. Bemis Department of Ecology, Evolutionary and Marine Biology, University of California, Santa Barbara, CA93106-9106, USA

Dr John Cast School of Pharmacy and Chemistry, Liverpool John Moores University, Byrom Street, Liverpool L3 3AF, UK

Mr Timothy C. Dintinger School of Pharmacy and Chemistry, Liverpool John Moores University, Byrom Street, Liverpool L3 3AF, UK

Dr K. Eulitz Center for Food Safety and Applied Nutrition, Food and Drug Administration, 200 C St SW, Washington DC 20204, USA

Dr William H. Gerwick College of Pharmacy, Oregon State University, Corvallis, OR 97331, USA

Professor Richard J. Hamilton School of Pharmacy and Chemistry, Liverpool John Moores University, Byrom Street, Liverpool L3 3AF, UK

Mr G.W. Hammond Micromass UK Ltd, Tudor Road, Altrincham, Cheshire WA14 5RZ, UK

Dr Inès Elizabeth Holzbaur Department of Food Science and Agricultural Chemistry, Macdonald Campus of McGill University, 21,111 Lakeshore Road, Ste Anne de Bellevue, Québec, Canada H9X 3V9

Dr Ashraf A. Ismail Department of Food Science and Agricultural Chemistry, Macdonald Campus of McGill University, 21,111 Lakeshore Road, Ste Anne de Bellevue, Québec, Canada H9X 3V9

Dr Robert S. Jacobs

Department of Ecology, Evolutionary and Marine Biology, University of California, Santa Barbara, CA93106-9106, USA

Dr Päivi Laakso

Raisio, Benecol Ltd, PO Box 101, FIN-21201, Raisio, Finland

Dr Peter Laggner

The Austrian Academy of Sciences, Institute of Biophysics & X-Ray Structure Research, Steyrergasse 17, A-8010 Graz, Austria

Dr Jennifer C. MacPherson

Department of Cell Biology, Cleveland Clinic Foundation, 9500 Euclid avenue, NC-10, Cleveland, OH 44195, USA

Dr Pekka Manninen

Institute of Biomedicine, Department of Pharmacology, University of Turku, FIN-20520 Turku, Finland

Dr Wouter L.J. Meeussen

Doornikse Steenweg 105, B-8580 Avelgem, Belgium

Dr M.M. Mossoba

Center for Food Safety and Applied Nutrition, Food and Drug Administration, 200 C St SW, Washington DC 20204, USA

Dr Antonio Nicodemo

Department of Food Science and Agricultural Chemistry, Macdonald Campus of McGill University, 21,111 Lakeshore Road, Ste Anne de Bellevue, Québec, Canada H9X 3V9

Professor Christopher J. Rhodes

School of Pharmacy and Chemistry, Liverpool John Moores University, Byrom Street, Liverpool L3 3AF, UK

Mr John A.G. Roach

Center for Food Safety and Applied Nutrition, Food and Drug Administration, 200 C St SW, Washington DC 20204, USA

Dr B.J. Rossell

Leatherhead Food Research Association, Randalls Road, Leatherhead, Surrey KT22 7RY, UK

Dr Jacqueline Sedman

Department of Food Science and Agricultural Chemistry, Macdonald Campus of McGill University, 21,111 Lakeshore Road Ste Anne de Bellevue, Québec, Canada H9X 3V9

Dr Vince P. Shiers

Leatherhead Food Research Association, Randalls Road, Leatherhead, Surrey KT22 7RY, UK

Dr James Todd

College of Pharmacy, Oregon State University, Corvallis, OR 97331, USA

Dr Frederick R. van de Voort

Department of Food Science and Agricultural Chemistry, Macdonald Campus of McGill University, 21,111 Lakeshore Road, Ste Anne de Bellevue, Québec, Canada H9X 3V9

Dr R. Alan Wheatley

Valley View, Canada Lane, Caistor, Market Rasen LN7 6RN, UK (Department of Chemistry, University of Hull)

Dr Andrew Young

School of Biological and Earth Sciences, Liverpool John Moores University, Byrom Street, Liverpool L3 3AF, UK

Mr Martin Yurawecz

Center for Food Safety and Applied Nutrition, Food and Drug Administration, 200 C St SW, Washington DC 20204, USA

Contents

4. NMR in conjunction with GC-MS and UV methods; a case study in marine lipids

97

J. C. MACPHERSON, D. L. BEMIS, R. S. JACOBS,
W. H. GERWICK AND J. TODD

8   Infrared spectroscopy of lipids: principles and applications   235
A. A. ISMAIL, A. NICODEMO, J. SEDMAN, F. R. VAN DE VOORT AND I. E. HOLZBAUR

9.   Electron spin resonance studies of lipids   270
C. J. RHODES AND T. C. DINTINGER

1   Introduction

John Cast and Richard J. Hamilton

1.1   Lipid nomenclature

Lipids are defined as derivatives of fatty acids and the definition includes
biosynthetically derived compounds. As such, lipids range from highly non-
polar substances such as saturated hydrocarbons to gangliosides, which are
insoluble in non-polar solvents but form micelles in aqueous solution.

   In oils and fats, the major components are triacylglycerols (**I**), which
are esters based on glycerol where R^1, R^2 and R^3 may be different fatty
acids. If only two of the hydroxyl groups in glycerol are esterified, a
diacylglycerol (**II**) is formed, there being two possible isomers. Finally, a
monoacylglycerol (**III**) has only one ester group, with two possible
isomers. Although much smaller in quantity in oils and fats, the
phosphoglycerides are very important in cell membranes. Most are
diacylglycerols in which the third hydroxyl group has been esterified with
phosphoric acid (**IV**). See Scheme 1a.

   By reaction of one of the (acidic) hydrogens of the phosphoric acid
group with other alcohols, the following glycerophospholipids are
formed: phosphatidyl choline (**V**), phosphatidyl ethanolamine (**VI**),
phosphatidyl serine (**VII**), phosphatidyl inositol (**VIII**) (Scheme 1b).

   Phosphoglycerols are based on glycerol, whereas the next type of lipid,
the sphingolipids, have a long-chain hydroxyl base (**IX**), sphingosine. The
fatty acids are attached to the base as an amide. Some sphingolipids have
sugar molecules attached to one of the hydroxyl groups in sphingosine.
Thus ceramides (**X**) are made up of a sphingosine molecule (**IX**) with
an acyl group attached at the amino group. Cerebrosides have an acyl
group attached to the amino group and R'' is a sugar, either glucose or
galactose. Cerebrosides (**XI**) are found in brain and ganglion cells.
Gangliosides (**XII**) have an acyl group attached to the amino group and
R'' is a sugar containing glucose, galactose and sialic acid (**XIII**). These
lipids are found in the myelin sheath of brain and nerve. Lastly, in
sphingomyelin (**XIV**), R'' is a phosphocholine group (Scheme 1c).

   Seven of the most common fatty acids found in combination with these
complex molecules to produce lipids, have the structures illustrated in
Table 1.1.

   The most abundant saturated fatty acids have an even number of
carbon atoms and include lauric acid, myristic acid, palmitic acid and
stearic acid. There is a short-hand representation of these acids which

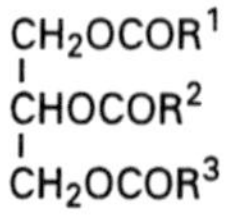

CH$_2$OCOR^1
|
CHOCOR^2
|
CH$_2$OCOR^3

(I)

CH$_2$OCOR^1　　　　　　　　　　CH$_2$OCOR^1
|　　　　　　　　　　　　　　　　　|
CHOCOR^2　　　　　　　　　　　CHOH
|　　　　　　　　　　　　　　　　　|
CH$_2$OH　　　　　　　　　　　　　CH$_2$OCOR^3

(II)

CH$_2$OCOR^1　　　　　　　　　　CH$_2$OH
|　　　　　　　　　　　　　　　　　|
CHOH　　　　　　　　　　　　　　CHOCOR^2
|　　　　　　　　　　　　　　　　　|
CH$_2$OH　　　　　　　　　　　　　CH$_2$OH

(III)

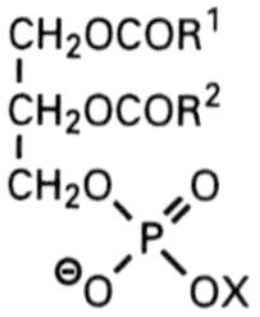

CH$_2$OCOR^1
|
CH$_2$OCOR^2
|
CH$_2$O＼　　＿O
　　　　　P
　⊖O＼　　＼OX

(IV)

Scheme 1a

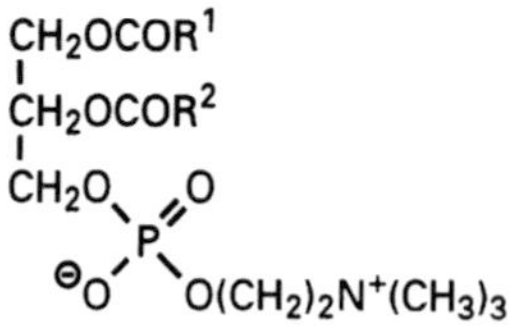

CH$_2$OCOR^1　　　　　　　　　　　CH$_2$OCOR^1
|　　　　　　　　　　　　　　　　　　|
CH$_2$OCOR^2　　　　　　　　　　　CH$_2$OCOR^2
|　　　　　　　　　　　　　　　　　　|
CH$_2$O＼　　＿O　　　　　　　　　　CH$_2$O＼　　＿O
　　　　P　　　　　　　　　　　　　　　　P
⊖O＼　　＼O(CH$_2$)$_2$N$^+$(CH$_3$)$_3$　　⊖O＼　　＼O(CH$_2$)$_2$NH$_3$$^+$

(V)　　　　　　　　　　　　　　　　　(VI)

CH$_2$OCOR^1　　　　　　　　　　　CH$_2$OCOR^1
|　　　　　　　　　　　　　　　　　　|
CH$_2$OCOR^2　　　　　　　　　　　CH$_2$OCOR^2
|　　　　　　　　　　　　　　　　　　|　　　　　OH OH
CH$_2$O＼　　＿O　　　　　　　　　　CH$_2$O＼　　＿O
　　　　P　　　　　　　　　　　　　　　　P
⊖O＼　　＼O(CH$_2$)CHCOO$^-$　　　⊖O＼　　＼O——⟨　　⟩——OH
　　　　　　　　　|
　　　　　　　　　NH$_3$$^+$　　　　　　　　　　　　OH OH

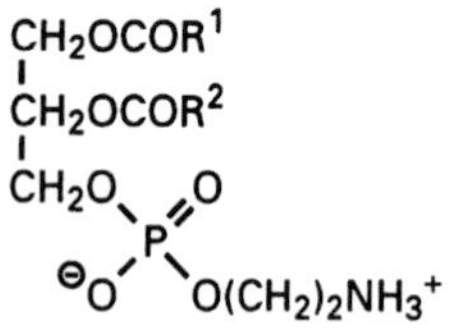

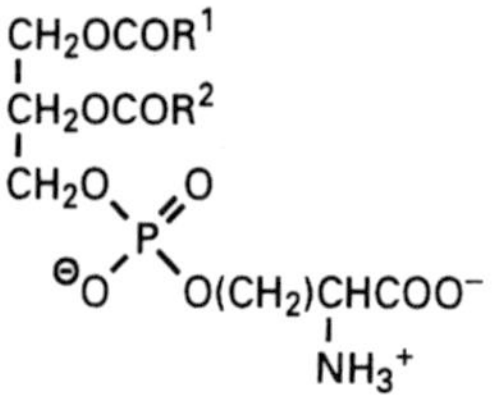

(VII) 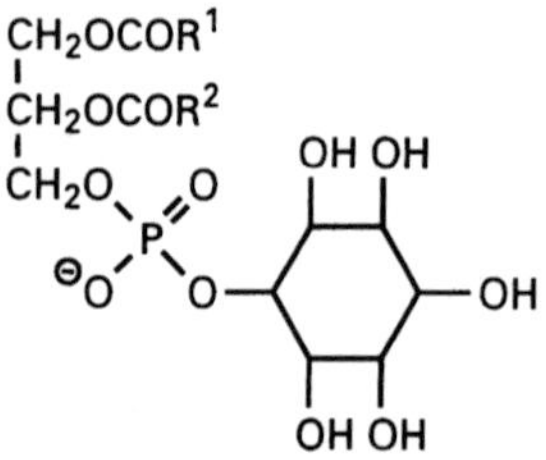　　　　　　　　　　　　　　　　(VIII)

Scheme 1b

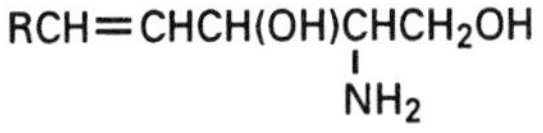

RCH=CHCH(OH)CHCH$_2$OH
|
NH$_2$

(IX)
Sphingosine

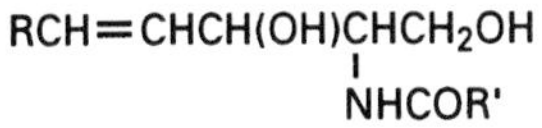

RCH=CHCH(OH)CHCH$_2$OH
|
NHCOR'

(X)

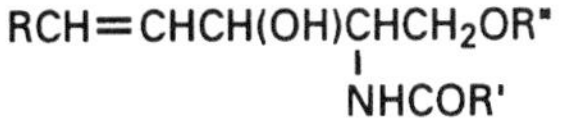

RCH=CHCH(OH)CHCH$_2$OR"
|
NHCOR'

(XI)
where R" is a sugar

RCH=CHCH(OH)CHCH$_2$OR"
|
NHCOR'

H$_2$C—OH
|
HC—OH
|
HC—OH

HO₂C, HO, HO, O, NHCOCH$_3$

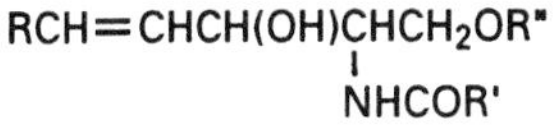

(XIII)
where R" is glucose, galactose
and sialic acid (XIII)

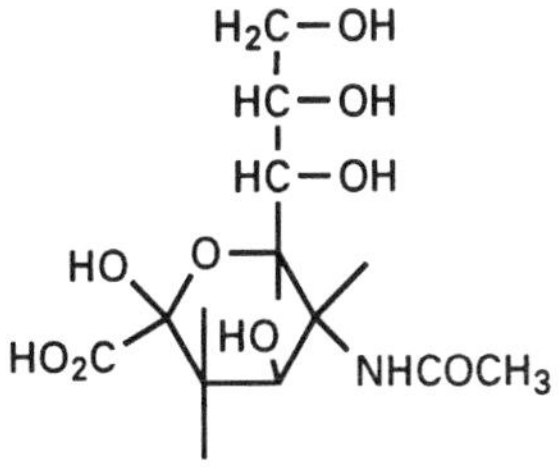

(XIII)
Sialic acid

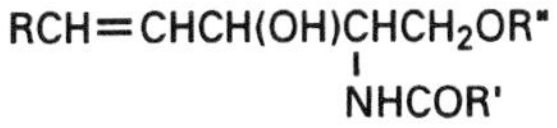

RCH=CHCH(OH)CHCH$_2$OR"
|
NHCOR'

(XIV)
where R" is phosphocholine

Scheme 1c

designates the number of carbon atoms followed by a colon and then the number of double bonds in the chain, e.g. 12:0 for lauric acid.

The most abundant unsaturated acid, in nature is oleic acid, whose shorthand representation is 18:1^{9c} where the superscript designates the position of the double bond from the carboxyl carbon atom and the configuration (*cis*) of the double bond.

1.2 Electromagnetic radiation

The investigation of the structure and the quantification of lipids has been greatly accelerated by the application of physical methods based upon

Table 1.1 Fatty acids commonly found in lipids

Trivial name	Designation	Systematic name	Structure
Lauric acid	12:0	Dodecanoic acid	
Myristic acid	14:0	Tetradecanoic acid	
Palmitic acid	16:0	Hexadecanoic acid	
Stearic acid	18:0	Octadecanoic acid	
Oleic acid	18:1^{9c}	Octadec-9cis-enoic acid	
Linoleic acid	18:2^{9c,12c}	Octadeca-9c,12c-dienoic acid	
Linolenic acid	18:3^{9c,12c,15c}	Octadeca-9c,12c,15c-trienoic acid	

electromagnetic radiation. The interaction of radiation with matter has characteristics that are dependent upon the nature of the radiation and of the structure, macroscopic and molecular, of the material. The radiation considered in this volume extends from the short wavelengths of the X-ray region, through ultraviolet, visible and infrared to the long wavelengths employed in nuclear magnetic resonance. (See Table 1.2).

Electromagnetic radiation is a form of energy involving both magnetic and electric fields, and can be described in terms of both wave and particulate properties. The relation between these two aspects is described by Planck's equation,

$$E = h\nu \qquad (1.1)$$

where E is the energy of the photon and ν is the frequency of the wave. Planck's constant, h, has the value $6.625 \times 10^{-34}\,\mathrm{J\,s}$. The energy in spectroscopic work is usually described in terms of frequency, wavelength or wavenumber.

The wavelength is the distance between identical parts of the wave, with symbol λ and the SI unit of the metre; frequency is the number of cycles of the wave passing a given point in unit time, with symbol ν and the unit is the reciprocal second ($\mathrm{s^{-1}}$) or Hertz (Hz). These quantities are related by the equation

$$c = \lambda\nu \qquad (1.2)$$

where c is the velocity of electromagnetic radiation.

A quantity generally employed in the infrared region is wavenumber, the reciprocal of wavelength, given the symbol $\bar{\nu}$ or, less commonly, σ. The usual spectroscopic unit of wavenumber is the reciprocal centimetre ($\mathrm{cm^{-1}}$).

$$\bar{\nu} = 1/\lambda \quad \text{and} \quad E = hc\bar{\nu} \qquad (1.3)$$

Energy considerations are of interest in spectroscopic studies and wavenumber has the virtues of being proportional to energy and measurable to the high accuracy possible with wavelength measurement.

The relationship between the attenuation of the incident radiation and the amount of absorbing material is expressed by the Beer–Lambert–Bouguer Law. The law applies to monochromatic radiation (Loudon, 1964; Wineforder, 1971) and holds throughout the whole frequency range of electromagnetic radiation, its form being modified according to the frequency and the absorbing medium. The most usual expression is that employed for absorption in the UV/visible and infrared region. In terms

Table 1.2 The electromagnetic spectrum. Frequently used units relating to different regions are indicated by bold entries

Region	Wavelength λ (nm)	Frequency ν (s^{-1})	Wavenumber $\bar{\nu}$ (cm^{-1})	Energy/mol E (J mol^{-1})
X-Ray	0.01–10	$3 \times 10^{19} - 3 \times 10^{16}$	$10^9 - 10^6$	$1.2 \times 10^{10} - 1.2 \times 10^7$
Far ultraviolet	**10–200**	$3 \times 10^{16} - 1.5 \times 10^{15}$	$10^6 - 5 \times 10^4$	$1.2 \times 10^7 - 6 \times 10^5$
Near ultraviolet	**200–400**	$1.5 \times 10^{15} - 7.5 \times 10^{14}$	$5 \times 10^4 - 2.5 \times 10^4$	$6 \times 10^5 - 3 \times 10^5$
Visible	**400–750**	$7.5 \times 10^{14} - 4 \times 10^{14}$	$2.5 \times 10^4 - 1.3 \times 10^4$	$3 \times 10^5 - 1.5 \times 10^5$
Near infrared	750–2500	$4 \times 10^{14} - 1.2 \times 10^{14}$	$\mathbf{1.3 \times 10^4 - 4 \times 10^3}$	$1.5 \times 10^5 - 4.8 \times 10^4$
Medium infrared	2500–25000	$1.2 \times 10^{14} - 1.2 \times 10^{13}$	$\mathbf{4 \times 10^3 - 400}$	$4.8 \times 10^4 - 4.8 \times 10^3$
Far infrared	25 000–300 000	$1.2 \times 10^{13} - 10^{12}$	**400–33**	$4.8 \times 10^3 - 4 \times 10^2$
Microwave	$3 \times 10^5 - 5 \times 10^8$	$10^{12} - 6 \times 10^8$	**33–0.02**	$4 \times 10^2 - 0.24$
NMR	$5 \times 10^8 - 5 \times 10^9$	$\mathbf{6 \times 10^8 - 6 \times 10^7}$	0.02–0.002	0.24–0.024

of the incident radiation intensity I_0, the transmitted intensity I, path length b and concentration c,

$$\log(I_0/I) = kbc \qquad (1.4)$$

where k is a constant depending upon the nature of the material and the frequency of the radiation; the quantity $\log(I_0/I)$ is defined as the absorbance, A.

When concentration is measured in moles per cubic decimetre and pathlength in decimetres, we write

$$A = \log(I_0/I) = \varepsilon cl \qquad (1.5)$$

where ε, the decadic molar absorptivity, has dimensions of $dm^2 mol^{-1}$.

Other quantities used to measure the absorptivity include $E_{1\,cm}^{1\%}$, the 'extinction coefficient', which is the absorbance of a solution of 1 g of substance in $100\,cm^3$ of solution of path length 1 cm. This extinction coefficient, $E_{1\,cm}^{1\%}$, has been employed in measurements where the relative molecular mass is not known, as with mixtures and many natural products.

For mixtures of substances and in the absence of interactions between them, Beer's law is additive, so that the total absorbance at any wavelength is the sum of the absorbances of the components at that wavelength; that is

$$A(\text{total}) = \sum \varepsilon cl \qquad (1.6)$$

The composition of an n-component mixture may be found if the absorbance of the mixture is determined at n wavelengths and the absorptivities are known at each of these wavelengths for each of the components (Bauman, 1962). This permits the construction of n simultaneous equations, the solution of which may be carried out rapidly by a computer program.

Beer's law is a limiting one and applies rigorously only if the refractive index remains constant. In practice the refractive index changes with concentration, but over the ranges usually employed the effect on linearity is negligible. The law assumes that there are no losses from the incident radiation apart from absorption; losses by reflection at cell faces or scattering, for example in solution, lead to departures from the law. There are several other causes of nonlinear relationships between absorbance and concentration (Lothian, 1963), and apparent deviations from Beer's law. One of the most common is the modification in species that may occur with changes in concentration. For example, if a species of greater absorptivity is produced as the concentration increases, the total absorbance will increase disproportionately and a positive deviation will be observed. There are some important instrumental causes of deviations (Goldring *et al.*, 1953). In dispersive instruments, mono-chromators employ slits to select the bandwidth of radiation measured; wide slits, which may be required to admit sufficient energy for the effective operation of the detector, pass radiation at absorptivities less than the peak value and so, at increasing concentrations, lead to negative deviations.

1.3   Ultraviolet/visible spectra

Since interest is largely at molecular level, some appreciation of bonding and orbitals is useful. Absorption of radiation in the ultraviolet and visible regions generally causes an increase in the energy of electrons within the molecule. Electrons can be assigned to quantised energy levels, or molecular orbitals. The form or shape of the orbitals depends upon the relative energies of electrons occupying them. The absorption of energy in this region involves the excitation of electrons from lower to higher molecular orbitals, and transitions between orbitals are governed by selection rules that determine the probability of the transition and so the intensity of the absorption band.

The types of orbitals encountered in most organic compounds can be represented as shown for those of acetaldehyde (Figure 1.1). Their relative energies are depicted in Figure 1.2.

Spectra, plots of the amount of radiation of a given energy absorbed or transmitted, may be presented by various parameters. In the UV/visible

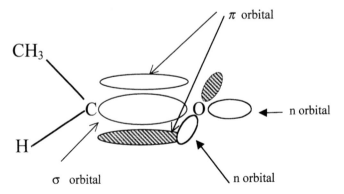

Figure 1.1   Depiction of the orbitals of acetaldehyde. Their relative energies are illustrated in Figure 1.2.

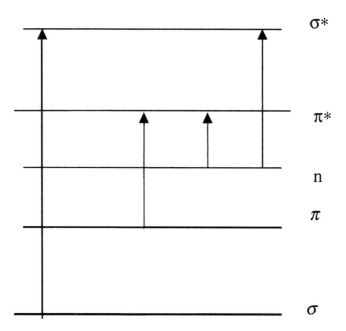

Figure 1.2 Depiction of the relative energies of the orbitals illustrated in Figures 1.1 and the transitions that can be observed between them.

region the most usual plot is of absorbance against wavelength; when it is desired to show bands of widely differing absorptivities, $\log A$ or $\log \varepsilon$ may be plotted on the ordinate. Infrared spectra are usually plots of transmittance versus wavenumber; transmittance is sometimes replaced by absorbance, especially for quantitative work.

In UV/visible spectroscopy, wavelength is still employed because, although disadvantageous from the point of view of energy, measurements in wavelength have formed the basis of sets of rules enabling prediction of λ_{max} in certain series of compounds, notably polyenes, α,β-unsaturated carbonyl compounds and some aromatic carbonyl compounds.

Since the molecular orbitals in conjugated compounds are dependent upon the essential structural features of the system, being modified to a small extent by the saturated groups adjacent to them, the UV/visible spectra are frequently characteristic of the conjugated system present in compounds and have served to identify such compounds. The term 'chromophore' was introduced to describe a group conferring colour on a compound and thence to groups absorbing in the UV/visible part of the spectrum, though the term is often employed for groups absorbing in any part of the electromagnetic spectrum.

An illustration of the effect of changes in chromophore may be found in the flavonoids, a group of substances sometimes found in association with oils and fats and giving colour to them. Markham (1982) has recorded the spectra of a series of similarly substituted flavonoids (Figure 1.3) which show absorption bands at increasing λ_{max} as the conjugation changes.

Rules for the calculation of λ_{max} of polyenes and conjugated unsaturated carbonyl compounds have been formulated by Woodward (1941), (Fieser and Fieser, 1949), and others. The Woodward–Fieser rules have proved valuable in the determination of the position and substitution patterns of the diene grouping within complex molecules, by the application of a base value to the unsaturated system and the addition of increments for other features. (see Table 1.3). A similar set of rules can be employed to predict the position of α,β-unsaturated carbonyl compounds (see Table 1.4). For the carbonyl compounds the position of λ_{max} depends upon the solvent; the tabulated values obtain for ethanol solutions.

Further discussion of the details of the organic aspects of electronic spectra may be found in texts listed in the bibliography.

Before the development of UV/visible spectroscopy as a quality control and investigative tool, the measurement of colour resulting from absorption in the visible region had become widely applicable and had given rise to a number of tests used extensively in the past. Some still employed include the Halphen and Baudouin tests.

The presence of cottonseed oil, an adulterant in consumable oils, may be determined by the Halphen test. Cottonseed oil gives a characteristic crimson colour when reacted with sulphur in carbon disulfide in the presence of amyl alcohol at 95°C. By using standards made up with amounts of cottonseed oil up to 20%, a semi-quantitative method can be provided for the measurement of adulteration. The colour is caused by cyclopropane/cyclopropene groupings in a fatty acid, so other fats containing this feature will give a positive Halphen test, e.g. Kapok oil or *Sterculia foetida* oil.

Sesame oil contains the antioxidants sesamol and sesamolin, which react in Baudouin's test. The oil is shaken with concentrated hydrochloric acid containing (1% w/v) sucrose (1 g/100 cm³ aqueous solution) and a brilliant crimson colour develops after 5 minutes. Some artificial colouring matters used in foods may also produce a similar colour under the test conditions. Rosin acids are sometimes found mixed with fatty acids; if their presence is suspected, the fats are heated with acetic anhydride, the mixture is cooled and concentrated sulphuric acid is carefully added; a fugitive reddish-violet colour indicates the presence of rosin acids. This is the basis of the Liebermann–Storch test. It should be

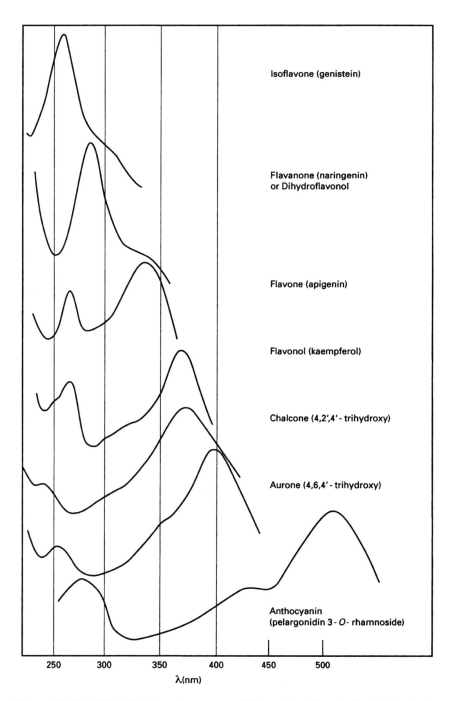

Figure 1.3 The UV/visible absorption spectra of flavonoids with equivalent hydroxylation substitution.

Table 1.3 Woodward–Fieser rules for π–π^* band of conjugated dienes and polyenes

Species	λ (nm)
Acyclic and heteroannular dienes	215
Increments	
Alkyl substituent or ring residue	5
Exocyclic double bond	5
Additional double bond in conjugation	30
Homoannular component	39

Table 1.4 Woodward–Fieser rules for π–π^* band of α,β-unsaturated carbonyl compounds

Species	λ(nm)
Acyclic or 6-membered ring ketones	215
5-membered ring ketones	202
Aldehydes	207
Acids and esters	195
Increments	
α-Alkyl substituent or ring residue	10
β-Alkyl substituent or ring residue	12
γ-and δ-Alkyl substituent or ring residue	18
Additional double bond in conjugation	30
Exocyclic double bond	5
Homoannular component	39

remembered that cholesterol also gives rise to colours when treated with sulphuric acid.

In addition to absorption of radiation, emission of radiation also has applications, mainly in quantitation. The initial excitation process may be either absorption of radiation, as in fluorimetric processes, or a chemical or biochemical reaction, as in chemiluminescence or bioluminescence.

1.4 Vibrational spectra

For a bond at a given electronic energy level, there is also a series of smaller energy levels, corresponding to vibrational energies of the bond. The energy differences between the ground and first excited vibrational levels correspond to radiation in the fundamental or middle infrared region. Absorption of infrared radiation causes an increase in the vibrational energy of bonds within the molecule without, in general, changing the electronic energy levels, and gives rise to a number of absorption bands characteristic of the molecule. In practice it is possible to associate many absorption bands with the vibration of particular bonds (Bellamy, 1975, 1980; Socrates, 1994), their position usually being measured as wavenumber.

Infrared spectroscopy is a widely used tool in routine work and investigative aspects in the lipid field, being relatively cheap and rapidly carried out with, for most applications, little sample preparation. The dispersive instruments have now been replaced almost entirely by Fourier transform (FT) instruments (Griffiths and de Haseth, 1986), except for some specialist applications, and their introduction has led to the more rapid acquisition of spectra and easier recording of spectra of very small samples. The necessary use of the computer to perform the fast Fourier transform of the interferogram facilitated the development of computerised spectral enhancement and quantitative measurement.

In early work spectra were, and most commonly today still are, obtained on liquids or on solids compressed in potassium bromide discs. Sampling techniques are now available to permit more convenient and rapid sampling. Attenuated reflectance devices, and in particular horizontal reflectance devices, enable spectra to be obtained on materials of low visible transparency and on aqueous solutions (Harrick, 1967). Diffuse reflectance lends itself to rapid sampling since it is necessary only to mix the substance with a non-absorbing support such as potassium bromide (Griffiths and Fuller, 1982). For these techniques, FTIR is a desirable procedure, since the radiation throughput is small, especially in the case of diffuse reflectance (DRIFT). FTIR has also made possible on-the-fly-scanning for gas chromatography – FTIR coupling.

Most of the characteristic functional groups lend themselves readily to qualitative and, in many cases, quantitative study by infrared spectroscopy. Commonly found groups include those listed in Table 1.5. Many of these groups are useful for quantitative work, and of especial importance is estimation of the *trans* and *cis* carbon – carbon double bonds.

In the near infrared region, absorption bands are mainly overtones and combination bands of vibrational modes of the middle or fundamental region. Instrumentation is generally designed to measure the diffuse reflected radiation and, although the technique does not lend itself easily to qualitative work, it has proved of great value in quantitative work, employing instruments, of both filter and monochromator types, with dedicated computing facilities and software (Williams and Norris, 1987; Osbourne and Fearn, 1988).

The discovery of the Raman effect (Raman and Krishnan, 1928) opened up an alternative way of examining the vibrational spectra of molecules. When a substance is irradiated by energy in the UV/visible region the scattered radiation contains, in addition to the incident

Table 1.5 Functional groups commonly detected in IR spectroscopy

Functional group	Wavenumber (cm^{-1})
Alkyl chains	2960–2870
Cyclopropyl	3020, 1050–1000
Alkenes terminal	
	3095–3075
cis	1660–1630, 730–650
trans	1680–1670, 980–865
Alkyne, terminal	3340, 700–610
Dienes	
conjugated	1650–1600
cis–trans	990–980, 968–950
trans–trans	990–984
Hydroxyl	
free	3650–3510
associated	3550–3200
Carboxylic acid monomer	
OH	3550–3500
C=O	1760
Carboxylic acid dimer	
OH assoc	3000–2500
C=O	1710
Aldehyde	
CH	2800, 2700
C=O	1725
Ketone C=O	1715
Epoxy	3050, 950–815, 880–750

frequency, radiation displaced from that frequency by quanta corresponding to vibrational frequencies. Whereas in the infrared region highly polarised bonds give rise to strong absorption, bonds that are highly polarisable produce strong scattering in a Raman spectrum. Thus some groups that give weak bands in the infrared, for example C—Br, C—I, C—S, C=C and aromatic systems, give strong bands in the Raman spectrum. Water, with polarised bonds of low polarisability, has low Raman scattering, so spectra can be recorded in aqueous solutions (Dollish *et al.*, 1974; Gardiner and Graves, 1989; Grasselli and Bulkin, 1991). Some of the problems associated with Raman spectra, in particular interference from fluorescence, are overcome with Fourier transform Raman spectroscopy (Hendra *et al.*, 1991), in which illumination of the sample is by near infrared radiation, so that the scattered radiation is in the infrared region, avoiding the fluorescence in the visible and making it possible to combine the instrument with FTIR.

When the incident radiation falls within an absorption band of the compound, resonance Raman scattering occurs (Koningstein, 1982), the scattering of some bands becoming considerably enhanced. Areas in which Raman and resonance Raman have been employed include carotenoids, lipid membranes, nucleic acids and metalloproteins.

1.5 Magnetic resonance spectra

The most widely used application of nuclear magnetic resonance spectroscopy, both proton and ^{13}C, in the lipid field is the determination of structure mainly by Fourier transform pulse NMR. This may involve complete structure elucidation, frequently in conjunction with infrared spectroscopy, or, in more limited work, identification or location of functional groups. Recently it has been shown that high-resolution NMR can be used to determine the composition of mixtures of lipids, rather as is done in the case of high-temperature GLC (Gunstone, 1993). Solids may pose problems in proton magnetic resonance, but here ^{13}C can provide useful data, and where this fails magic angle spinning (Levy *et al.*, may be applicable, as in the determination of oil in whole seeds (Haw and Maciel, 1983).

In addition to information from nuclear spin transitions, measured by chemical shift, and the further information from spin–spin coupling, relaxation studies as the nuclei return to their equilibrium energy levels can give information of diffusion processes and hence of the phases present, for example ratios of liquid to solid in fats. Work in this area of wide-line NMR does not require the expensive instrumentation common in high-resolution work.

Some of the structural features of fatty acids that can be identified by ^1H NMR include protons of alkane, alkene, alkyne, cyclopropyl, cyclopropenyl, epoxide and carboxyl groups and the position of the alkene group; with higher frequency instruments (220 MHz and above), *cis* and *trans* double bonds can be distinguished and ^{13}C NMR permits in many cases the assignment of bands to most carbon atoms in the molecule, and facilitates the distinction of *cis* from *trans* isomers.

1.6 Mass spectrometry

The early work in mass spectrometry, associated with the names of Aston, J.J. Thomson and Dempster, was in the areas of physics and physical chemistry, and the application to organic chemistry was initiated by Conrad in 1930 (de Hoffmann *et al.*, 1996). The accessibility of commercial instruments from 1940 onwards led to the development of cleavage patterns for organic molecules, a concept that has proved valuable in the elucidation of the structure of organic compounds. The burgeoning of methods of ionisation in addition to electron impact, such as chemical ionisation, field ionisation and desorption (FI and FD), ionic bombardment (SIMA; Rose and Johnstone, 1987), laser ionisation and desorption (LIMA and LD; Hillenkamp, 1983) and fast atom bombardment (FABMS; Martin *et al.*, 1982), along with alternative mass analysers such as time of flight, quadrupole and cyclotron resonance analysers, have made it possible to examine a wide variety of materials including labile compounds and those of high molecular mass.

Thus, matrix-assisted laser desorption/ionisation with time-of-flight mass analysis (MALDI-TOF) was developed by Karas and Hillenkamp (1988) to study proteins. Fast atom bombardment produces molecular ions from polar and ionic compounds, so that compounds of molecular mass corresponding to m/z values of 100 000 have been structurally identified (Caprioli, 1990). Chapman and Goni (1994) list many mass spectrometric studies covering fatty acids, wax esters, monoacylglycerols, diacylglycerols, triacylglycerols, phosphoglycerides, glycolipids and sphingolipids.

1.7   X-Ray spectroscopy (crystallography)

In the X-ray field, most interest in lipids has centred on X-ray diffraction. In early work, Müller (1927, 1929) analysed the crystal structure of stearic acid and nonacosane.

Malkin and his group, from studies on the long spacings of ethyl esters of fatty acids, were led to a recognition of polymorphism (Malkin, 1954). Subsequently, many investigations of acyl glycerols have been carried out by X-rays (reviewed by Chapman, 1965, and Shannon, 1989).

Some of the intractable lipid components early examined by X-ray diffraction were long-chain waxes from plant surfaces and Kreger (1948) showed that these produced X-ray interferences with small diffraction angles from which accurate 'estimates' of the chain length could be made. Kreger and Schambert (1956) then obtained X-ray diagrams of C_{12}, C_{14}, C_{16}, C_{18} and C_{20} alcohols and of C_{16}, C_{17}, C_{18}, C_{22} fatty acids as well as of those of the esters produced from the C_{12} alcohol and linseed oil fatty acids. Some of the earliest work on the use of X-rays includes investigations of waxes by Shearer (1931) and work with long-chain alcohols by Müller (1927, 1929).

1.8   Chromatography – spectroscopy hyphenated techniques

The linking of separative methods with spectroscopic procedures, leading to hyphenated techniques—a term coined by Hirschfeld (1980)—has proved of considerable value in the lipid field. The transfer of an eluate from a capillary gas chromatography column to a mass spectrometer may be effected directly with a wide column or with a packed column using a jet separator.

The methods introduced for LC-MS have been various and include moving belt (Mcfadden, 1979), monodisperse aerosol generation interface (MAGIC; Willoughby and Browner, 1984; Bowers, 1989), particle beam, (PB; Willoughby and Browner, 1984), thermospray (TSP; Blakely and Vestal, 1983; Arpino, 1989), atmospheric pressure chemical ionisation (APCI; Bruins et al., 1987), electrospray (ESP; Bruins et al., 1987), the latter particularly useful for relative molecular masses of about 100 000. An example of the use of thermospray-MS is by Kirn and Salem (1989), who employed it to determine the structures of hydroperoxy derivatives of polyunsaturated fatty acids. Gas chromatography has been linked to FTIR by a light pipe system (Azarraga, 1980; Gurka et al., 1987) or cryogenic methods (Bourne et al., 1990). Liquid chromatography requiring methods removing the solvent is a greater challenge and many methods have been used (Fuoco et al., 1989), including MAGIC (Robertson et al., 1988). Some of the polar lipids that can now be analysed by LC-MS and GC-MS are reported by Myher and Kuksis (1995) and Karlsson (1998).

References

Arpino, P. (1989) *Mass Spectrosc. Rev.*, **8** 35.

Azarraga, L.V. (1980) *Appl. Spectrosc.*, **34** 224.

Bauman, R.P. (1962) *Absorption Spectroscopy*, Wiley, New York.

Bellamy, L.J. (1975) *The Infrared Spectra of Complex Molecules*, vol. 1, 3rd edn, Chapman and Hall, London.

Bellamy, L.J. (1980) *The Infrared Spectra of Complex Molecules*, vol. 2, 2nd edn, Chapman and Hall, London.

Blakely, C.R. and Vestal, M.L. (1983) *Anal. Chem.*, **55** 750.

Bourne, S., Haefner, A.M., Norton, K.L. and Griffiths, P.R. (1990) *Anal. Chem.*, **62** 2448.

Bowers, L.D. (1989) *Clin. Chem.*, **35** 1282.

Bruins, A.P., Covey, T.R. and Henion, J.D. (1987) *Anal Chem.*, **59** 2642.

Caprioli, R.M. (1990) *Anal. Chem.*, **62** 477A-485.

Chapman, D. (1965) *The Structure of Lipids*, Methuen, London.

Chapman, D. and Goni, F.M. (1994) in *The Lipid Handbook*, 2nd edn (eds F.D. Gunstone, J.L. Harwood and F.B. Padley), Chapman and Hall, London.

de Hoffmann, E., Charette, J. and Stroobant, V. (1996) *Mass Spectrometry, Principles and Applications*, Wiley, Chichester.

Dollish, D.R., Fateley, W.G. and Bentley, F.F. (1974) *Characteristic Raman Frequencies of Organic Compounds*, Wiley Interscience, London.

Fieser and Fieser (1949) *Natural Products Related to Phenanthrene*, Rheinhold, New York.

Fuoco, R., Pentoney, S.L. Jr. and Griffiths, P.R. (1989) *Anal. Chem.*, **61** 2212.

Gardiner, D.J. and Graves, P.R. (1989) (eds) *Practical Raman Spectroscopy*, Springer-Verlag, Berlin.

Goldring, L.S., Hawes, R.C., Hare, G.H., Beckman, A.O. and Stickney, M.E. (1953) Anomalies in extinction coefficient measurements. *Anal. Chem.*, **25** 869.

Grasselli, J.G. and Bulkin, B.J. (1991) *Analytical Raman Spectroscopy*, Wiley, New York.

Griffiths, P.R. and de Haseth, J.A. (1986) *Fourier Transform Infrared Spectrometry*, Wiley, New York.

Griffiths, P.R. and Fuller, M.P. (1982) in *Advances in Infrared and Raman Spectroscopy*, vol. 9 (eds R.J.H. Clark and R.E. Hester), Heyden, London.

Gunstone, F.D. (1993) *Advances in Lipid Methodology* (ed. W.W. Christie), The Oily Press, Dundee.

Gurka, D.F., Titus, R., Griffiths, P.R., Henry, D. and Giorgetti, A. (1987) *Anal. Chem.*, **59** 2362.

Harrick, N.J. (1967) *Internal Reflection Spectroscopy*, Interscience Division, Wiley, New York.

Haw, J.F. and Maciel, G.E. (1983) *Anal. Chem.*, **55** 1262.

Hendra, P.J., Jones, C. and Warnes, G. (1991) *Fourier Transform Raman Spectroscopy— Instrumentation and Chemical Applications*, Ellis Horwood, Chichester.

Hillenkamp, F. (1983) in *Proceedings of 2nd International Conference of Ion Formation from Organic Solids* (ed. A. Benninghoven), Springer-Verlag, New York.

Hirschfeld, T. (1980) *Anal. Chem.*, **52** 297-312A.

Karas, M. and Hillenkamp, F. (1988) *Anal. Chem.*, **60** 2299-2301.

Karlsson, A.A. (1998) in *Lipid Analysis in Oils and Fats.* (ed. R.J. Hamilton), Blackie Academic and Professional, Glasgow.

Kirn, H.Y. and Salem, N. (1989) *Prostaglandins*, **37** 105.

Koningstein, J.A. (1982) *Introduction to the Theory of the Raman Effect*, Reidel, Dordrecht.

Kreger, D.R. (1948) *Rec. Trav. Botan. Neerl.*, **41** 603-736.

Kreger, D.R. and Schamhart, C. (1956) *Biochim. Biophys. Acta*, **19** 22-44.

Levy, G.C., Lichter, R.L. and Nelson, G.L. (1980) *Carbon-13 Nuclear Magnetic Resonance*, 2nd edn, Wiley, New York.

Lothian, G.F. (1963) *Analyst*, **88** 678.

Loudon, G.M. (1964) *J. Chem. Educ.*, **41** (7) 391.

Malkin, T. (1954) *Prog. Chem. Fats and Other Lipids*, **2** 1.

Markham, K.R. (1982) *Techniques of Flavanoid Identification*, Academic Press, London, pp 1-49, 72-99.

Martin, S.A., Costello, C.E. and Biemann, K. (1982) *Anal. Chem.*, **54** 2362.

Mcfadden, W.H. (1979) *J. Chromatogr. Sci.*, **17** 2.

Müller, A. (1927) *Proc. Roy. Soc.*, London, **A114** 542-546.

Müller, A. (1929) *Trans. Faraday Soc.*, **25** 347-348.

Myher, J. and Kuksis, A. (1995) *INFORM*, **6** 1068-1072.

Osbourne, B.G. and Fearn, T. (1988) *Near-infrared Spectroscopy in Food Analysis*, Longmans, New York.

Raman, C.V. and Krishnan, K.S. (1928) *Nature*, **121** 501.

Robertson, R.M., de Haseth, J.A., Kirk, J.D. and Browner, R.F. (1988) *Appl. Spectrosc.*, **42** 1365.

Rose, M.E. and Johnstone, R.A.W. (1987) *Mass Spectrometry for Chemists and Biochemists*, Cambridge University Press, Cambridge.

Shannon, J. (1989) Ph.D. thesis, Liverpool Polytechnic.

Shearer, G. (1931) *J. Chem. Soc.*, 315; *Proc. Roy. Soc. (London)* (1925) **A108** 655-666.

Socrates, G. (1994) *Infrared Characteristic Group Frequencies*, 2nd edn, Wiley, London.

Williams, P. and Norris, K. (1987) *Near-Infrared Technology in the Agricultural and Food Industries*, American Association of Cereal Chemists, Inc., St Paul, MN.

Willoughby, R.C. and Browner, R.F. (1984) *Anal. Chem.*, **56** 2626.

Wineforder, J.D. (ed.) (1971) *Spectrochemical Methods of Analysis*, Wiley Interscience, New York.
Woodward, R.B. (1941) *J. Am. Chem. Soc.*, **63** 1123.
Woodward, R.B. (1942) *J. Am. Chem. Soc.*, **64** 72.
Woodward, R.B. (1942) *J. Am. Chem. Soc.*, **64** 76.

Bibliography

Ardrey, R.E. (1994) *Liquid Chromatography Mass Spectrometry*, VCH, Weinheim.
Breitmaier Eberhard (1993) *Structure Elucidation by NMR in Organic Chemistry, A Practical Guide*, Wiley, Chichester.
Brown, S.S. (1980) *An Introduction to Spectroscopy for Biochemists*, Academic Press, London.
Budzikiewicz, H., Djerassi, C. and Williams, D.H. (1964) *Structure Elucidation of Natural Products by Mass Spectrometry*, vols I and II, Holden-Day, San-Francisco.
Budzikiewicz, H., Djerassi, C. and Williams, D.H. (1964) *Interpretation of Mass Spectra of Organic Compounds*, Holden-Day, San-Francisco.
Cooper, J.W. (1991) *Spectroscopic Techniques for Organic Chemists*, Wiley, New York.
Diehl, B.W.K. (1998) in *Lipid Analysis in Oils and Fats* (ed. R.J. Hamilton), Blackie Academic and Professional, Glasgow.
Field, L.D., Sternhell, S. and Kalman, J.R. (1996) *Organic Structures from Spectra*. 2nd edn, Wiley, Chichester.
Fleming, I. and Williams, D.H. (1989) *Spectroscopic Methods in Organic Chemistry*, 4th edn, McGraw-Hill, Maidenhead.
Harris, R.K. (1994) *Nuclear Magnetic Resonance Spectroscopy*, Longman Scientific and Technical, Harlow.
Hans-Otto Kalmowski, Berger, S. and Braun, S. (1991) *Carbon-13 NMR Spectroscopy*, Wiley, Chichester.
Levy, G.C., Lichter, R.L. and Nelson, G.L. (1980) *Carbon-13 Nuclear Magnetic Resonance Spectroscopy*, 2nd edn, Wiley Interscience, New York.
Pavia, D.L., Lampman, G.M. and Kriz, G.S. (1996) *Introduction to Spectroscopy*, 2nd edn, Saunders College Publishing, Harcourt Brace College Publishers, FL.
Rao, C.N.R. (1974) *Ultraviolet and Visible Spectroscopy*, Chemical Applications, 3rd edn, Butterworth, London.
Scott, A.I. (1964) *Interpretation of the Ultraviolet Spectra of Natural Products*, Pergamon Press, Oxford.
Silverstein, R.M., Bassler, G.C. and Morrill, T.C. (1991) *Spectrophotometric Identification of Organic Compounds*, 5th edn, Wiley, New York.
Stothers, J.B. (1972) *Carbon-13 NMR Spectroscopy*, Academic Press, New York.
Wehrli, F.W., Marchand, A.P. and Wehrli, S. (1988) *Interpretation of Carbon-13 NMR Spectroscopy*, 2nd edn, Wiley, Chichester.
Yergey, A.L. *et al.* (1989) *Liquid Chromatography Mass Spectrometry:Techniques & Applications*, Plenum, New York.

2 Atomic spectroscopy for heavy metal determination in edible oils and fats

G.W. Hammond, V.P. Shiers and J.B. Rossell

2.1 Introduction

Atomic spectroscopy, which incorporates both atomic absorption spectroscopy (AAS) and atomic emission spectroscopy (AES) specifically, and inductively coupled plasma spectroscopy (ICP), are techniques commonly used in the oils and fats industry to determine concentrations of metallic contaminants. These metals contaminate oils and fats from a variety of sources.

All edible fats are produced by natural processes. Vegetable oils are produced by agriculture, animal fats are produced from the carcass or milk fat of animals that have usually been fed on agricultural crops, while fish oils are produced from fish harvested at sea. A metal may be present in the soil where an agricultural crop is grown as a result of its natural presence or by introduction to the soil via agrochemicals, fertilisers, sewage, refuse or industrial wastes. In Australia, it was established (McGowan and Petch, 1993) that phosphate rocks used in the production of phosphate fertiliser contained trace quantities of cadmium. These residues contaminated the soil where the fertiliser had been used and cadmium was then bioaccumulated by safflower plants, leading to eventual contamination of safflower seeds, 11 of 14 samples being above the maximum permitted concentration. Other sources of contamination may be environmental. It is known that fish may become contaminated as a result of living in contaminated water; the infamous Minamata Disease, in which several Japanese people died, was caused by the use of fish that had been contaminated with methylmercury discharged into the sea by a chemical plant (Sakamoto *et al.*, 1996). Metals may also be introduced during processing, transport or storage, or could be residues of processing aids, such as hydrogenation catalysts.

In the majority of cases, metals reduce the quality of an oil. Several metals, especially iron and copper, are known to catalyse the oxidation of oils and fats (Swern, 1982). Even in ancient Roman times, it was recognised that holding olive oil in copper vessels was likely to shorten its shelf-life (Patterson, 1992). Other metals may be toxic, while a third class may impede processing. Since any method of determination of these metal contaminants must be reliable, at least at the concentration at which the metal is a problem, but preferably at even lower levels, it is relevant first of all to review the concentrations of metals at which their harmful effect becomes apparent.

Swern (1982) reported the concentrations of metals at which the oxidative stability of lard is reduced by 50%. These concentrations are given in Table 2.1. Some catalytic oxidation will probably be apparent at concentrations still lower than these; for instance, Swern (1982) reports that $0.02\,\text{mg kg}^{-1}$ of copper will reduce the flavour stability of margarine; this is a level only one-fifth of the concentration found to reduce the rate of oxidation of lard by 50%. However, the data in Table 2.1 do give a general guide to concentrations at which any analytical method for the metal should be reliable.

The situation with regard to the critical concentration of toxic metals is less clear. While it is known that metals such as mercury, cadmium and lead are toxic, there are few recommendations in the literature specifying maximum permitted levels in edible oils. There is a draft European Union regulation (III/5125/95/Rev. 3 of March 1997) stating that the maximum

Table 2.1   Critical metal concentrations (mg kg^{-1} oil)

	Based on oxidation (mg kg^{-1})[a]	Based on toxicity (mg kg^{-1})	Authors' estimate (mg kg^{-1})[b]	Based on processing (mg kg^{-1})
Lead		0.1[c]		
Copper	0.5			
Copper	0.02[d]			
Iron	0.6			
Manganese	0.6		2.0	
Chromium	1.2		0.1	
Nickel	2.2		0.25	
Vanadium	3.0		4.0	
Zinc	19.6			
Aluminium	50.0			
Cadmium			0.03	
Mercury			0.02	
Molybdenum			0.07	
Selenium			0.8	
Tin			56	
Sodium				0.4
Phosphorus[f]				2[e]

[a]Swern (1982) and Gordon (1990).
[b]Authors' estimate based on permitted total weekly intake in comparison with that of lead (see text).
[c]EU limit (Draft Commission regulation III/5125/95/Rev. 3 of March 1997), and FAO/WHO (1993).
[d]Level needs to be reduced to $0.02\,\text{mg kg}^{-1}$ to preserve margarine flavour (see Swern, 1982).
[e]Minimum 2.0, maximum $20.0\,\text{mg kg}^{-1}$ for best stability in soya-bean oil (List and Erickson, 1980).
[f]Phosphorus, although not a metal, is an important element determined by AA and included in this chapter.

permitted amount of lead in an edible oil should be $0.1\,\text{mg}\,\text{kg}^{-1}$. This corresponds to the limit in Codex Alimentarius specifications on edible oils and fats (FAO/WHO, 1993). Any method for the determination of lead must therefore be accurate at least at this concentration, but preferably at a still lower level.

There are no recommendations about permitted levels of the other hazardous metals in oils and fats. The present authors therefore made estimates on the following basis. The WHO has set a maximum weekly intake value for lead of $25\,\mu\text{g}\,\text{kg}^{-1}$ body weight (WHO, 1996). It has also set a maximum weekly intake for mercury of $5\,\mu\text{g}\,\text{kg}^{-1}$ body weight. The authors therefore reasoned that, if the maximum permitted weekly intake of mercury is 20% of that of lead, it is appropriate to seek a method for the determination of mercury that will be reliable at 20% of the limit of detection (LOD) sought in the method for lead, i.e. about $0.02\,\text{mg}\,\text{kg}^{-1}$. On this same basis, concentrations at which we should have reliable methods for the determination of other metals have been estimated to give the data in Table 2.1. It is not intended that this should be taken as a definitive list of concentrations at which the toxic effects become apparent, since toxicity will depend on the amount of oil consumed in comparison with other components of the diet, but rather as a guide to the concentrations at which analytical methods should be reliable.

List and Erickson (1980) reviewed the influence of lecithin on the stability of soya-bean oil and came to the conclusion that the amount of lecithin should lie between 60 and $600\,\text{mg}\,\text{kg}^{-1}$ for maximum flavour stability of the oil.

As the common method for the determination of lecithin is to determine the concentration of phosphorus and then multiply by the factor 30 (AOCS, 1996), this equates to an optimum concentration of phosphorus of $2\text{--}20\,\text{mg}\,\text{kg}^{-1}$.

Alkali refining of crude oil introduces sodium soaps, which then contaminate the oil. Hot-water washing is the conventional and very effective method of removing soap. This is important as any residues are absorbed by bleaching earth and hydrogenation catalysts, the efficiency of the bleaching or hydrogenation process being thereby reduced. It is therefore recommended that the soap content should be reduced to below $5\,\text{mg}\,\text{kg}^{-1}$ (Hui, 1996), which corresponds to a sodium content below $0.4\,\text{mg}\,\text{kg}^{-1}$.

By these means the critical concentrations of sodium and phosphorus at which an analytical method should be reliable were drawn up (Table 2.1). As a whole, the table indicates that it may sometimes be more important to ensure a low metal content of an oil in order to make sure that it is not a hazard to health than to prevent oxidation, whilst in other cases the opposite is true.

2.2   Traditional analytical approaches to the determination of metals

There are several analytical approaches to the determination of metals at the levels indicated in Table 2.1. These include colorimetry, polarography/voltammetry, atomic absorption spectroscopy (AAS) and inductively coupled plasma-emission spectroscopy (ICP). Colorimetric methods relate loosely to spectral properties of the oil in that a colour is developed by reaction with a colour-forming reagent and the intensity of the colour is measured at a specific wavelength, usually in a spectrometer or colorimeter, and the intensity of the colour at that wavelength is used to determine the concentration of the metal. Polarography, in contrast, has no relationship to spectral properties.

It is appropriate to review the experimental techniques used in the traditional methods in order to appreciate fully the advantages of atomic spectroscopy.

The so-called 'traditional methods' are mainly colorimetric and polarographic. In both cases, it is necessary to convert the sample to an aqueous extract containing the metal. This may be carried out by dry ashing, but here there is a danger of losing some of the volatile metals such as mercury, lead or cadmium. The present authors therefore prefer wet ashing. AOAC Official Method 945.58 (AOAC, 1990) for cadmium in food, for instance, specifies a typical and reliable wet ashing procedure. This involves digesting a known amount, usually about 5 g, of the food such as an oil or fat with 10 ml of a 50/50 mixture of sulfuric and nitric acids. If the sample tends to char rather than oxidise evenly, 5–10 ml of additional sulfuric acid may be added. The digestion is continued with addition of further 10 ml quantities of nitric acid until the evolution of sulfur trioxide indicates that the digestion is complete. The solution is cooled and 15 ml of saturated ammonium oxalate solution are added and the mixture is again heated until fumes are evolved. This product, or related products produced by slight modifications of the technique, may then be used for the colorimetric or polarographic determination of the metals in question.

This wet oxidation process is quite laborious, especially if a number of samples needs to be digested. The procedure can also be dangerous if not carried out according to the instructions. It is, for instance, possible to generate oxidised compounds such as organic nitrates, which can decompose violently if heated at too high a temperature. One aspect of the approved procedures is to heat the mixture until fumes of sulfur trioxide are evolved, which demonstrates that nitrates have been destroyed. Automatic systems have therefore been introduced, which alleviate the tedium of the wet oxidation process and, as they are automated, do not need constant attention and there is less temptation for junior staff to attempt to take short cuts. The apparatus is also therefore safer than manual oxidation in a busy laboratory. The automatic Gerhardt unit is shown in Fig. 2.1. This not only has timed sequences for the various stages of the digestion but also has an acid trap at the top of the digestion unit designed to trap any volatile metals that could otherwise be lost. It is therefore claimed to give more reliable results with volatile metals such as mercury, lead or zinc.

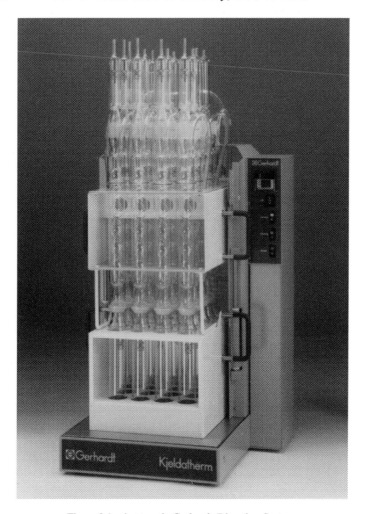

Figure 2.1 Automatic Gerhardt Digestion System.

2.2.1 Colorimetric methods

2.2.1.1 Iron and copper
The standard ISO/BSI method for the determination of iron and copper (BSI, 1976) involves digestion of the fat with nitric acid, extraction with

light petroleum to remove unreacted organic material, and dilution of the acidic aqueous phase, an apparently more straightforward procedure than the AOAC one.

In the case of copper (Ramsey, 1960; AOAC, 1990 method 960.40), diethyldithiocarbamate is used as the colorimetric reagent. The residue from the wet ashing/extraction is filtered, if necessary, and diluted with water to 100 ml. An aliqot of 25 ml is treated with citrate–EDTA solution and brought to a pH level of 8.8 by addition of thymol blue indicator and slow addition of ammonia solution until the indicator shows the appropriate green-blue colour. Then, 1 ml of diethyldithiocarbamate solution is added and the copper diethyldithiocarbamate compound is extracted with 15 ml of carbon tetrachloride. The absorbance of the carbon tetrachloride extract is measured at the peak maximum near to 400 nm, and the amount of copper is determined by reference to a calibration curve. If it is suspected that bismuth or tellurium is present, and may have interfered, the carbon tetrachloride solution is returned to a separator and 10 ml of 5% potassium chloride solution are added. If the carbon tetrachloride solution becomes colourless, these two interfering metals are absent. If colour remains and the presence of interfering metals is indicated, a second 25 ml of red (i.e. copper- containing) diethyldithio-carbamate solution is treated with two 10-ml portions of $1 \, mol \, l^{-1}$ sodium hydroxide solution, which decomposes the bismuth and tellurium compounds, the copper complex being stable under these conditions. The carbon tetrachloride solution is drained and again used for the determination of the absorbance at the peak maximum near to 400 nm. The blank value is reported to be $1 \, \mu g$ copper (Ramsey, 1960), which corresponds to a detection limit of $0.2 \, mg \, kg^{-1}$, or 0.2 ppm, with the specified starting weight of 5 g.

The closely related American Oil Chemists Society method Da 31-58 (AOCS, 1996) gives a value for the repeatability of $0.03 \, mg \, kg^{-1}$ but a reproducibility of $0.4 \, mg \, kg^{-1}$ on samples containing $3 \, mg \, kg^{-1}$.

In a second method for the determination of copper (British Standard 684, section 2.16; BSI, 1976), a diluted aqueous extract is treated with a solution of zinc dibenzyldithiocarbamate and filtered. The optical density of the filtrate is then determined and compared with that of a calibration curve, from which the amount of copper in the original sample can be determined. This procedure can determine as little as $5 \, \mu g$ of copper, which, based on the starting sample weight of 25 g, is $0.2 \, mg \, kg^{-1}$, the same detection limit as for the first method mentioned.

Neither of these colorimetric methods for copper can achieve the limit of detection needed to ensure that copper is at a lower concentration than the $0.02 \, mg \, kg^{-1}$ specified by Swern (1982) to preserve margarine flavour (Table 2.1).

With iron (BS 684, Section 2.17; BSI, 1976) the aqueous extract is prepared as in the case of the BSI method for copper above, and sulfur dioxide solution is added. The diluted aqueous solution is then titrated with sodium acetate buffer solution until neutral to Congo red paper, when 1,10-phenanthroline solution is added and the optical density of the solution is determined. The iron content is determined by reference to a calibration curve. The limit of detection is much the same as that of the BSI method for copper.

Several colorimetric methods rely on colour reactions of dithizone (diphenylthiocarbazone, $C_{13}H_{12}N_4S$) with the metal, different metals being determined by use of the different relative stabilities and/or solubilities of the dithizonates at different pH values. Clifford and Wichmann (1936) give an interesting review of the properties of dithizone, and the importance of ensuring that the reagent is pure before use in colorimetry.

2.2.1.2 Cadmium

In the case of cadmium (Shirley et al., 1949; AOAC, 1990, method 945.58) the determination of the metal in the residue from wet ashing involves addition of citric acid, dilution to 25 ml, addition of thymol blue indicator and adjustment to pH 8.8 by slow addition of ammonia solution followed by quantitative dilution to 125 ml. Aliquots are extracted with portions of dithizone–chloroform solution until the chloroform is clear, to extract all metals. Copper, mercury and almost all nickel and cobalt are removed by washing the chloroform solution with dilute hydrochloric acid, which decomposes the dithizone compound of cadmium, all metals except cadmium and zinc (if present) being retained in the chloroform layer, which is discarded. The aqueous layer is then rendered alkaline, and extracted with further portions of dithizone in chloroform solution. Cadmium and zinc (if present) dithizonates are converted to chlorides by shaking with hydrochloric acid solution, and the aqueous phase is again rendered alkaline. Exactly 25 ml of dithizone in chloroform solution are added; the mixture is shaken for exactly 1 minute; the layers are then permitted to separate for exactly 3 minutes and the organic layer is filtered through a plug of cotton wool. The first 5 ml are discarded and the remainder is used to fill an absorption cell, which is used to measure the absorbance at 510 nm. The amount of cadmium is determined by reference to a calibration plot.

Recoveries of about 0.1 µg were obtained (corresponding to a detection limit of $0.02 \, mg \, kg^{-1}$ on a 5 g starting weight) on cadmium added after the ashing stage. However, it is admitted that some cadmium could be lost during ashing or on the leaching of cadmium from a dry ash residue (Shirley et al., 1949).

2.2.1.3   Lead

The dithizone method for lead (Clifford and Wichmann, 1936; AOAC, 1990, method 934.07) has a proviso (note K) that complete digestion is accomplished with difficulty. It is recommended (method 963.21C) that the sample is digested in a Kjeldahl flask with 25–50 ml nitric acid followed by cautious addition of 40 ml sulfuric acid and then warming. The oxidising conditions are maintained by adding further quantities of nitric acid whenever the mixture turns black or darkens. The digestion is continued until all organic matter is destroyed and fumes of sulfur trioxide are evolved. After this, 100 ml of water are added and enough hydrochloric acid to dissolve any calcium sulfate in the residue. Any anhydrous silica or barium sulfate is pulverised with a flat-ended stirring rod and the mixture is filtered. Lead sulfate (if present) is dissolved in hydrochloric acid and the mixture on the filter is leached with 10–20 ml hot hydrochloric–citric acid solution, followed by 10–20 ml hot ammonium acetate solution. The flasks and filter are finally rinsed with hot water.

Lead is separated by precipitation of sulfide by cooling the aqueous solution, adding citric acid and adjusting to pH 3.0–3.4 with bromophe-nol blue indicator by addition of ammonium hydroxide solution. Copper sulfate is optionally added, as a co-precipitant, and hydrogen sulfide is passed through the solution for 3–5 minutes until the solution is saturated. The solution is immediately filtered.

The sulfides are dissolved, without washing, in 5 ml of hot nitric acid. The original flask is washed with hot water and the washings are combined with the nitric acid solution, which is then boiled to remove hydrogen sulfide. The product is transferred to a 200-ml separator and treated with citric acid solution, rendered alkaline with ammonia (pH > 8.5 with thymol blue indicator), and treated with 5 ml of 10% potassium cyanide solution. Lead is then extracted with 20-ml portions of dithizone–chloroform reagent until the extract shows no sign of the red lead–dithizone complex. The extracts are drained into a separator containing 25 ml of 1% nitric acid and shaken, and the chloroform is drained into a second separator containing 1% nitric acid. This is shaken and the chloroform is discarded. The nitric acid solutions are combined, filtered, adjusted to 50 ml with 1% nitric acid, and treated with 10 ml of ammonium cyanide solution. The solution is shaken with a measured amount of dithizone reagent, when a red colour develops. The absorbance of the chloroform layer is measured at 510 nm and the amount of lead is determined by reference to a calibration curve. The limit of detection of this method is dependent on the blank value, and it is claimed (Clifford and Wichmann, 1936) that blank values as low as 1 μg can be obtained with attention to detail. This corresponds to a limit of detection of lead of $0.2\,mg\,kg^{-1}$ or 0.2 ppm for a 5 g starting weight. This method is not, therefore, sufficiently sensitive for enforcement of EU legislation, which calls for a maximum concentration of lead of $0.1\,mg\,kg^{-1}$ (Table 2.1).

2.2.1.4 Mercury

Since mercury is more volatile than other metals, the critical step is the digestion. AOAC method 952.14 (Klein, 1952; AOAC, 1990) specifies special apparatus and stipulates that, because of the volatility of mercury compounds, careful heating during the digestion and sample preparation is essential. Digestion must be almost complete otherwise residual organic matter may combine with mercury and prevent or hinder extraction with dithizone. On the other hand, oxidising material in the digest must be destroyed, otherwise the dithizone itself is oxidised and the mercury is not quantitatively extracted. The acidity of the final sample must be adjusted to about $1\,mol\,l^{-1}$ by addition of ammonium hydroxide (it should not be over $1.2\,mol\,l^{-1}$; nor should it become alkaline, to avoid formation of mercury complexes).

The less than $1.2\,mol\,l^{-1}$ acidic sample is transferred to a separator and dithizone–chloroform solution is added. The mixture is shaken and the chloroform layer drained into a separating funnel. The extraction of the sample with dithizone solution is repeated until the dithizone–chloroform extracts remain green, indicating that all the mercury has been extracted. The extracts are combined in a separating funnel containing a mixture of hydrochloric acid and ammonium hypochlorite. The mixture is shaken for 1 minute and allowed to separate. The chloroform layer is drained into $0.1\,mol\,l^{-1}$ hydrochloric acid solution, and sodium thiosulfite solution is added. The chloroform layer is drained and discarded to remove copper dithizonate. Sodium hypochlorite solution is added to decompose the mercury thiosulfate complex and to oxidise excess thiosulfate.

Ammonium hypochlorite solution is added and the mixture shaken vigorously. All hypochlorite must be reduced, as it would otherwise oxidise dithizone to a yellow form, which would be recorded as mercury. The aqueous solution is extracted with chloroform, which is discarded to leave a colourless aqueous product. Acetic acid solution is added and the solution is extracted with a calculated volume of dithizone in chloroform solution. The red chloroform layer is separated, filtered and used to measure the absorbance at 490 nm. The amount of mercury is determined by reference to a calibration curve. A ring test of the method (Klein, 1952) found a mean deviation between analysts of $0.0208\,\mu g\,g^{-1}$ on a sample with a mean concentration of $0.34\,\mu g\,g^{-1}$. On the basis that the limit of detection is twice the deviation, the limit of detection of this method is $0.04\,\mu g\,g^{-1}$.

2.2.1.5 Zinc

Zinc may be also determined with dithizone (Alexander and Taylor, 1944; Taylor and Alexander, 1945; AOAC, 1990, method 944.09). The wet-ashed aqueous extract is diluted to 40 ml and two drops of methyl red and 1 ml of copper sulfate (scavenger agent) are added. Hydrochloric acid is added until the solution is $0.15\,mol\,l^{-1}$ with respect to this acid, and the pH is 1.9–2.1. Hydrogen sulfide is passed through the solution to precipitate copper, cadmium, bismuth, tin, antimony, mercury and silver as sulfides. These are filtered off and the filtrate is thoroughly washed with dilute hydrochloric acid followed by distilled water. The combined filtrate and washings are boiled to remove hydrogen sulfide, and the mixture is then treated with bromine water. The solution is again boiled until free from bromine. The solution is neutralised to phenol red with ammonium hydroxide solution and then made slightly acid with hydrochloric acid solution. The solution is then made up to a standard volume containing $0.2–1.0\,\mu g\,l^{-1}$ zinc.

Twenty ml of this resulting solution is treated with aqueous ammonium citrate and dimethylglyoxime solutions and an α-nitroso-β-naphthol–chloroform solution (to combine with any nickel and cobalt), and the resulting chloroform layer is rejected. The aqueous layer is washed with chloroform to complete removal of nickel and cobalt.

The aqueous layer should at this stage have a pH of 8.0–8.2. It is treated with dithizone/chloroform solution and additional chloroform. The chloroform layer is quantitatively removed, and is treated with $0.04\,mol\,l^{-1}$ hydrochloric acid to transfer zinc to this aqueous layer, which is then treated with ammonium citrate solution and carbon tetrachloride. A calculated quantity of dithizone solution is added and the absorption of the chloroform–carbon tetrachloride solution measured at 540 nm. The amount of zinc is determined by reference to a calibration plot.

The main problem with this method is the reduction of the blank value. Alexander and Taylor (1944) report that zinc contamination due to extraction from Pyrex glassware was a persistent problem. In a later paper, these researchers say that it is impossible to overemphasise the importance of proper cleaning of glassware and the precautions that must be observed in guarding against contamination (Alexander and Taylor, 1945). By careful attention to detail, including immersion of all glassware in boiling nitric acid followed by thorough rinsing, they were able to reduce the blank value to $0.23\,\mu g$ of zinc. This amounts to a limit of detection of $0.1\,mg\,kg^{-1}$ on a 23 g sample, or $0.5\,mg\,kg^{-1}$ (0.5 ppm) on a 5 g sample.

2.2.1.6 Manganese

In contrast with the above metals, the determination of manganese (AOAC, 1990, method 921.02) is relatively straightforward, involving oxidation to permanganate ion. There are no interfering metals, so complicated separation techniques are not needed.

An extract of the food is treated with ammonium oxalate to precipitate calcium oxalate, which may then be titrated with permanganate to determine calcium. To the filtrate is added a measured amount of potassium iodate or periodic acid, and the solution is boiled until no further colour development takes place. The absorbance is measured at 530 nm and the amount of manganese is determined by reference to a calibration curve.

In a collaborative trial (Mitchell, 1921), an average difference of 0.03% Mn_3O_4 (equivalent to 216 mg kg^{-1} Mn) was obtained, giving a rough estimate of the limit of detection of about 200–250 ppm, much higher than the critical concentration mentioned in Table 2.1.

2.2.1.7 Phosphorus

A newcomer to the field may think that there is no problem with the colorimetric determination of phosphorus, as there are standardised methods published by the American Oil Chemists Society (AOCS, 1996, method Ca 12-55) and the International Union of Pure and Applied Chemistry (IUPAC, 1987, method 2.423). Unfortunately, however, it is not quite so straightforward. It was unanimously agreed by the IUPAC Oils and Fats Committee that parts of the colorimetric method for phosphorus were poorly written and technically ambiguous (Pike, 1995). It was therefore decided to withdraw the IUPAC method in favour of the alternative colorimetric AOCS method. At the same time, the American Oil Chemists Society realised that its own method (Ca 12-55) relied upon the use of carcinogenic hydrazine sulfate as a reducing agent. The AOCS therefore listed its own method for withdrawal. It is possible that both organisations thought that analysts could use the alternative method, not realising that this too had been withdrawn. At that time, there were AA methods (e.g. IUPAC 2.423), but these stipulated that the AA procedure should be calibrated against a sample of known phosphorus content and that this should be determined by the (now withdrawn) colorimetric method. The Leatherhead Food Research Association was therefore commissioned to develop a colorimetric method free from hazardous reagents and to confirm that the method had the required sensitivity and reliability.

A literature survey revealed a promising Dutch method (NEN 6349). This employed special hand-made calibrated glass sample tubes and a matching metallic heating block. These were not easily available in the UK (or, it transpired, in the Netherlands). The method was therefore modified slightly to use standard graduated glass flasks and a muffle furnace for heating. These modifications were quite satisfactory (Cooke *et al.*, 1996). The ISO Oils and Fats Committee (ISO TC34/SC11) is considering the method for approval. If no problems arise, it should be available in published form as ISO 10540 by the time of publication of this book.

The Dutch method, revised as a result of work carried out at the Leatherhead Food Research Association (Cooke *et al.*, 1996), specifies that the sample should first be melted and rendered homogeneous. In particular, any sediment should be incorporated fully into the oil. This is because many crude oils contain phosphatides such as lecithin, which separate out from damp oils on standing. As phosphatides are rich in phosphorus, any failure to incorporate a phosphatide-rich sediment fully would have a profound influence on the magnitude of the final result for phosphorus content, an aspect not taken into account in some method texts. A borosilicate glass tube, calibrated at 15 ml, with a drawn-out stoppered neck, and containing about 30 g of magnesium hydroxycarbonate, is weighed. Ten to fifteen drops of the sample are added from a Pasteur pipette and the tube is re-weighed to gain the weight of sample used. The tube is heated to $350°C$ until the sample is carbonised to a black mass and then at $550°C$ until completely white (about 4 hours). The tube is allowed to cool and the white residue is dissolved in 2 ml of hydrochloric acid and boiled. Sodium hydroxide solution (0.6 ml) is added to the cooled solution followed by 5 ml of a reducing solution comprising metol, sodium sulfite and sodium bisulfite and 2.5 ml of sulfate–molybdate reagent. The tubes are stood in the dark for 20 minutes and then filled to the 15 ml mark with sodium acetate solution. The absorbance of the solution is then measured against a blank at 720 nm. The limit of quantification is about $1 \, \text{mg kg}^{-1}$.

2.2.2 Polarography/voltammetry

A few methods for the determination of low concentrations of metals rely on the technique of polarography or voltammetry. This is an electrochemical technique in which the metal is reduced at a small electrode. The electrical potential at which reduction takes place varies from metal to metal; this enables some discrimination. More than one metal can therefore be determined at the same time, but the flexibility here is somewhat limited. Conditions in the electrochemical cell are arranged so that the cathode at which the metals, or one of several other species, are reduced has a small surface area. Because of this, the electrode suffers a 'concentration polarisation' (hence the name 'polarography') in which the current which passes is limited by the diffusion of the metal ions to the electrode from the bulk of the solution. This is in turn dependent on the concentration of metal ions in the solution. A sweeping electrical potential is generally applied and a wave-form volts/amps curve is obtained (thus the alternative name 'voltammetry'). The voltage at which a species is reduced relates to its nature, and the current developed relates to its concentration. In anodic stripping voltammetry, materials such as lead, cadmium, copper and mercury can be concentrated on the microelectrode, usually a dropping mercury electrode, by electrochemical deposition before the process is reversed to strip the material from the electrode. This can give extraordinary sensitivity down to $10^{-10}\,\text{mol}\,\text{l}^{-1}$ and less (Bersier, 1987). The technique is, therefore, well suited to the determination of low concentrations of metals in aqueous solutions. Readers interested in the background theory and the various degrees of sophistication of this technique are referred to standard textbooks on the subject, such as that by Pomeranz and Meloan (1982).

However, it has been contended that polarography/voltammetry is an under-utilised analytical method (Bersier, 1987). Thus in the 15th edition of the 'Official Methods of Analysis of the Association of Official Analytical Chemists' (AOAC, 1990), there are only five methods that utilise anodic stripping voltammetry for the determination of metals in food, and these all relate to the determination of lead and cadmium in different foods. Method 982.23 (AOAC, 1990) describes the determination of cadmium and lead in food, and includes a specific digestion process in which an extra nitrate oxidation is included to convert any tin present to Sn(IV), which does not interfere with the determination of lead in the anodic stripping procedure since its oxidation potential is sufficiently far removed from that of lead (Capar, 1987).

This method specifies that the sample should be dry-ashed with potassium sulfate and nitric acid at 500°C, taking all precautions to avoid contamination of the samples or reagents. At least three control blanks should be prepared and these should include any extra water or nitric acid added during the digestion. Five to ten grams of sample are weighed into the ashing vessel and 5 ml of potassium sulfate ashing solution is added together with any additional water to ensure that the sample is fully submerged. The sample is agitated with a glass rod to mix thoroughly and the mixture is then heated to dryness under a glass cover. The residue is then heated in a furnace at 500–550°C for at least 4 hours and cooled. A white ash should result.

The sides of the vessel are washed with water; 2 ml of nitric acid is added and the sample is again heated to dryness, after which it is again heated at 500°C for 30 minutes. The nitric acid treatment is repeated until a white ash is obtained. To this 1 ml of water and 5 ml of nitric acid are added and the mixture is stood for 5 minutes until all residue dissolves. Heating in this dilute acid solution is avoided as far as possible to minimise conversion of any tin to the Sn(II) form. The solution is then quantitatively diluted to 50 ml in a volumetric flask and allowed to stand so that any precipitate of silica can settle to the bottom of the flask. Then, 5 ml of clear supernatant liquid are pipetted from the top of the flask and mixed with 5 ml of an electrolyte solution containing acetic acid, sodium acetate and tartaric acid. The solution is purged with nitrogen and the dropping mercury electrode of the polarograph is adjusted to a predetermined rate. The solution is electrolysed at -0.8 V vs a standard calomel electrode for (typically) 1–2 minutes to deposit metals on the electrode electrolytically. A linear sweep voltage is applied and the wave height of the current curve obtained is measured at -0.62 for cadmium and -0.45 for lead. The estimated quantification limits for a 10 g starting weight are 0.005 ppm for cadmium and 0.010 ppm for lead (Gajan et al., 1982), much better (lower) than in the colorimetric methods discussed earlier.

2.2.3 Summary of traditional methods

In general, the majority of metals can be determined by traditional techniques at concentrations sufficiently low for the assessment of triglyceride oil quality. One exception is manganese, where the conversion to permanganate and determination of this colorimetrically has a very high limit of detection. Furthermore, this method has not been revised or improved for over 50 years. With most of the other metals the main problem is lack of specificity. In order to attain sufficiently low limits of detection it is necessary to use colour-forming reagents of high colour-forming ability. There are only a few of these available. One of the best is dithizone, but this reagent forms coloured compounds with several metals. In order to avoid interference from another metal species, and thus over-reporting, and yet avoid jeopardising full recovery of the target analyte, and thus under-reporting, it is necessary to employ elaborate separation techniques. These separation techniques are tedious and employ toxic reagents such as the combination of hydrogen sulfide and potassium cyanide used in the colorimetric method for lead. The procedures may introduce unwanted experimental error and therefore detract further from the appeal of the colorimetric methods. A related situation applies with polarography and related electrochemical techniques where overlap of reduction potentials can be avoided only by preliminary separation steps. Polarography is in any case an under-utilised procedure as discussed earlier (Bersier, 1976). The big advantage of atomic spectroscopy is that the metals can be separated by means of their different atomic spectral characteristics. This makes it unnecessary to carry out preliminary separations and thus avoids any attendant errors. Atomic spectroscopy also has the necessary sensitivity to provide very low limits of detection. The various attributes of atomic spectroscopy will therefore be explained in greater depth, including theory and methodologies.

2.3   Introductory theory of atomic spectroscopy

The term 'atomic spectroscopy' includes several related techniques of elemental identification and, provided appropriate standards are used, quantification. The most common forms of atomic spectroscopy are atomic absorption (AAS), atomic emission (AES), atomic fluorescence (AFS) and, more recently, inductively coupled plasma emission (ICP). Of these, AAS, AES and ICP (which is a type of AES) are most relevant to metal analysis in edible oil samples. An overview of the theory, concentrating on these techniques, is presented below as an introduction to the subsequent part of the chapter dealing with the methods and practical applications.

2.3.1   Energy of electronic transitions

The wave–particle duality concept shows that, depending upon how electromagnetic radiation is observed, it may exhibit both wave and particle properties. Einstein postulated that the energy of radiation is quantized and made up of packets of energy that became known as photons. The energy of these photons depends upon their frequency. Planck's relationship describes the energy of a photon $E = h\nu$, where E is the energy of the photon in joules, h is Planck's constant and equal to 6.626×10^{-34} Js, and ν is the frequency of the photon. Depending upon the energy of incident photons, the atom or molecule will be excited in different ways. For example, ultraviolet and visible light will result in electronic changes, i.e. electronic or atomic spectroscopy; infrared radiation will result in changes in the vibrational spectroscopy of molecules; and microwave radiation will result in changes in the rotational spectroscopy of molecules. Of interest here is electronic spectroscopy of atoms and more particularly of atomised metals, i.e. metals in the vapour form, achieved by heating the sample in a furnace to a temperature of the order of 2000°C. Since electrons within atoms absorb or emit photons at energies equal to the energy difference between the atomic energy levels, the energy of a photon absorbed by an atom during an electronic transition is comparable with the energy difference between these atomic energy levels. This is demonstrated by rearranging Planck's relationship using $\nu = c/\lambda$, where c is the speed of light in a vacuum. Planck's relationship can then include wavelength of light, λ, giving, $E = hc/\lambda$ or $E \propto 1/\lambda$. The energy of the electronic transition is therefore inversely proportional to the wavelength of light absorbed or emitted. As the series of related energies required for the electronic transitions that occur between the ground and excited electronic states in each atom are specific to that atom type, so too are the series of spectral lines of elements. This in turn shows that each element will absorb or emit light at specific wavelengths that will be related to those energy level transitions between ground and excited states. The transitions may also occur between different excited states. In general, this provides much better selectivity than with the traditional techniques discussed earlier, and as it is normally possible to find a spectral line that does not overlap that of any other element, complicated separation techniques are not needed. Fig. 2.2 shows in very basic form different electronic transitions

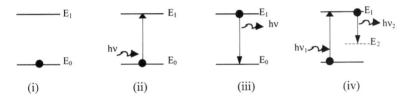

Figure 2.2 Simplified depiction of electronic energy levels.

resulting in absorption, emission and fluorescence. If a spectrometer is 'tuned' to a single wavelength, related to a spectral line of the analyte element then light absorbed at that wavelength should indicate the presence of that element. However, while this is in principle straightforward, care should be taken to ensure that there is no interference from another atom at that specific wavelength. This is because, although the series of electronic transitions may be unique for each element, there is the possibility that an individual energy of transition—and therefore wavelength—may be similar for more than one atomic species. In this case it is necessary to assess whether the interfering atom may be present in the sample. If this is the case, then an alternative energy or wavelength may need to be selected.

In Fig. 2.2, (i) shows two energy levels in an atom; an electron occupies the lower energy level (E_0) and the higher energy level (E_1) is unoccupied in the ground state. Diagram (ii) shows atomic absorption; a photon interacts with the ground-state atom with an electron in energy level E_0 and, because the energy of the photon, $h\nu$, is the same as the energy difference between the two energy levels (i.e. $E_1 - E_0 = h\nu$), the photon is absorbed by the atom and the electron is promoted to the higher level, E_1. Diagram (iii) shows atomic emission; the atom is already in an excited state and, by emitting a photon of energy $h\nu$ that is again the same as the energy difference between the two energy levels, the electron reverts to level E_0 and, in this simplified case, the atom is restored to its ground state. In (iv) is shown a very simplified case of atomic fluorescence; a photon of energy $h\nu_1$ interacts with the ground-state atom with an electron in the energy level E_0 and, as with atomic absorption, the photon is absorbed and the electron is promoted to the higher level, E_1. In atomic fluorescence, the electron almost instantaneously reverts to a lower energy level, emitting a photon of energy $h\nu_2$, producing fluorescence. Here, the initial absorption is of greater energy than the subsequent fluorescence, i.e. $h\nu_1 > h\nu_2$.

The energy differences between electronic levels in atoms are therefore characteristics of the individual atoms. While there may be some occasional overlap here, the energy difference, or the energy (and therefore wavelength) of the photon to be absorbed, is selected in the atomic spectrometer program to identify the specific metal analyte in question. However, it is usual that, as well as identification, quantification is also required. This can be achieved by applying the Beer–Lambert law, which states that $A = \varepsilon c l$, where A is the absorbance at the energy (or wavelength) of the determination, ε is the absorption coefficient, formerly known as the extinction coefficient, with units of $(\text{concentration} \times \text{length})^{-1}$, c is the concentration and l is the path length of light through the sample. Since ε and l are effectively constant at a single wavelength, the concentration can be determined directly from the measured absorbance. This is done by comparing absorbances of samples with the absorbances of standard solutions of known concentration of the analyte. Care should of course be taken to ensure that analyte concentrations in the samples and the standards are within the linear range for the calibration (see section on calibration and quality control below).

2.3.2   Absorbance and transmittance

The term absorbance is often used in spectroscopy when discussing electronic transitions and excited states. However this may also be expressed as transmittance. For example, a light source has an intensity, I_0, and the relevant detector detects light at an intensity I_1 (see Fig. 2.3). The amount of absorbed light can easily be seen to be $I_0 - I_1$. The transmittance, T, is defined as I_1/I_0 and the percentage transmittance, $\%T$, is defined as $100 \times T$. The relationship between transmittance and absorbance is $A = \log_{10}(I_0/I_1)$. Of course, modern spectrometers are usually highly sophisticated and so all these calculations will be carried out automatically. However, it is often important to investigate the raw absorption or transmittance data that are obtained in order to ensure that the correct and expected intensities are being observed, based upon previous results with standard materials and reference solutions.

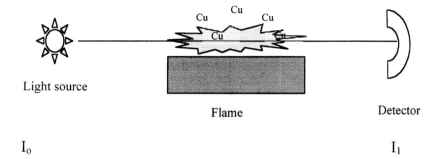

Figure 2.3 Light path of a flame atomic absorption spectrometer.

Taking a real example, we have a vegetable oil sample and we want to determine its copper content by atomic absorption spectroscopy. We suspect that the level is approximately 100 ppb, but we want to confirm the exact level as close as possible. At these concentrations atomic absorption with a flame is insufficiently sensitive and so a stabilised temperature platform furnace (STPF) is used. Later in the chapter, a method for the determination of copper at 100 ppb and below is described. We install the copper hollow-cathode lamp into the instrument and key in the correct wavelength that is associated with the main copper electronic energy transition, at 324.7 nm. We also ensure that the slit width is correct. After analysing a number of copper standard solutions, usually three plus a blank, we analyse the sample and then compare the peak area of the sample with peak areas of standards using a calibration curve to obtain the copper concentration in the sample.

2.3.3   Peak broadening

One effect that may be observed is that the peaks may not always be very narrow and well defined. This effect is known as broadening, and by far the most common cause of peak broadening is called Doppler broadening. This very common effect is well known with sound waves: when the wavelength of the sound of an approaching fire engine siren appears shorter than it really is, and, when it has passed and is travelling away from the observer, it appears to have a longer wavelength. The same is observed with atoms in atomic spectroscopy. If an atom is approaching the detector when a photon is absorbed, then the light wave will appear to have a shorter wavelength. If, however, the atom is moving away from the detector when the photon is absorbed, then the wavelength of the absorbed light will appear longer. The resultant is that the peak will appear broader than may be expected. This can be expressed in frequency terms as follows (equation 2.1):

$$v_1 = v\left(1 - \frac{v_1}{c}\right)^{-1} \qquad (2.1)$$

where v is the actual frequency of the transition (in this case absorption), v_1 is the frequency at which the absorption is observed to occur, v_1 is the velocity of the atom travelling towards the detector in the spectrometer, and c is the speed of light. Then, by applying the Maxwell distribution, the resultant range of values for v_1 produces line broadening described by equation 2.2 and the resulting peak is Gaussian-shaped:

$$\Delta v = \frac{v}{c}\left(\frac{2kT\ln 2}{m}\right)^{1/2} \qquad (2.2)$$

While the velocity is expressed relative to the speed of light, the ratio is still significant at the temperature (typically 2000°C) needed to vaporise/atomise the metal atoms in question.

Another cause of line broadening may be pressure broadening, where exchange of energy between atoms through collisions results in broadening.

It is possible to reduce or remove the effects of Doppler and pressure broadening, but there is another form of broadening that cannot be removed, and that is the natural line broadening. This is due to the fact that, while the absorption of a photon is very fast, the lifetime of the resultant excited state is still sufficiently long to ensure that the effects of the Heisenberg uncertainty principle are relevant. This means that it is not possible to know exactly the lifetime and the energy of the excited state, which results in the slight uncertainty with respect to the energy of the transition, and therefore the so-called natural line broadening.

For the interested reader there are many texts that go into the subject of atomic absorption spectroscopy theory in considerably greater detail than is required here. These texts concentrate on the fundamental theory, including selection rules, angular momentum quantum numbers, electronic transitions, partial energy level diagrams, etc. However, here we will concentrate on the two important types of atomic spectroscopy—absorption and emission—for the purpose of trace metal analysis in edible oils. This will include the conventional absorption and emission spectrometers, as well as that for ICP, which is an AES-type instrument.

2.3.4   Atomic absorption spectroscopy (AAS)

In many cases a hollow-cathode tube light source containing the analyte metal is used to generate light at the specific wavelengths. It is also possible to use multi-element lamps. These enable a number of metals to be analysed using the same lamp, although the required wavelengths and slit widths need to be selected for each analyte element. This means that a sample must be analysed sequentially for each metal. As we will see later, this is beginning to be superseded by the inductively coupled plasma emission spectrometer (ICP), which can analyse for several metals simultaneously given the correct instrumental settings.

As described above, atomic absorption relates to the electromagnetic light that is absorbed when an electron in an atom is excited from its ground state energy level to a higher energy level. In practice there are a few (normally only two or sometimes three) intense wavelengths of light that are concentrated upon in atomic spectroscopy experiments for each element. Light at these specific wavelengths is generated from the appropriate hollow-cathode tube sources. During the analysis procedure, the sample undergoes a staged heating programme, ultimately reaching its atomisation temperature, typically in the order of 2000°C. The steps of this heating programme will be described in greater detail later in this chapter. The volatile elemental atoms then enter the light path of the instrument and absorb the light at the specific wavelength, resulting in excited-state atoms. It is this absorption of light that is measured and used to quantify the presence of the specific metal analyte.

2.3.5 Atomic emission spectroscopy (AES)

Just as an element will absorb light at the wavelength corresponding to the electronic transition as defined by Planck's relationship, so will light be emitted at those wavelengths when an atom returns to a lower energy state from a higher one. This is the basis of atomic emission spectroscopy (see Fig. 2.2). The high temperatures from the flame, furnace or inductively coupled plasma provide sufficient energy to excite atoms into higher energy levels. The atoms subsequently decay back to the ground state and, in doing so, emit light. As has already been described above, these transitions are between distinct energy levels and therefore the resulting emission lines are narrow, despite the broadening influences mentioned earlier. If there are several elements present in the analyte, the resulting spectra can be crowded and a high-resolution spectrometer is required to resolve the individual spectra and thereby identify all the atom types present. This is in fact the advantage of AES over AAS: that because all atoms are excited simultaneously, they may also be detected and identified simultaneously.

2.3.6 Inductively coupled plasma emission spectrometry (ICP)

The inductively coupled plasma emission system is a relatively new addition to the range of atomic spectrometers. However, because of its sensitivity (particularly the hyphenated technique, ICP-MS) and ability to carry out rapid simultaneous multi-element determinations, it is coming to be seen as a necessary tool in the analytical laboratory. The principle of the emission is the same as with conventional AES, although the atomisation is carried out at a considerably higher temperature, typically 7000°C. Here the sample is nebulised and carried in the flow of plasma gas, usually argon. The plasma torch consists of concentric quartz tubes, with the inner tube containing the sample aerosol and argon, and the outer tube containing a flowing gas for cooling purposes. A radio frequency generator, usually 1–5 kW at 27 MHz, produces an oscillating current in an induction coil around the tubes. This creates an oscillating magnetic field, which in turn produces an oscillating current in the ions and electrons of the argon gas plasma. It is this plasma that acts as the excitation source for the analyte atoms. The main disadvantage of ICP is also the reason for its success: the high temperature of the plasma. At these temperatures there is a high possibility that other electronic transitions, or interferences, will also be excited and thereby included in the resulting emission spectra. The spectrometer therefore requires highly sophisticated hardware and software to eliminate these potential interferences. Because of this, the cost of these instruments can also be significantly higher than that of a basic AAS or conventional AES instrument.

A full description of the theory of AA, AES and ICP is given in a comprehensive review by Jackson and Lu (1998) and in reports of the Royal Society of Chemistry Analytical Methods Committee (1998).

2.4 Methods and procedures for the determination of metals in oils

2.4.1 Atomic absorption/emission spectroscopy

There are two methods published by the British Standards Institution in BS 684 Methods for determining heavy metals in fats and oils by graphite furnace atomic absorption spectrophotometry:

- BS 684: Section 2.18:1995 (ISO 8294: 1994). Determination of copper, iron and nickel contents, and
- BS 684: Section 2.21: 1995 (ISO 12193: 1994). Determination of lead content.

Of the metals determined by these methods, it is usually the presence of copper and iron that is considered important because of their oxidative nature when present above certain concentrations in the oils (see

Table 2.1). For the determination in oil of the bioaccumulating toxic metal cadmium, there is a draft IUPAC method that is a good basis for analysis (IUPAC, 1996, private communication).

Other standard methods for the determination of metals in oils are not available; however, there are many non-standardised methods for determining these metals in foods. These methods generally specify some form of wet or dry ashing of the sample that is not always necessary for the determination of metals in oils. The only common exceptions to this are for mercury and arsenic determinations. These metals are extremely volatile and so the sample must be wet acid-digested first as described earlier in the section on traditional analyses. There are many methods published for the total determination of mercury and arsenic in food, but if routine cheap analyses are to be provided then a suitable method is the Digestion of Food Samples for Total Mercury Determination (Hon Way *et al.*, 1985).

It is important to note that mineral oils are also routinely analysed using ICP in industry, although the full benefits that may be gained from the use of ICP have not yet been fully realised by the food industry.

2.4.2   Graphite furnace absorption spectroscopy

2.4.2.1   Sampling and sample preparation
Oil and fat samples will be in one of two forms, either a solid fat or a viscous oil. Fats tend to have a low melting point and raising the temperature to 60°C will melt most fats and allow subsequent mixing by agitation or shaking.

The laboratory sample to be analysed should be derived from the bulk material by a suitable means of sub-sampling. The sub-sample must be representative of the whole, and in extreme circumstances it may be necessary to take several sub-samples and report the result as an average.

Most oils and fats require many different analyses to be carried out in order to characterise them fully. The results for some of these tests, such as the colour and peroxide value of an oil, may change as a result of oxidation if the sample is heated. Therefore, the temperature of the samples should be raised with caution or the sample divided so that heat-sensitive tests use an alternative sub-sample.

For graphite furnace analyses, oil samples generally present no problem and can be delivered directly into the furnace tube by way of an automated pipette, usually an integrated autosampler. If the sample is a solid fat, then heating to 60°C is necessary, which is achieved either by heating in an oven and removing directly before analysis or using a heated autosampler.

Alternatively, flame AA and ICP analyses, where the sample is aspirated into the atomisation chamber, generally require the sample to have a low viscosity. In this case the sample is mixed with a known amount of solvent, such as hexane or cyclohexane. If flame AA is being carried out, it may be preferred to use a less flammable solvent such as octan-2-ol.

2.4.2.2 Calibration and quality control

It is important with all trace metal determinations to use volumetric flasks and other glassware that are scrupulously clean. If the glassware has previously been used for oils or metal solutions it should be washed using hot water and detergent to remove the oil and rinsed with distilled and deionised water. All glassware should then be either soaked preferably in 10% (v/v) concentrated nitric acid for several hours or alternatively rinsed with hydrochloric acid–water (1:1) to remove any metal ions from the surface of the glassware. Finally, the glassware should be rinsed thoroughly with distilled and deionised water to remove the acid solution and then dried in an oven before the oil samples are added.

The metals present in the samples are in an organic matrix, making aqueous standards unsuitable. Organometallic standards are available from many sources, e.g. Alpha Products or Johnson Mathey. These organometallic standards are carried in an organic solvent (e.g. xylene) that readily dissolves in oils. Standards should be prepared freshly from stock solutions because at low concentrations the stability of the metal ion solutions is poor owing to adsorption onto the surfaces of their container. If solutions are to be kept for periods of time, they should ideally be transferred to disposable plastic containers. In any case, the stock solutions should be made fresh every few days, or when it is noticed that the expected absorption for the standards is not being achieved.

Blank oils are needed for the preparation of standards and dilution of samples, where necessary. Refined vegetable oils are usually adequate and can be purchased from retail outlets. Some applications may be more specific and a blank oil of a specific variety may be required. The metal content of the retail oil sample should be checked and, if the absorption level is unacceptable, indicating a significant quantity of analyte metal, then purification through an alumina column will remove most trace impurities from the oil (BS method). The amount of metal, if any, in the blank is determined by using previous data of the specific absorption of a known mass of the metal under investigation.

Calibration of an atomic absorption spectrometer is carried out using external calibration for AA (if emission is being studied it is possible to calibrate by iterative methods). Three or four concentrations of a metal are prepared using the organometallic standard and the blank oil. The samples are analysed using the appropriate method for each metal. It is important to note that if too concentrated a standard is used then the limit of the spectroscopic technique will be exceeded. The concentrations selected should produce a linear calibration curve, but polynomial calibrations are not uncommon and most computer software packages are able to cope with these. Generally, for best accuracy, it is advisable to know approximately the result you expect and calibrate over as small a range as possible, above and below the expected concentrations.

A calibration is only valid for the life of the current graphite furnace tube. The tube will deteriorate over time and the expected decrease in absorbance means that regular quality control including replicate samples and regular reanalysis of standards is needed, with recalibration if necessary. This will establish when the tube should be replaced and ensure that the results are always valid. The next tube and calibration may be slightly different from the previous one, but as data are compiled statistics on the calibration curve can be collected; if the quality control measurements do not meet a set criterion, a new tube should be installed and recalibrated. The samples run immediately prior to this recalibration should then be repeated. The frequency with which the tube needs to be replaced is dependent on the metal analysed and the severity of the technique employed. If the temperature is taken to the limits of graphite furnace work (2700°C) in order to atomise the sample, the tube will not last as long as if the temperature is raised to only 2200°C.

2.4.2.3 Sample introduction

Normally for flame AA and ICP the use of solvents and dilution of the samples increases the analysis set-up time. The use of a graphite furnace with autosampler (Fig. 2.4) can reduce the required staff time and also the hazards caused by flammable solvents, and also only a small amount of sample is required (typically 10 μl per determination). The biggest drawback when using graphite furnace for atomisation is the volatility of metals at high temperatures and the subsequent loss of sample. The volatility of the metals can be reduced by the introduction of matrix modifiers. A typical graphite furnace instrument with autosampler is shown in Fig. 2.4.

2.4.2.4 Graphite tubes

A note of caution is necessary when analysing oil samples. Although these are fairly viscous at room temperature, the viscosity is dramatically reduced when the sample is heated only gently. When high temperatures are employed to remove carbon during pre-treatment, the sample can migrate to the ends of the graphite tube and contaminate the quartz lens windows. To overcome this problem L'Vov platforms are always used for analysis of oils at the Leatherhead laboratory. The sample is injected onto a graphite platform that has a shallow reservoir moulded into it. The

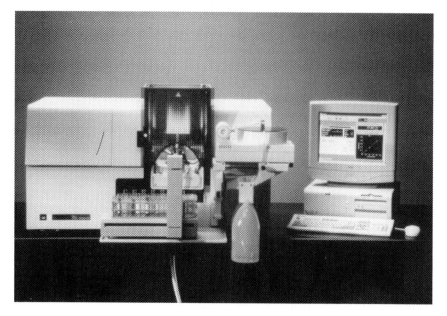

Figure 2.4 Typical graphite furnace with autosampler.

sample then remains in the reservoir as the heating takes place. When using a L'Vov platform it is necessary to alter standard programmes because the heat at the centre of the L'Vov platform is generally lower than when the sample is conventionally heated on the walls of the tube. Also, the heating process takes longer, so programme times may need to be increased according to the analysis. If the sample is not delivered correctly onto the platform, the sample may not be pretreated properly which can result in smoke production during atomisation. This invalidates the result by artificially increasing the light absorption.

If the oil or fat being heated has a high water content, spitting can occur as soon as the sample is heated above the boiling point of water. This results in loss of sample and contamination of the quartz lens. This necessitates disassembly of the atomisation chamber, which must be thoroughly cleaned before the next sample is analysed. It is not always apparent how much water is present in the samples and so a drying step should be routinely included. A drying step of about 30–60 seconds at 120°C should suffice.

As discussed above, when analysing samples by graphite furnace AA, a pretreatment temperature of about 700°C is necessary. However, at these high pretreatment temperatures some metals are very volatile, which results in loss of material and under-estimation of the metal content.

Examples of such metals are arsenic, mercury, cadmium and lead. In the case of cadmium and lead, matrix modifiers can be employed to prevent loss due to volatilisation. This is not possible with arsenic and mercury and they must be wet-acid digested, as discussed in the section on traditional analyses, before analysis by either flame AA or ICP.

For copper, iron and nickel determinations, there is no need for a matrix modifier because the boiling points of these metals are high and so they do not volatilise to a significant extent at 700°C. The samples are injected directly from an autosampler into the graphite furnace tube, preheated to remove carbon and then atomised. One problem associated with iron is that it may be adsorbed onto the surface of the graphite furnace tube and so be atomised at the same time as the iron present in the sample. Also, if the sample contains a reasonable amount of iron, this too may become adsorbed onto the surface of the tube and contaminate the next sample. To remove iron from the tube's surface, niobium is injected several times and atomised; this coats the tube and helps prevent iron binding to the surface. Even with the niobium treatment, if the level of iron measured is higher than around 500–1000 ppb, then it is very likely that there may be carry over to the next sample. If these levels are expected, it is best to include a blank air sample (i.e. nothing injected) between real samples to allow all iron to be removed between samples. While this may be a point to consider for most metals, it seems to be a particular problem for iron.

When determining lead, lecithin can be used as a matrix modifier. The sample is mixed with an equal quantity of 2% (w/v) lecithin in cyclohexane in a glass vial. The standards are prepared in a similar way and this allows the exact mass of sample to be known. Checks on the validity of the method can be made to ensure that an adequate amount of matrix modifier is being employed.

It might be thought that, because lead and cadmium are in the same group of the periodic table, lecithin may also be employed as the matrix modifier for cadmium. However, studies have shown that this is not the case but that platinum(II) chloride can be used instead. In the referenced method for cadmium determination, it is recommended that the matrix modifier should be prepared in an aqueous solution. This is not always practical for the analysis because of the nature of the sample. Thorough drying of the matrix modifier is necessary before the addition of the oil sample, which slows analysis time, and, if the aqueous phase is not dried efficiently, spitting problems will be apparent, with loss of sample. Also, some autosamplers are not able to cope with multiple-tasking for individual samples. As an alternative, the platinum(II) chloride can easily be dissolved in methanol at the same concentrations mentioned in the referenced method. The methanol can then be added at the same time as the sample and dried in the usual manner. This prevents spitting and shortens analysis time, and can be a big help if many samples are to be analysed. It is recommended to pretreat the tube once with platinum(II) chloride in a blank oil to maintain a constant platinum level for all determinations.

2.4.2.5 *Wet acid digestion for mercury and arsenic determinations*

The nature of these metals makes their determinations inherently difficult. They are volatile even at room temperature and so, once the temperature is raised above 100°C, most of the metal is lost. The traditional means of preparing these samples involves long digestion processes, refluxing the sample and scrubbing the gases produced through an acidic solution to trap the volatile metals. This, however, is not always suitable for rapid multiple determinations. One method in the literature proposes a digestion using a mixture of three acids: hydrochloric, nitric and sulfuric. The hydrochloric acid is added first, followed by the nitric acid, and this mixture is allowed to settle before the addition of the sulfuric acid. The presence of the two acids before the sulfuric is to allow any mercury or arsenic ions to form a stable complex. The sample is then heated for an hour on a steam bath before extraction and removal of the oil phase with hexane. The metal ions are then present in the aqueous phase and are made up to volume in a flask before being analysed with a hydride generator. The metal is reduced to its ground state hydride by the addition of either stannous chloride or sodium borohydride and the resulting gas is collected in a cell and emission spectroscopy is carried out. In general, sodium borohydride is used in preference to stannous chloride because the glassware requires less stringent cleaning between determinations. Stannous chloride is very persistent and is not easily removed by simple rinsing with distilled and deionised water.

2.4.3 *Inductively coupled plasma spectroscopy*

Although more sample may be necessary and the addition of solvent is required compared with simple graphite furnace techniques, ICP is potentially a very useful technique for the analytical chemist determining heavy metals in food. ICP allows multiple metal determinations to be carried out from a single sample aspiration after the instrument has been calibrated for all the metals of interest. This allows fast turnover of samples.

Other benefits of ICP include the fact that the technique is linear over a much wider calibration range and, because of the higher temperatures involved, there are fewer interferences experienced than with flame AA. ICP is therefore a very robust method for determining heavy metals and will be used more as the instrumentation improves and becomes more readily available to the analytical chemist through reducing costs.

2.5 Summary

The reasons for analysing for metals in oils include identifying the presence of oxidative catalysts; ensuring the absence of toxic metals; and for good manufacturing practice. The levels at which metals catalyse oxidation or become a safety problem are outlined in this chapter. In addition there is an introductory outline to the theory behind the atomic spectroscopic techniques. For the interested reader, there are numerous texts that concentrate on atomic spectroscopy in particular and spectroscopy in general. Finally, we have outlined a number of different methods that are appropriate for the analysis of the most common trace metals that may be present in edible oils. It should be noted here that while the standard reference methods are a useful starting point, they normally need to be modified and optimised for an individual laboratory's equipment. Once this is done and the method is validated and the appropriate quality measures are in place, the analyses may commence with confidence.

References

Alexander, O.R. and Taylor, L.V. (1944) Improved dithizone method for the determination of zinc in foods. *J. Assoc. Off. Anal. Chem.*, **27** 325-331.

AOAC (1990) *Official methods of analysis of the Association of Official Analytical Chemists*, 15th edn, Association of Official Analytical Chemists, Arlington, VA.

AOCS (1996) *Official methods and recommended practices of the American Oil Chemists Society*, 4th edn, AOCS, Champaign, IL. Method Ca 12-55(93).

Bersier, P.M. (1987) Do polarography and voltammetry deserve wider recognition in official and recommended methods? *Anal. Proc.*, **24** 44-49.

BSI (1976) British Standard 684, sections 2.16 and 2.17 "Determination of copper and iron—colorimetric method". British Standards Institution, London.

BS 684: Section 2.18:1995 (ISO 8294: 1994) Determination of Copper, Iron and Nickel.

BS 684: Section 2.21: 1995 (ISO 12193: 1994) Determination of Lead Content.

Capar, S.G. (1987) General referee reports Committee on Residues. *J. Ass. Off. Anal. Chem.*, **70** 295-296.

Clifford, P.A. and Wichmann, H.J. (1936) Dithizone methods for the determination of lead. *J. Assoc. Off. Anal. Chem.*, **19** 30-156.

Cooke, M.V., Shiers,V.P., Farmer, M.R. and Rossell, J.B. (1996) *Colorimetric Determination of Phosphorus in Oils*, Leatherhead Food RA, Research Report No. 735.

FAO, WHO (Food and Agriculture Organization/World Health Organization of the United Nations) (1993) *Codex Alimentarius Specifications on Fats Oils and Related Products*, vol. 8, FAO/WHO Food Standards Programme, FAO, Rome.

Gajan, R.J., Capar, S.G., Subjoc, C.A. and Sanders, M. (1982) Determination of lead and cadmium in foods by anodic stripping voltammetry, I. Development of method. *J. Ass. Off. Anal. Chem.*, **65** 970-977.

Gordon, M.H. (1990) The mechanism of antioxidant action in vitro, in *Food Antioxidants* (ed B.J.F. Hudson), Elsevier Applied Science, London, pp 1-18.

Hon Way, L., Go, D., Fedczina, J. and Dallins, J. (1985) Digestion of food samples for total mercury determination. *J. Ass. Off. Anal. Chem.*, **68** (5) 891-893.

Hui, Y.H. (1996) Refining and bleaching, in *Bailey's Industrial Oil and Fat Products*, 5th edn, (ed Y.H. Hui), Wiley-Interscience, New York, p 164.

IUPAC (1987) *Standard Methods for the Analysis of Oils, Fats and Derivatives*, 7th revised and enlarged edn, The International Union of Pure and Applied Chemistry.

Jackson, K.W. and Lu, S. (1998) Atomic absorption, atomic emission, and flame emission spectrometry. *Anal. Chem.*, **70** 363R-383R.

Klein, A.K. (1952) Report on mercury. *J. Ass. Off. Anal. Chem.*, **35** 537-542.

List, G.R. and Erickson, D.R. (1980) Storage, handling and stabilisation, in *Handbook of Soy Oil Processing and Utilisation* (eds. D.R. Erickson, E.H. Pryde, O.L. Brekke, T.L. Mounts and R.A. Falb), American Soybean Association, St Louis, MO, pp 267-353.

McGowan, R. and Petch, J. (1993) *Victorian Produce Monitoring—Results of Residue Testing 1992/93*, Department of Food and Agriculture, Victoria, Australia. ISSN 1036-1227.

Mitchell, J.H. (1921) Report on inorganic plant constituents—Manganese. *J. Assoc. Off. Anal. Chem.*, **4** 393-394.

Patterson, H.B.W. (1992) *Bleaching and Purifying Oils and Fats—Theory and Practice*, American Oil Chemists Society, Champaign, IL.

Pomeranz, Y. and Meloan, C.E. (1982) Voltammetry (polarography), in *Food Analysis Theory and Practice*, revised edn, AVI, Westport, CT, p 205.

Ramsey, L.I. (1960) Report on metals, other elements and residues in foods. *J. Ass. Off. Anal. Chem.*, **43** 695-700.

Royal Society of Chemistry—Analytical Methods Committee (1998) Evaluation of analytical instrumentation Parts I & II. Atomic absorption spectrophotometers. *Analyst*, **123** 1407-1414, 1415-1423.

Sakamoto, M., Nakano, A. and Akagi, H. (1996) Evaluation of the significance of methyl mercury contamination in red blood as indicator of human fish consumption. *Foods and Food Ingredient Journal of Japan*, **167** 109-115.

Shirley, R.L., Benne, E.J. and Miller, E.J. (1949) Cadmium in biological materials. *Anal. Chem.*, **21** 300-303.

Sonntag, N.V.O. (1982) Reactions of Fats and Fatty Acids in *Bailey's Industrial Oil and Fat Products*, vol. 1, 4th edn, (ed. D. Swern), Wiley, New York, p 152.

Swern, D. (ed.) (1982) *Bailey's Industrial Oil and Fat Products*, vol. 1, 4th edn, Wiley, New York.

Taylor, L.V. and Alexander, O.R. (1945) Report on zinc. *J. Assoc. Off. Anal. Chem.*, **28** 271-277.

WHO (1996) *Trace Elements in Human Nutrition and Health*, World Health Organisation, Geneva.

3 Lipid chemiluminescence

R. Alan Wheatley

Although there are no known chemiluminescent reactions that are specific to lipids, progress has occurred by the ingenious application of existing — usually well established — chemiluminescence phenomena to the solution of problems in lipid analysis. Some areas are well developed. The scope of the chapter includes lipids derived from mevalonate; the quest for steroid determinations has frequently been successful and includes interesting assays based on a reaction between an inorganic oxidant and an inorganic reductant, not a superficially promising scenario for lipid analysis. Assays for lipid peroxidation products are also well represented. On the other hand, the reluctance of triacylglycerols to cause the emission of chemiluminescent signals is fortunate, given their ubiquity in biological samples. The applications described here are set out by analyte category in Table 3.1, which, by indicating the section(s) in which each method is covered, gives a more direct route to the specific items of information in the chapter.

It is the objective of this chapter to provide an appreciation of the potential role of chemiluminescence in the study of lipids. Mindful that the reader may not have any previous acquaintance with chemiluminescence techniques, the fundamentals have been covered and space is allocated to description and explanation of the instrumentation. In addition, practical details are provided in selected cases of applications of the technique to lipid analysis.

3.1 What is chemiluminescence?

Luminescence is the emission of light due to transitions of electrons from molecular orbitals of higher energy to those of lower energy, usually the ground state or lowest unoccupied molecular orbitals. Such transitions are referred to as relaxations. Figure 3.1 shows four electronic energy levels S_0, S_1, S_2 and T_1 and the possible transitions between them. S_0 represents the ground state, while S_1, S_2 and T_1 represent higher-energy excited states; S_0, S_1 and S_2 are singlet states in which all the electrons form pairs of opposed spins whereas T_1 is a triplet excited state, in which not all electrons are paired off in that way. Each energy level is subdivided into a number of vibrational states, each characterised by an amount of vibrational energy that accompanies the potential energy of the electrons occupying the orbitals.

Table 3.1 Chemiluminescence assays dealt with in the text

Analyte	CL system	Section	Reference
Fatty acids	Luminol	3.4.3	Naslund et al. (1990)
	Peroxy–oxalate	3.6.4	Honda et al. (1985b)
	Peroxy–oxalate	3.6.4	Grayeski and DeVasto (1987)
Unsaturated fatty acids	Ultraweak CL	3.11	Das et al. (1997)
Prostaglandins	Luminol	3.4.4	Neuport et al. (1996)
Lipid hydroperoxides	Luminol	3.4.5	Wieland et al. (1990b)
	Luminol	3.4.5	Matthäus et al. (1994)
	Luminol	3.4.5	Pettersen (1994)
	Luminol	3.4.5	Yamamoto et al. (1987); Frei et al. (1988)
	Luminol electroCL	3.5	Sakura and Terao (1992a)
	Ultraweak CL	3.11	Singh et al. (1996a)
Peroxidation carbonyls	Luciferase	3.7	Wieland et al. (1990a)
	Peroxy–oxalate	3.6.5	Mann and Grayeski (1990)
	Dinitrophenylhydrazine	3.9	Townshend and Wheatley (1998c)
Antioxidants	Luminol	3.4.6	Metsa-Ketela (1990)
	Luminol	3.4.6	Whitehead et al. (1992); Maxwell et al. (1993)
	Luminol	3.4.6	Kricka et al. (1993)
	Luminol	3.4.6	Kim et al. (1993)
	Luciferase	3.7	Mashiko et al. (1993)
Choline phospholipids	Luminol	3.4.7	Bissé et al. (1990)
	Luciferase	3.7	Dukhovich et al. (1990)
Cholesterol	Peroxy–oxalate	3.6.6	Rigin (1978)
	Lucigenin	3.8	Sasamoto et al. (1995)
Bile acids	Luminol	3.4.8	Maeda et al. (1993)
	Luciferase	3.7	Lekhakula et al. (1991)
	Peroxy–oxalate	3.6.6	Imai et al. (1989); Higashidate et al. (1990)
	Sulfite	3.10	Psarellis et al. (1994)
Sex hormones	Peroxy–oxalate	3.6.6	Nozaki et al. (1988)
	Peroxy–oxalate	3.6.6	Imai et al. (1989); Higashidate et al. (1990)
	Sulfite	3.10	Syropoulos et al. (1990)
	Sulfite	3.10	Deftereos and Calokerinos (1994)
	Lucigenin	3.8	Sato et al. (1996)
Corticosteroids	Peroxy–oxalate	3.6.6	Imai et al. (1989); Higashidate et al. (1990)
	Peroxy–oxalate	3.6.6	Koziol et al. (1984)
	Sulfite	3.10	Koukli and Calokerinos (1990)
	Sulfite	3.10	Deftereos and Calokerinos (1994)
Lipoproteins	Luminol	3.4.5	Zimmermann et al. (1993)
	Luminol	3.4.9	O'Kane et al. (1992)
	Luminol	3.4.9	Kessler et al. (1994)
	Luminol	3.4.9	Smith et al. (1997)

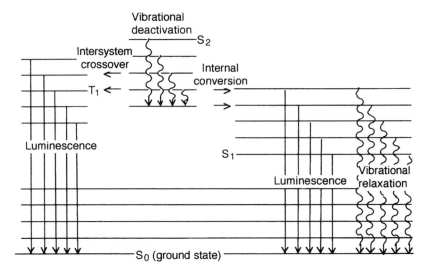

Figure 3.1 Jablonski diagram showing four electronic energy levels S_0, S_1, S_2 and T_1 with their vibrational fine structure and the transitions between them that affect luminescence.

Luminescence is classified according to the excited state that gives rise to it and to the source of the energy that caused the excited state to be populated with electrons. The promotion of electrons to an excited state is called excitation. In many cases, this is brought about by absorption of visible or ultraviolet radiation. In such a case, if the luminescence arises because electrons are relaxing from a singlet excited state to a singlet ground state then it is called *fluorescence*, and generally occurs within 10^{-11} to 10^{-5} s. The transition is very fast because it involves no reversal of electron spin. If, however, it arises due to relaxation from a triplet excited state, then the luminescence is called *phosphorescence*, which generally occurs within 10^{-4} to 100 s. If the excitation is the result of energy released in a chemical reaction, the luminescence is called *chemiluminescence*. A subset of chemiluminescence phenomena occurring in the biosphere as a result of enzyme-catalysed reactions are called *bioluminescence*. *Electrochemiluminescence* is another distinct subset of chemiluminescence phenomena, made up of those reactions in which the excited species is produced at an electrode during electrolysis.

Before luminescence occurs, there is a non-radiative loss of energy (due to collisions) as the excited state relaxes to a lower vibrational state while remaining at the same electronic energy level. This type of transition is called *vibrational deactivation*; it has to occur even more rapidly than fluorescence and typically occurs within 10^{-12} s of excitation. Therefore the luminescence involves the emission of photons of lower energy (higher wavelength) than would otherwise be the case. Another possible transition is *internal conversion*, in which an electron transfers from a lower vibrational state of a higher electronic energy level to a higher vibrational state of a lower electronic energy level, without any significant gain or loss of energy; such a transition, $S_2 \rightarrow S_1$, is shown schematically in Figure 3.1. In *intersystem crossing*, internal conversion would involve also reversal of the spin of the electron in a transition from a singlet to a triplet state, which would then give rise to phosphorescence; the transition $S_2 \rightarrow T_1$ in Figure 3.1 is of this type. Finally, luminescence is not inevitable. The energy of excitation can be dissipated as heat by interaction with the solvent; this is referred to as *vibrational relaxation*. Solvents with widely spaced vibrational levels can accept large quanta of energy and therefore, by promoting vibrational relaxation, extinguish the luminescence; this is known as *quenching*.

The strength of the emission compared with the number of molecules in the excited state is called the *quantum yield* (Φ_F). The quantum yield of fluorescent emission can be calculated as shown in equation (3.1) and while the quantity Φ_F should be the same in a chemiluminescence phenomenon as in the fluorescence phenomenon involving the same excited state, it can only be calculated in the same way by conducting a separate fluorescence experiment, for chemiluminescence does not depend on the absorption of photons.

$$\Phi_F = \frac{\text{No. of photons emitted}}{\text{No. of photons absorbed}} \tag{3.1}$$

The strength of the emission compared with the number of reactant molecules is called the *chemiluminescence quantum yield* (Φ_{CL}).

$$R \xrightarrow{\;\Phi_C \cdot \Phi_E\;} P^* \xrightarrow{\;\Phi_F\;} P + h\nu \tag{3.2}$$

$$\Phi_{CL} = \Phi_C \cdot \Phi_E \cdot \Phi_F \tag{3.3}$$

Equation (3.2) shows a reaction in which Φ_C is the proportion of reactant molecules (R) that are converted to product (P) and Φ_E is the proportion of product molecules that are formed in the excited state (P*). Equation (3.3) shows that Φ_{CL} is a product of three yields, one of which is Φ_F. Φ_{CL} varies from 10^{-8} for Cl–MgBr to 0.88 for firefly luciferin. Φ_{CL} for luminol is restricted by a small Φ_F for the phthalate dianion product.

Because Φ_{CL} depends on Φ_F, it would be reasonable to suppose that chemiluminescence would be affected by substitution in product molecules in the same way as is fluorescence. In that case, electron donors, e.g. —OH, —OMe, —NH$_2$, —CN, would increase Φ_F and electron acceptors, e.g. —NO$_2$, —SO$_3$H, —Cl, —COOH, would decrease Φ_F. Alkyl groups would have no tendency either to increase or to decrease luminescence. Φ_F is increased by rigid planarity of molecules (which facilitates π-bond delocalisation) and by conjugation; conjugation also causes a bathochromic shift in the emission wavelength. Such generalisations must be used with great caution, for the only product species to which they can apply are the ones that give rise to the chemiluminescent emission; it is by no means obvious what this is in any particular case.

3.2 Chemiluminescence literature

3.2.1 General considerations

Early reviews of the chemiluminescence of organic compounds by Gundermann (1965) and McCapra (1966) pointed out that this is generally weak ($\Phi \sim 0.05$, cf. $\Phi \sim 1$ for bioluminescence) and was detectable only because $\sim 10^3$ photons cm^{-3}s^{-1} could be measured; detection of single photons is now feasible. Reactions were identified as having enthalpy changes of between 160 and 300 kJ mol^{-1} (corresponding to emission at about 400–750 nm). In a more recent review, Robards and Worsfold (1992) defined minimum conditions for a visible chemiluminescent reaction as an exothermicity of at least 180 kJ mol^{-1} and a pathway to an excited state that relaxes with the loss of a photon (see Figure 3.1). Such reactions occur mainly in the chemistry of aromatic compounds, e.g. luminol, lucigenin, lophine, aryl oxalates, pyrogallol, phthalocyanins and most of the luciferins of bioluminescent organisms.

The emitter is regarded as a product molecule in an electronically excited state, and the concentration of such molecules will depend on the equilibrium between states and on the structural similarity between the emitter and the transition state of the reaction. Where it occurs, chemiluminescence is diminished by radiationless transfer, external quenching, absorption of emitted light or the transfer of energy to impurities.

Applications of chemiluminescence frequently depend on exploiting the spectrofluorimetric properties of substances (Psarellis et al., 1994). If the analyte is a fluorophore, chemi-excitation by energy transfer can be applied as in peroxy–oxalate chemiluminescence; or if the analyte is the reductant in an inorganic oxidation, then a fluorophore may be used as a sensitiser. Peroxy–oxalate, luminol and bacterial luciferase chemistry are the most versatile chemiluminescence systems, the latter attaining a detection limit of 10^{-21} mol in a determination of β-D-galactosidase

(Tanaka and Ishikawa, 1986). Consideration of these and other reaction systems that have proved applicable to lipids will constitute the bulk of this chapter. Major general mechanisms involved in chemiluminescent reactions are (a) peroxide decomposition to form carbonyl compounds, (b) electron transfer and (c) production of singlet oxygen, especially dimers. These will be discussed as they become relevant to our theme. Chemiluminescent reactions have such diverse applications as in acid–base indicators for titrations in opaque solutions and short-wavelength radiometry.

3.2.2   Other chemiluminescence reviews; a brief guide

There has been a recent review (Anderson et al., 1997) of clinical and biomedical analysis, citing 166 publications in the period 1 October 1994 to 1 October 1996 dealing with applications of chemiluminescence. An earlier review (Kricka, 1991) cites 109 sources and emphasises the sensitivity, speed, simplicity and safety of the techniques. A bibliography (Kricka and Stanley, 1996) covers the second half of the 1995 literature; throughout the decade, there has in the same journal been a succession of half-yearly bibliographies from these authors. Chinese work on electro-chemiluminescence over 30 years has been separately reviewed (Wang et al., 1991), with 20 references covering basic principles and analytical applications; developments and applications of chemiluminescence in China during the period 1976–88 are the subject of a later review (Geng, 1992) with 37 references. One brief review (Nakano, 1996) covering 12 references on photon emission and disease includes lipid peroxidation and another (Styrov et al., 1991) has six references on the principles of chemiluminescence and applications including the analysis of solid surfaces and determinations of gases in mixtures and in vacuo. A review (Yoshimura et al., 1992) with 57 references covers methods for determining lipid peroxidation and active oxygen species (superoxide and hydroxyl radicals, hydrogen peroxide and singlet oxygen).

The review by Robards and Worsfold (1992) has already been mentioned; it lists 249 references (19 of which are directly related to lipid chemiluminescence) and covers the main reaction systems and instrumentation as well as dealing with applications to a wide range of analytes. The perspective provided by Chandross (1994) is much briefer (24 references) and more selective, but the extension of the discussion to electrochemiluminescence will be of interest to readers of this chapter. Townshend's (1990) thorough review (99 references) gives particular emphasis to the important chemiluminescence systems based on luminol oxidation and on the peroxy–oxalate reaction.

3.3 Instrumentation

Emission intensity is proportional to the concentration of either an intermediate or a product in an electronically excited state and this depends on the rate of the reaction producing it. Analytical detection of chemiluminescence usually involves no wavelength selection and signals are heavily dependent on the mixing rate of reactants. Detection limits are determined mainly by background emission from contaminants and side products rather than instrumental limitations. Selectivity is achieved by on-line treatments rather than processing of the signal, which has little fine-structure (Robards and Worsfold, 1992).

3.3.1 Flow luminometry

Flow systems are commonly used. In these cases the signal depends on flow rate, tubing dimensions, order of reagent addition and flow-cell volume; immobilised reagents and stopped-flow techniques are sometimes useful. A suitable flow injection manifold is shown in Figure 3.2 and is essentially as described by Townshend (1990). The flow-cell, positioned immediately after the mixing point in the manifold, should be of relatively large volume so that a high proportion of the total emission enters the detector; optimisation will favour conditions that lead to emission occurring during the passage of the sample through the flow-cell. These techniques have been applied successfully to determinations of a wide variety of substances.

A flow luminometer (Figure 3.3) is described by Faizullah and Townshend (1985). It consists of a Teflon T-piece (to which Teflon tubing was joined by flanged screw-in connections), a flat coil flow-cell

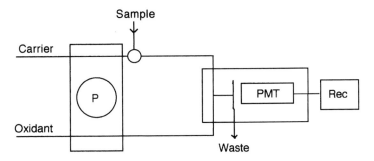

Figure 3.2 Flow injection manifold for measuring oxidative chemiluminescence. PMT, photomultiplier tube; Rec, chart recorder.

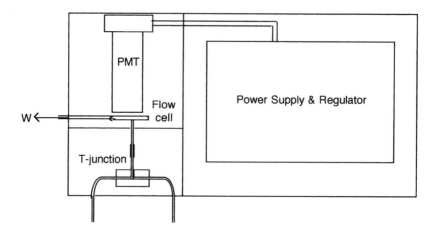

Figure 3.3 Internal layout of a flow luminometer. PMT, photomultiplier tube; W, waste. The diagram is not to scale.

(Figure 3.4) and a photomultiplier tube (EMI model 9844B). The luminometer is housed in a stainless steel box made light-tight by sealing all joints and holes with black adhesive tape. To prevent light-piping, the Teflon inlet and outlet tubing is sheathed with opaque material.

Earlier studies (Burguera and Townshend, 1979; Burguera et al., 1980; Wheatley, 1983) anticipated the design of the flow-cell. This is illustrated in Figure 3.4 and is made from a 30 cm length of 0.8 mm i.d. glass tubing, 25 cm of which is in the form of a flat coil, internal volume 125 μl, held close to the window of the photomultiplier tube and backed by a reflective aluminium plate (to enhance the delivery of emitted light to the detector). The coil has inlet and outlet tubes which are inserted into tight PVC connection sleeves to make unions with the Teflon manifold. The

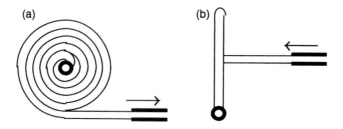

Figure 3.4   (a) Front elevation (looking along the direction of inflow) and (b) side elevation of the flat coil flow-cell. (Approximately × 2; the arrows show the direction of flow.)

flow-distance between the T-junction confluence and the entrance to the flat coil flow-cell is about 5 cm, representing a dead space of about 25 µl. A two-phase flow-cell has also been devised that permits analytes to be dialysed into the reagent with immediate measurement of emitted light (Mullin and Seitz, 1984).

The bialkali photocathode in the photomultiplier tube has maximum response between 300 nm and 500 nm and detection is facilitated by its large area (compared with the modest dimensions of the whole tube, 94 mm×52 mm o.d.). The luminometer also has a 1500 V integrated power supply and voltage regulator. Unless there is an inner sealed space for the photomultiplier and its connections, the *luminometer must be opened only in darkness and 30 min after disconnection from the mains electrical supply.*

Hemmi *et al.* (1995) have reported the use of a photodiode as a low-cost substitute for a photomultiplier for quality control in the food industry. Their chemiluminescence detector was constructed by assembling a silicon photodiode (Hamamatsu Photonics S1227-1010, photosensitivity 0.42 A/W at 720 nm) into a stainless steel body, 0.9 mm from a flow cell of ~78.5 µl capacity. The layout of the detector is shown in Figure 3.5. The photodiode was connected to a ±15 V regulated DC power supply and received light through a 10 mm diameter quartz window. The device was used to detect hydrogen peroxide resulting from the enzymic oxidation of L-lactate using immobilised L-lactate oxidase; results of very good precision gave a linear calibration from 1×10^{-7} mol dm^{-3} to 1×10^{-3} mol dm^{-3}, with a detection limit of 8×10^{-8} mol dm^{-3} and a sampling rate of 60 h^{-1}. There was a correlation coefficient of 0.992 between determinations using this detector and determinations using an established HPLC assay with an ultraviolet absorption detector.

A flow-cell suitable for construction by the experimenter that has been used for investigations using luminol electrochemiluminescence is

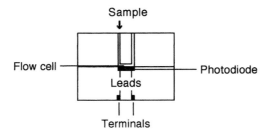

Figure 3.5 Schematic diagram of a photodiode-based chemiluminescence detector (Hemmi *et al.*, 1995).

described by Sakura and Imai (1988). The sample–reagent mixture passes into spiral coil of 0.6 mm i.d. transparent polythene, within which is inserted the platinum wire working electrode, 0.5 mm i.d. × 10 cm. The outlet from the spiral coil feeds a pool contained in a PTFE cell, 17 mm diameter, where the end of the electrode is connected to the source of the applied voltage. The auxiliary electrode (platinum wire, 0.5 mm i.d. × 10 cm) and the reference electrode (Ag/AgCl) are also immersed in the pool, from which the sample–carrier–reagent mixture passes to waste. The polythene spiral coil is situated next to a window of a photomultiplier, so that light measurement is continuous. The cell, coil and photomultiplier tube are contained within a PTFE sleeve, which should be light-tight. A submicrolitre flow detector has also been described (Arora et al., 1997); it has been used for tris(2,2'-bipyridyl)-ruthenium(II) electrochemiluminescence, but not so far for luminol electrochemiluminescence.

3.3.2 Batch luminometry

Typical of earlier work is Buergera and Townshend's (1979) use of a pulse technique in which the light-emitting reaction occurred when the reactants were directly pipetted into a glass cell in a light-tight compartment. More typical of modern batch luminometry is the adaptation and automation of the luminol chemiluminescence assay for the Dynatech ML 3000 microplate luminometer (Kricka et al., 1993). A 20 µl sample and 100 µl prepared signal reagent are multiple-pipetted within 2 min into each of the 96 wells. Time course of the chemilumines-cence from each of 96 wells can be displayed simultaneously. In a chemiluminescent antioxidant assay, the sample quenches the chemilu-minescence from luminol mixed with enhancer and horseradish peroxidase. The time until the emission is restored to 10% of its initial value is measured automatically. Also automated by the microtitre plate luminometer (Hastings et al., 1989) is the determination of ATP chemiluminescence at 80 fmol dm^{-3} detectivity, used to study microbial growth, bactericides, chemotaxis, sperm viability and antitumour drugs. Rapid chemiluminescence cell counts can be conducted at a detectivity for yeast cells of 8.8×10^4 (Kricka, 1993).

3.3.3 Microcomputer applications

A different design of luminometer (Matthäus et al., 1994) consists of a black box, within which samples are positioned by a step motor for luminol delivery or luminometry with a photomultiplier tube. A schematic diagram is shown in Figure 3.6. The luminol dispenser

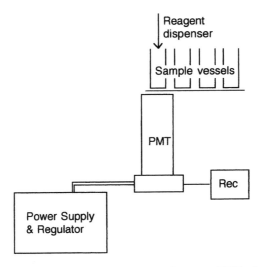

Figure 3.6 Internal layout of an automated batch luminometer. PMT, photomultiplier tube; Rec, data recorder. The diagram is not to scale.

consisted of a compressed-air-driven syringe fed from a reagent reservoir via an electronically controlled three-way valve. The sample vessels were held in a plate that was driven by a step motor to present each vessel in turn to the window of the photomultiplier. The data recorder was a computer interfaced via an analogue-to -digital converter; it was also programmed to output control signals to the syringe, the three-way valve and the step motor.

A computerised luminometer is described (Bochev *et al.*, 1992) that simultaneously performs integrations and/or spectral measurements of chemiluminescence from up to six samples, together with data processing and analysis of the shape and magnitude of the kinetic chemilumines-cence graphs (slopes, signal maxima and minima, area under curve, etc.). Within a light-tight chamber, six cuvettes in a rotating holder are maintained at a temperature ($\pm 0.1°C$) in the range 25–45°C. Filters, housed in a separate rotating disc beneath the cuvette holder, are used for wavelength selection; the integral signal is measured if no filter is used. In other respects, the internal layout is as shown in Figure 3.6. Light from each cuvette in sequence passes through the selected filter, immediately beneath which is placed the window of the detecting photomultiplier (EMI 9635 QB, 220–600 nm or EMI 9658 R, 300–700 nm). The photo-multiplier is fan-cooled to reduce the dark current. Signals pass from the photomultiplier, through an amplifier/amplitude discriminator to the microcomputer (Apple IIe); these operate as a photon-counting module.

The luminometer has been used to evaluate activated leukocyte metabolism by measuring the luminol chemiluminescence induced by the release of reactive oxygen species. The time-course of iron(II)-catalysed peroxidation in phospholipid liposomes has also been studied simultaneously at a range of wavelengths.

There is a report (Lin *et al.*, 1991) of a control and measuring device for electrochemiluminescence interfaced with an Apple II microcomputer, with programming in BASIC and in 6502 assembly language. The device carries out data acquisition and processing, controls scanning and rest times and performs automatic electrode washing.

Photon counting has been carried out (Scott, 1994) by using a vacuum isolated photomultiplier (160–850 nm) at low temperature for stable noise. The instrumentation allowed simultaneous temperature measurement as the sample is heated by converting a thermistor output voltage to frequency.

3.3.4   Charge-coupled device (CCD) applications

Vacuum tube devices, originally developed for astronomy, have been adapted for use in chemiluminescence and bioluminescence imaging (Ingle *et al.*, 1994). The main advantage of these photon-counting cameras lies in their ability to study time-dependent phenomena over a very wide range of light levels and to produce quantitative data on the time course of luminescence. Images are distinguished from noise by electronic amplification using microplate intensifiers held at high vacuum between two fibre-optic blocks. An image projected onto the front surface is transmitted to the intensifier by one of the blocks; the second transfers to a CCD camera the enhanced image from the phosphor screen, which converts to photons the electron avalanche produced in the intensification process. Each original photon event has, in the enhanced image, a Gaussian profile, which restricts the system resolution. Therefore, a high-performance microprocessor calculates the centroid of the pulse to relocate the event more precisely. Digital brightness control monitors each pixel and when in any frame this reaches a pre-set level the tube can be switched off in less than 1 μs while still maintaining an accurate stream of counting rate data to the host computer. This protects the intensifier from the very bright point sources of light that are typical of luminescence phenomena; it also affords protection against misuse of the device.

The Image Research bioluminescence image quantifier (BIQ) system consists of an image intensifier and a CCD detector (Mitchell, 1991). It detects single photons of energies corresponding to visible wavelengths, and displays an image of chemiluminescence or bioluminescence from an array of 250 000 points. It is useful for reading microtitre plates, gels and blots and can also be operated at TV frame rates for the observation of dynamic processes, e.g. changes in intracellular calcium.

There is a report (Mueller-Klieser *et al.*, 1991) of photon-counting video imaging using a charge-coupled device to give spatial resolution, e.g. of ATP/lactate distribution in human melanoma tissue. Metabolite distribution in tissue has been studied (Mueller-Klieser and Walenta, 1991) using a frozen layer and a viscous reagent solution to minimise diffusion.

Video enhanced contrast differential interference contrast microscopy has been used (Suzaki, 1995) simultaneously with fluorescence in a dual imaging system to observe the phagocytosis–exocytosis sequence in a single neutrophil. The relationship between exocytosis and transient cellular calcium levels has been explored. Superoxide radicals and other oxidants from stimulated neutrophils have been detected by chemiluminescence at a detectivity of 59 amol per count.

3.3.5 Instrumentation reviews

Ninety commercially available luminometers and imaging devices for chemiluminescence and bioluminescence have been surveyed (Stanley, 1992a); photographs of the instruments were published separately (Stanley, 1992b). A further survey covered 200 commercially available kits and reagents (Stanley, 1993). Stanley (1997) has also produced the latest (at the time of writing) in a series of regular updates of these surveys. There is a bibliography (Kricka and Stanley, 1992) with 113 references covering luminometry and other instrumentation for chemiluminescence and bioluminescence. A review (Ugarova, 1991) of Russian literature on chemiluminescence and bioluminescence includes published work on instrumentation. A short review in Japanese (Ueno, 1991) cites 10 papers in which photographic detection is reported, covering a variety of chemiluminescence systems.

3.4 Oxidation of luminol

One of the most widely used chemiluminescent systems involves the oxidation of luminol, 3-aminophthalhydrazide, by oxygen or hydrogen peroxide in the presence of base in aprotic solvents (e.g. dimethyl sulfoxide, dimethylformamide). The reaction also occurs in water or alcohols, but a supplementary oxidant (such as ferricyanide, hypochlorite or persulfate), or an enzyme (e.g. microperoxidase, horseradish peroxidase, cytochrome *c*), a metal ion, pulse radiolysis or sonic waves is needed (Roswell and White, 1978). 4-Aminophthalhydrazide (isoluminol), is also chemiluminescent on oxidation by oxygen or hydrogen peroxide.

A number of substances are reported to be enhancers of luminol chemiluminescence. Among these, 4-aminophenol has been widely used (Whitehead et al., 1992) to enhance the horseradish peroxidase-catalysed chemiluminescent oxidation of luminol or isoluminol by hydrogen peroxide. 4-Phenylboronic acid has been more recently reported (Kricka and Xiaoying, 1995) in the same role. This enables the enzyme to be determined at a detectivity of 76 amol.

Townshend (1990), dealing especially with solution chemiluminescence in flow injection analysis, has reported determinations of hydrogen peroxide, other oxidants and substances reacting with them, including several lipids. McCapra (1987) has reviewed potential applications in biosensors.

3.4.1 Mechanism

The postulated mechanism for the luminol reaction (Roswell and White, 1978) proceeds in aprotic solvents via the luminol dianion and then either the radical ion or the azaquinone (Scheme 3.1), which itself is

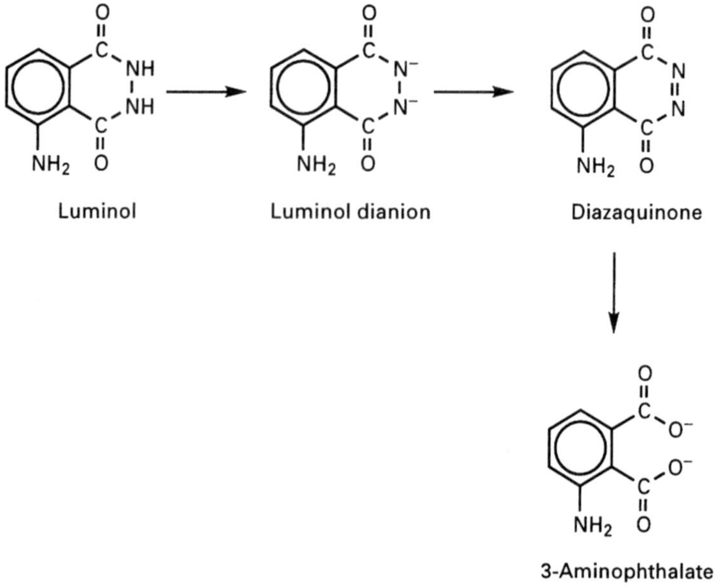

| Luminol | Luminol dianion | Diazaquinone |

3-Aminophthalate

Scheme 3.1 Oxidation of luminol.

chemiluminescent and has the same emitter as luminol; subsequent steps may involve the superoxide ion as luminol chemiluminescence is quenched by superoxide dismutase. The monoanion leads to the radical anion in protic solvents. A triplet state adduct of oxygen and the luminol dianion or, alternatively, a peroxide of luminol have been proposed as plausible transition states (McCapra, 1966). In microperoxidase-catalysed isoluminol oxidation, the monoanion is postulated (Yamamoto *et al.*, 1987, and references therein) to form its semiquinone as a result of hydrogen abstraction by radicals generated by peroxide bond cleavage. Semiquinone radicals then react with molecular oxygen to form super-oxide, which reacts with more semiquinone to produce isoluminol endoperoxide, which decomposes to 4-aminophthalate with light emission. The mechanism of the enzyme-catalysed reactions has been discussed by Kim *et al.* (1990).

The excited product in the luminol reaction, yield > 50%, is 3-aminophthalate dianion (Gundermann, 1965; McCapra, 1966), which emits at $\lambda_{max} = 424$ nm, its fluorescence spectrum being identical to the chemiluminescence spectrum. The quinonoid form ($\lambda_{em} = 510$ nm) is favoured by aprotic solvents. In mixed solvents, there is usually less 425 nm emission in chemiluminescence than in fluorescence; this seems to depend on the fraction of ion-pairs, which in chemiluminescence is determined by the transition state rather than the ground state equilibrium. In mild oxidation conditions, acid aldehydes can be formed instead of 3-aminophthalate dianions; these may be produced in a dark reaction that branches from the chemiluminescent mechanism before the emitting dianion is produced (Roswell and White, 1978).

3.4.2 Other hydrazides

Phthalhydrazides, analogues of luminol, are often strongly chemiluminescent when oxidised by oxygen or hydrogen peroxide and, although there are so far no reports of their application to lipid analysis, some of them have real potential. Gundermann (1965) points out that electron-donating substituents, especially at the 3-position, enhance chemiluminescence unless the group is large enough to experience steric hindrance at the 3-position from the adjacent carbonyl group, in which case electron delocalisation is impeded and chemiluminescence is reduced. There is, however, no steric hindrance of electron-donating substituents at the 4-position and these do not reduce chemiluminescence, e.g. 4-diethylaminophthalhydrazide emits more strongly than luminol and over a wider pH range. Polysubstituted phthalhydrazides are strongly fluorescent but not necessarily strongly chemiluminescent. 7-Dialkylaminonaphthalene-1,2-dicarbohydrazides emit 2–3 times more strongly than luminol but

$4'$-dialkylaminostilbene-2,3-dicarbohydrazides   only   two-thirds   as
strongly as luminol in their optimised conditions (40% dimethyl
sulfoxide). Alteration to the luminol heterocyclic ring abolishes the
chemiluminescence. Unsubstituted phthalic hydrazide is chemiluminescent only in aprotic solvents, where there can be intermolecular energy
transfer from the excited phthalate dianion to hydrazide monoanion
(Roswell and White, 1978). Linear monoacylhydrazides are chemiluminescent (White *et al.*, 1967) but less so than luminol, and the mechanism
appears to be different.

3.4.3   Fatty acids

Naslund *et al.* (1990) describe a rather complex assay for fatty acids, in
which enzymic reactions generate hydrogen peroxide. Fat cells are
incubated at 37°C in a shaking water bath with Krebs–Ringer phosphate
buffer, bovine serum albumin (BSA), adenosine deaminase, glucose and
ascorbic acid; free fatty acids released are bound to the BSA. The
principle of the assay is first to release the fatty acids from albumin
conjugates by treatment with sodium dodecyl sulfate and then to combine
them with coenzyme A by incubation with the synthase and ATP. The
acyl-coenzyme A so formed is subsequently oxidised by molecular oxygen
in the presence of the enzyme acyl-CoA oxidase, with the production of
hydrogen peroxide; this is then detected by its chemiluminescent reaction
with luminol in the presence of horseradish peroxidase and diethylene-
triaminepentaacetic acid. Ascorbic acid inhibits acyl-CoA synthase, so is
removed by addition of ascorbate oxidase at the synthesis stage.
Unreacted coenzyme A would interfere with the luminol reaction and
so must be removed before light measurement by addition of *N*-
ethylmaleimide. Compared with a calibration using hydrogen peroxide
standards, recoveries of fatty acids of carbon number 10 to 18 were
$> 90\%$, with the exception of linolenic acid. Detectivity was $5 \, \mu mol \, dm^{-3}$
on a sample of $6.25 \, \mu l$ (0.03 nmol) and the analytical range was 0.05–
5 nmol. The results correlate ($r = 0.997$) with spectrophotometric detection using the NEFA-C test kit, for which detectivity was 2 nmol. The
cost per assay was about 50 cents.

3.4.4   Prostaglandins

A sensitive and reliable chemiluminescent immunoassay for prostaglandin E_2 has been devised (Neuport *et al.*, 1996) for use with 96-well
microtitre plates. This is based on a competitive reaction between mouse
monoclonal anti-PGE_2, free antigen and solid-phase bound antigen. The

plates are coated with PGE_2 conjugated with bovine serum albumin; analyte pre-incubated with antibody is added. Residual free antibody is captured by solid phase bound PGE_2–BSA conjugate. Monoclonal antibody captured on the plate is determined by biotinylated antibody and a complex of avidin and biotin-labelled horseradish peroxidase, which catalyses luminol oxidation in the presence of 4-iodophenol enhancer. The assay is valid over the range 10–$50\,000\,\mathrm{pg\,ml^{-1}}$ and quantifiable above $100\,\mathrm{pg\,ml^{-1}}$ ($2\,\mathrm{pg/cell}$) for buffer or above $150\,\mathrm{pg\,ml^{-1}}$ for plasma. Intra-day relative standard deviations range from 3.2% to 8.9% for buffer and from 4.2% to 17.7% for plasma; corresponding inter-day values are 2.9% to 19.8% for buffer and 3.6% to 21.2% for plasma. The assay can be used for biological fluids such as plasma or suction blister fluid.

3.4.5   Lipid hydroperoxides

This section describes a highly successful field of application for luminol chemiluminescence, in which four strands can be detected in the reported work. Suitable methods for assessing the quality of food are sought in some papers, while in others the studies are of more general biochemical interest, such as enzyme activity or lipid membranes. The peroxidation of lipoproteins, associated with many diseases including cancer and heart disease, is a third line of interest, while among the applications of luminol chemiluminescence to high pressure liquid chromatography (HPLC) detection, determinations of lipid peroxides in blood are a major concern.

Lipid peroxides in oxidised foods can be detected by chemilumines-cence using luminol (Matthäus et al., 1994). A 1 ml sample (1 g fat or oil in 10 ml 2:1 acetone–ethanol) is reacted with 1 ml of a reagent comprising $0.7\,\mathrm{mmol\,dm^{-3}}$ luminol, $3.85\,\mathrm{\mu mol\,dm^{-3}}$ haemin and $11.8\,\mathrm{mmol\,dm^{-3}}$ sodium carbonate. Chemiluminescence is measured for $79\,\mathrm{s}$ and integrated. Using trioctanoylglycerol containing $100\,\mathrm{mmol\,kg^{-1}}$ methyl linoleate as the model system, peroxide concentration was linear over the range 0.65–$19\,\mathrm{mmol\,kg^{-1}}$ fat and the detection limit was $0.01\,\mathrm{mmol\,kg^{-1}}$ fat. Interference occurred in the presence of EDTA and the antioxidants β-carotene, cysteine and α- and β-tocopherol. Results correlated fairly well with peroxide determinations by a range of other methods for relatively short storage periods ($< 50\,\mathrm{h}$). After prolonged storage, the results correlated most closely with peroxide determinations by iron(III) thiocyanate and by oxygen uptake. The method was applied, after suitable calibration, to commercial soybean and grapeseed oils that had been oxidised by storage for various times at 50°C. The assay looks useful as a screening method for the early detection of deterioration of food.

Pettersen (1994) used luminol as a sensitiser of hypochlorite-induced chemiluminescence of fish oils in a method suitable for evaluating rancidity; $0.2 \, \mu mol \, dm^{-3}$ luminol was found to enhance the signal by a factor of 7.2. In optimised conditions, $50 \, \mu l$ oil was emulsified with $500 \, \mu l$ 2-methylpropan-2-ol to which $10 \, \mu l$ $13 \, \mu mol \, dm^{-3}$ luminol had been added. Samples in disposable cuvettes were presented from a carousel. Chemiluminescence was recorded for 3 min after the addition of $105 \, \mu l$ aqueous sodium hypochlorite. The luminometer, model LKB 1251 was maintained at 25°C and the cell compartment and cuvette headspace were flushed with nitrogen, because the presence of air was found to elevate the light emission; signal intensity was also found to be temperature-dependent. Measured chemiluminescence increased during the time of storage and was negatively correlated with the sensory quality of the oil as judged by a trained panel. Because of earlier evidence that chemiluminescence of this kind is not correlated with peroxidation value as measured by other methods, Pettersen concludes that the light emission is caused by several molecular species that might include secondary oxidation products, some of which (viz., the aldehydes and ketones) are the main contributors to the off-flavours of oxidised oils.

Luminol has been used (Kondo et al., 1994) to investigate the activity of lipoxygenase on linoleic acid. The chemiluminescence reagent comprised $0.8 \, nmol \, dm^{-3}$ cytochrome c and $3 \, nmol \, dm^{-3}$ luminol in $0.2 \, mol \, dm^{-3}$ borate buffer (pH 9.0). A 2 ml portion of this reagent was added to 100 i.u. soybean lipoxygenase in $50 \, \mu l$ $0.2 \, mol \, dm^{-3}$ borate buffer (pH 9.0) and $0.95 \, ml$ $0.02 \, mol \, dm^{-3}$ sodium linoleate. The mixture was incubated at 30°C and chemiluminescence was measured every 10 s for 150 s or 300 s. Reaction rates were determined by measuring the gradient over the 60 s period following each reading. Calibration for lipoxygenase was linear up to 100 international units. Cytochrome c increased the intensity of the chemiluminescence by a factor of over 50. More than $2 \, \mu g$ tocopherols per assay interfered.

Gumuslu et al. (1997) have reported on the effect of detergents on emission intensity, which is important in chemiluminescence studies involving lipid membranes. Chemiluminescence was detected by means of a liquid scintillation counter in single-photon mode from a reaction mixture comprising haemoglobin, luminol, 2-methylpropane 2-hydroperoxide and one of the following detergents: digitonin, Triton X–100, taurocholic acid, cetylpyridinium chloride and sodium dodecyl sulfate. All detergents caused a diminished maximum chemiluminescent signal and retarded the appearance of the maximum. The most prominent diminution was from $50 \, mg \, dl^{-1}$ or $100 \, mg \, dl^{-1}$ digitonin; $5 \, mg \, dl^{-1}$ sodium dodecyl sulfate caused the least prominent diminution and this was hence the preferred detergent and final concentration.

Oxidative modification of lipids and proteins due to reactive oxygen species are postulated to contribute to cancer. Zimmermann *et al.* (1993) isolated low-density lipoprotein (LDL) from blood by coating onto tubes and determined the extent of its peroxidation by incubating with luminol and measuring the total emitted light during 10 s intervals. Compared with healthy subjects, cancer patients have an elevated 'peroxidation state' of LDL, which is possibly an indication of immune defence against the cancer cells, due to proliferation or necrosis. This is further raised 10 days after surgery but declines somewhat over 6 months, except that for relapsing patients it declines to preoperative level only.

Chemiluminescence has been used extensively to detect substances separated by chromatography. Pre- and post-column derivatisation with a chemiluminescence reagent or a post-column reactor are popular strategies (Kricka, 1993). Worsfold (1990) has described the basic principles of flow-injection analysis and HPLC in combination with chemiluminescence detection. An HPLC assay for lipid hydroperoxides with picomole detectivity is based on the microperoxidase-catalysed oxidation of isoluminol (Yamamoto *et al.*, 1987; Frei *et al.*, 1988). The optimised post-column detection system consisted of a Kratos mixer in which the eluent $(1\,\text{ml\,min}^{-1})$ was merged with pumped flow $(1.86\,\text{ml\,min}^{-1})$ of the chemiluminescence reagent $(1\,\text{mmol\,dm}^{-3}$ isoluminol and $25\,\mu\text{g\,ml}^{-1}$ microperoxidase in 70% methanol–30% borate buffer, pH 10), before entering a mixing coil $(92\,\mu\text{l})$ and then passing to a $10\,\mu\text{l}$ flow cell in an LC fluorimeter, which was used as a photon detector with the excitation source turned off. Hydroperoxides of different parent lipids could be identified by the use of retention time data. The assay also determined hydrogen peroxide, but not the endoperoxide (dioxetane) prostaglandin H_2. Antioxidants give negative peaks due to the quenching of background luminescence. Because the HPLC separation prevented interference from antioxidants, the method was suitable for biological samples such as blood plasma.

Since it is otherwise difficult to separate hydroperoxides from plasma antioxidants in a single chromatographic step, plasma was separated into an aqueous methanol phase (subsequently gel filtered before chromatography) and a hexane phase (evaporated and then redissolved in methanol–2-methylpropan-2-ol, the HPLC mobile phase). When lipid hydroperoxides were added to plasma, the hexane phase contained those of triacylglycerols, cholesterol and cholesterol esters, which have detection limits of about $0.03\,\mu\text{mol\,dm}^{-3}$; the aqueous methanol phase contained those of cholesterol, fatty acids and phospholipids, with detection limits of $0.01\,\mu\text{mol\,dm}^{-3}$. The recoveries of the parent lipids and of antioxidants were between 95% and 103%, while those of lipid hydroperoxides were much lower, probably owing to binding with plasma proteins. This was particularly so in the case of cholesterol 7-hydroperoxide (37% total in both phases). Spiked hydrogen peroxide could not be recovered from plasma, probably because of the latter's catalase activity; however, if the hydrogen peroxide is added after the extraction step, recovery is 72%, compared with 91% from phosphate buffered saline (in which detectivity is $0.25\,\mu mol\,dm^{-3}$). Neither hydrogen peroxide nor lipid peroxides were detectable in human plasma; peaks that chromatographically mimic these analytes are artefacts produced during the extraction process.

Ubiquinol-10 is detected in plasma at a similar retention time to cholesterol ester hydroperoxides; the assay may therefore be sensitive to other aromatic diols. Selectivity is gained by repeating the determination after reduction of the sample with sodium tetrahydridoborate or triphenylphosphine or tin(II) chloride. Only those peaks that fail to appear after reduction can be attributed to lipid hydroperoxides.

Wieland *et al.* (1990b) report an HPLC assay for lipid hydroperoxides in butanol extract of blood using post-column chemiluminescent oxidation of luminol in the presence of cytochrome c; detection limits varied from 50 to $200\,nmol\,dm^{-3}$ depending on the nature of the peroxidised lipid and calibration was linear up to $10\,\mu mol\,dm^{-3}$. Because lipophilic antioxidants and coenzyme Q gave negative peaks, interference from these sources (which occurs if ultraviolet absorption at 233 nm is used for detection) was avoided; unoxidised triacylglycerol gave no peak. The assay was also applied to LDL that had been oxidised by Cu^{2+} for 24 h; results correlated well with assay by ultraviolet absorption and there was a major peak corresponding to linoleic acid hydroperoxide that accounted for 80% of the peroxides present.

In another HPLC assay for lipid hydroperoxides (Wadano *et al.*, 1991), the chemiluminescence catalysts microperoxidase and cytochrome c were immobilised on suitable gels, e.g. Affi-prep 10, which were then packed in columns $10\,cm \times 4\,mm$. The mobile phase, 9:1 methanol–chloroform, was merged with the reagent, $1\,\mu g\,ml^{-1}$ luminol in $50\,mmol\,dm^{-3}$ borate before passing through the enzyme reactor.

3.4.6   Antioxidants

Antioxidants are of interest to lipid chemists as their study, as well as being interesting in its own right, is often the best approach to investigating lipid peroxidation phenomena. Metsa-Ketela (1990) determined the total radical-trapping antioxidants in plasma (TRAP) by measuring the duration of extinction of the chemiluminescent luminol reaction initiated by the free radicals generated on thermal decomposition of 2,2-azobis(propane-2-amidinium) chloride (ABAP). A $20\,\mu l$ sample was injected into an assay mixture of 450 µl oxygen-saturated 100 mmol dm^{-3} phosphate-buffered (pH 7.4) saline (PBS), 50 µl 400 mmol dm^{-3} ABAP in PBS, 25 µl 160 mmol dm^{-3} sodium linoleate and 50 µl 10 mmol dm^{-3} luminol at 37°C when the chemiluminescence had stabilised. The reaction took place in a cuvette on a temperature-controlled carousel; luminescence readings were automatically accessed to a computer. The presence of linoleate enhances the intensity of chemiluminescence (by preventing the termination of radical chains), but did not affect the duration of extinction, which had a linear relationship with the amount of antioxidant. Aqueous trolox (6-hydroxy-2,5,7,8-tetramethylchroman-2-carboxylic acid), a phenolic antioxidant, was used as a standard; TRAP values for human blood plasma were expressed as equivalent concentrations of trolox.

Whitehead et al. (1992) and Maxwell et al. (1993) used an assay for antioxidant capacity in biological fluids for the detection of oxidative stress in human disease. It was based on radical scavenging by antioxidants which interferes with p-aminophenol-enhanced luminol chemiluminescence. The reagent is made up from tablets containing luminol, enhancer and oxidant together with an assay buffer. A 20 µl sample, which must be fresh, was added to 100 µl of a solution of horseradish peroxidase (as a conjugate with immunoglobulin G), 100 µl made-up assay reagent and 800 µl deionised water. The time to 10% recovery of initial light output was linearly related to aqueous antioxidant concentration. Aqueous trolox was used as a standard; uric acid and ascorbic acid in aqueous solutions are stoichiometrically equivalent to trolox but they are incompletely measured in serum. Bilirubin and sulfydryl antioxidants (proteins, glutathione) in serum are seriously under-represented in the assay, but a separate qualitative assessment of the recovery of light emission yields some information about their activity in serum. In particular, the maximum intensity of the recovered chemiluminescence is inversely proportional to the total protein concentration. Deproteination of the serum sample loses approximately 40% of the measured antioxidant activity; the time course of the chemiluminescence in the absence of protein resembles that from an aqueous solution as in Figure 3.7(A), rather than from whole serum as in (B).

Normal serum antioxidant capacity was determined to be 447.1±60.2 µmol dm^{-3} for males and 380.9±80.0 µmol dm^{-3} for females (healthy students), sex differences being attributable to those in uric acid levels; total antioxidant in semen was about twice the serum level. The total antioxidant capacity of cerebrospinal fluid is low; samples of it can therefore be assayed undiluted. An acceptably low within-batch assay imprecision of ∼2% was reported. The assay can detect acute or chronic

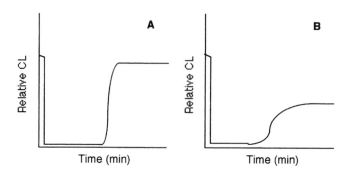

Figure 3.7   Time course of chemiluminescence during an assay of (A) an aqueous solution of trolox and (B) a 1:10 diluted human serum sample. Adapted from Whitehead *et al.* (1992).

depletion of antioxidants, which are the first line of defence against reactive oxygen species.

The chemiluminescent antioxidant assay of Maxwell *et al.* has been automated using the Dynatech ML 3000 microplate luminometer (Kricka *et al.*, 1993). The working solution of horseradish peroxidase is 0.125 mg dm^{-3} in Tris (0.1 mol dm^{-3}, pH 8.6) and is stored at −20°C; 50 μl of this solution is added to 1 ml Kodak Amerlite signal reagent in the dark; 100 μl aliquots are added within 2 min by multiple pipette to 20 μl sample in each of the 96 wells. The time taken is measured for the light emission to be restored after quenching to 10% of its initial intensity. The time course of the chemiluminescence from each of 96 wells can be displayed simultaneously.

The assay was applied to comparisons of antioxidants (trolox, albumin, transferrin, ceruloplasmin, uric acid, ascorbic acid, bilirubin, cholesterol and glucose) and to the determination of antioxidant levels in diabetics and in nutrient-depleted patients. Delay time is typically 15 min and has a linear relationship with the concentration of most antioxidants; results cannot be compared between runs. Measured antioxidant levels decline for 2 h at room temperature and 4°C and are then stable for up to 6 days.

A chemiluminescence method for screening antioxidants has been reported (Kim *et al.*, 1993). Ten microlitres of horseradish peroxidase (1 μg/ml), antioxidant (1 mg/ml) and signal reagent (luminol + H$_2$O$_2$ + enhancer) were incubated for 10 min in a 96-well plate; chemiluminescence was measured by an Amerlite analyser (integrated luminescence reader and data processor). Antioxidant activity was measured as the percentage inhibition of the chemiluminescence. Light emission diminished in the decreasing order: δ-tocopherol, ascorbic acid, hydroxyquinoline, propyl gallate and butylated hydroxytoluene, the light emission diminishing as antioxidant concentration rises in each case. Peroxidation of phosphatidylcholine liposomes (as measured by the thiobarbituric acid test) was inhibited in a similar order. In a further study (Zhang *et al.*, 1996), the order of ability to scavenge phenoxide radicals (in the horse-radish peroxidase + luminol + 4-iodophenol + perborate system) was consistent with the ability to suppress lipid peroxidation (in rat liver microsomes). Unrelated results were obtained on assay in a superoxide ion generating system.

3.4.7   Choline phospholipids

One direction in which luminol chemiluminescence may be applied analytically is for the determination of hydrogen peroxide produced by enzymic oxidations. Assays of this kind enjoy the specificity of the enzyme used. A chemiluminescent enzyme method has been reported (Bissé *et al.*, 1990) for the determination of serum choline-containing phospholipids (PL). This method is simpler and quicker than other methods involving solvent extraction and digestion of the PL followed by specific determination of inorganic phosphorus. It links luminol chemiluminescence with three enzyme-catalysed reactions, as shown in equation (3.4.).

$$PL \xrightarrow{\text{PD}} PA + \text{choline} \xrightarrow{\text{COD}} \text{betaine}$$
$$+H_2O_2 \xrightarrow[\text{luminol}]{\text{POD}} 3 - \text{aminophthalate} + h\nu \tag{3.4}$$

The chemiluminescent reagent is freshly made up from equal volumes of $2 \, \text{mmol dm}^{-3}$ luminol, $11 \, \text{U ml}^{-1}$ choline oxidase (COD) and $9.5 \, \text{U ml}^{-1}$ peroxidase (POD). The serum is diluted 1:600 with Tris-HCl buffer (pH 8.5) and sonicated for 10 min to homogenise. A $300 \, \mu\text{l}$ sample is incubated at 37°C for 20 min with $100 \, \mu\text{l}$ $22.3 \, \text{U ml}^{-1}$ phospholipase D (PD); then $100 \, \mu\text{l}$ of the chemiluminescent reagent is injected. Chemiluminescence was recorded against time and was found to increase with pH up to 8.5; pH dependence seems to be less pronounced at low phospholipid concentration. Chemiluminescence increases linearly with time of incubation of phospholipid with phospholipase D and with amount of choline oxidase used in the assay; the signal increases with temperature up to 37°C. Comparison with results from Boehringer–Mannheim phospholipid kit using human serum gave $r = 0.941$ ($n = 35$). Chemiluminescence calibration was linear up to $5 \, \text{mmol dm}^{-3}$ phospholipid and the relative standard deviation (within or between runs) varied from 2.2% to 4.0%.

3.4.8   Steroids

Another enzyme-linked chemiluminescence assay is a flow injection determination of bile acids (Maeda *et al.*, 1993). The sample is injected into a carrier of $0.2\,mmol\,dm^{-3}$ nicotinamide–adenine dinucleotide in $0.05\,mol\,dm^{-3}$ pyrophosphate buffer (pH 9.5) flowing at 1.0–$1.2\,ml\,min^{-1}$, passing through a column of 3α-hydroxysteroid dehydrogenase immobilised on glass beads. The carrier is merged with a channel of 1-methoxy-5-methylphenazinium methylsulfate flowing at $0.55\,ml\,min^{-1}$. The merged flow passes through a reaction coil before merging again with a channel of $0.2\,mmol\,dm^{-3}$ isoluminol and $1\,\mu mol\,dm^{-3}$ microperoxidase in $0.8\,mol\,dm^{-3}$ phosphate buffer (pH 9.5) also flowing at $0.55\,ml\,min^{-1}$. In an alternative approach, bile acids can be derivatised (Kawasaki *et al.*, 1985) with aminobutylethylisoluminol; the derivative luminesces on hydrogen peroxide oxidation.

3.4.9   Lipoproteins

The determination of peroxidation in LDL has already been considered in section 3.4.5. Most lipoprotein chemiluminescence assays combine aspects of protein chemistry (especially separation techniques) with a light-emitting detection reaction, which in most cases so far is luminol oxidation.

Pre-β high-density lipoprotein can be separated from α-lipoprotein by agarose gel electrophoresis and then determined by immunoblotting with chemiluminescence (O'Kane *et al.*, 1992). The sample is transferred onto a poly(vinylene difluoride) membrane by capillary blotting and successively incubated with sheep anti-human apolipoprotein A antiserum (1 h, 37°C) and rabbit anti-sheep immunoglobulin–horseradish peroxidase conjugate (30 min, 37°C). The membrane is then developed by Western blotting and HRP-catalysed 4-aminophenol-enhanced chemiluminescence is recorded on film and quantitated by transmission densitometry at 600 nm. The assay has a linear calibration from 8.1 to $180\,mg\,dm^{-3}$; the relative standard deviation is 7% at $22.1\,mg\,dm^{-3}$ and 4.9% at $44.3\,mg\,dm^{-3}$.

Immunoluminometric assays for lipoprotein (a) and a series of apolipoproteins have been described (Kessler *et al.*, 1994). Polystyrene spheres are coated with antibody specific for apolipoprotein and with biotinylated antibody and conjugated streptavidin, 7-[*N*-(4-aminobutyl)-*N*-ethylamino]naphthalene-1,2-dicarbohydrazide. Catalase and alkaline hydrogen peroxide are added and chemiluminescence measured. Results correlate tolerably well with a nephelometric method ($r = 0.65$ to 0.888) and there are advantages over nephelometry except for apolipoprotein

C-II. The detection limit is $< 5 \, \text{mg} \, \text{l}^{-1}$. The relative standard deviation is 2.9% to 5.9% intra-assay ($n = 15$ to 144) and 6.2% to 12.2% inter-assay ($n = 15$ to 24). The typical calibration range extends up to $800 \, \text{mg} \, \text{dm}^{-3}$.

An assay of human apolipoprotein B_{48} using a commercially available antibody is reported (Smith et al., 1997) that requires no delipidation of samples and requires only one overnight ultracentrifugation step. 4-Aminophenol-enhanced luminol chemiluminescence is used to visualise the proteins on a Western blot and is 10 times more sensitive than Coomassie blue staining after polyacrylamide gel electrophoresis in the presence of sodium dodecyl sulfate.

3.5 Electrochemiluminescence of luminol

Electrochemiluminescence is sensitive because of the short emission lifetime and the small electrode surface area. The reaction pathways lend themselves to control of the emission by varying the applied voltage or the electrode selected. Other distinct advantages are the economic use of reagents and applicability to neutral aqueous solutions such as biological fluids, in contrast with conventional luminol chemiluminescence in strongly alkaline or non-aqueous solutions.

An assay of lipid hydroperoxides has been devised (Sakura and Terao, 1992a) using electrochemiluminescence at a vitreous carbon electrode. An outline of the electrochemical oxidation of luminol is shown in Scheme 3.2. The model analyte was a 1:1 mixture of methyl linoleate hydroperoxides in 30% aqueous acetonitrile solution at pH 7.4. The optimum applied voltage was 0.7 V; over the range 0.5–1.0 V luminol alone is oxidised. Loss of one electron by the monoanion gives diazasemiquinone, which dismutes to produce diazaquinone (Sakura, 1992); this reacts quantitatively with the analyte, briefly emitting light ($\lambda_{max} = 440 \, \text{nm}$, decay 0.38 s) without interfering signals being generated. Below 0.4 V, there is no chemiluminescence.

At applied voltages above 1.0 V, the analyte itself is oxidised (and the emission becomes complicated ($\lambda_{max} = 430$–$440 \, \text{nm}$, decay 0.5 s) because the $-NH_2$ of diazaquinone is oxidised, forming $- NH_2^{\cdot +} - NH^{\cdot}$ and $-NH^+$, which react with the oxidation products of methyl linoleate hydroperoxide, giving an interfering signal. At 1.2 V, the intense chemiluminescence can be suppressed by superoxide dismutase to the value obtained at 0.7 V. This, together with the pattern of absorbance changes, establishes the oxidation reaction as a conversion of analyte to conjugated triene together with superoxide, the latter being responsible for the additional luminescence.

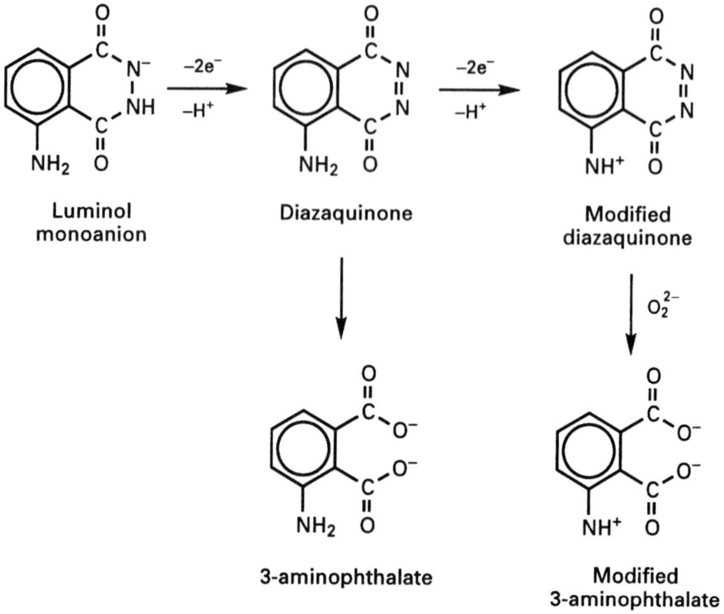

Scheme 3.2   Electrochemical oxidation of luminol.

The detection limit for the lipid hydroperoxide is 0.3 nmol at $S/N = 1.5$ in the optimised conditions, which include the use of phosphate buffered (pH 7 to 9) 30% acetonitrile as carrier, flowing at $0.3 - 0.5 \, ml \, min^{-1}$; the optimum luminol concentration is $10–20 \, \mu mol \, dm^{-3}$. There was no interference from uric acid and ascorbic acid in amounts similar to the analyte concentration, but there was from much greater amounts; α-tocopherol enhances the chemiluminescent signal. When the glassy carbon electrode was replaced by platinum, over the range 0.5–1.0 V both luminol and methyl linoleate hydroperoxide are oxidised, giving out light ($\lambda_{max} = 430 - 460 \, nm$, decay 0.38 s). The detection limit for methyl linoleate hydroperoxide is 0.1 nmol at $S/N = 2.5$. With either electrode, the optimum conditions are the same. There was no emission from methyl hydroxyoctadecadienoate, a major reduction product of linoleic acid hydroperoxide, in either method. A comparison (Sakura and Terao, 1992b) of electrochemiluminescence and amperometric detection of lipid hydroperoxides showed that both were very sensitive methods.

A further development (Atwater et al., 1997) of luminol electrochemiluminescence is described as 'reagentless' flow injection analysis. Addition of luminol and adjustment of pH are achieved by in-line crystal beds. It has so far been applied only to hydrogen peroxide determination.

3.6 The peroxy–oxalate reaction

Peroxy–oxalate chemiluminescence (PO-CL) was first observed by Chandross (1963) as a very weak bluish-white emission from oxalyl chloride, Cl·CO·CO·Cl, oxidised by hydrogen peroxide; a similar blue emission occurs from related oxalyl peroxides (Gundermann, 1965). Much more intense emission is obtained in the reaction between aryl oxalates and hydrogen peroxide in the presence of a fluorophore. Applications of this reaction to lipid analysis are reported in major reviews of chemiluminescence by Townshend (1990) and by Robards and Worsfold (1992). In a thorough review including 96 references, Kwakman and Brinkman (1992) have discussed liquid chromatographic applications of the peroxy–oxalate reaction.

3.6.1 General experimental procedure

Because in PO-CL analysis, the analyte is an added fluorophore to which energy is transferred, the various applications have much in common and it is useful to arrive at a general experimental procedure for reactions of this kind. The rate of PO-CL depends especially on pH and the presence of a nucleophilic base catalyst for ester hydrolysis; a typical protocol would involve the use of 0.25–$10 \, \text{mmol dm}^{-3}$ bis(2,4,6-trichlorophenyl) oxalate (TCPO) or bis(2–nitrophenyl) oxalate (2NPO) to detect 10–$1000 \, \text{mmol dm}^{-3}$ hydrogen peroxide, the catalyst being imidazole or triethylamine. The concentration of oxalate affects the intensity, but not the kinetics, of the signal. Steijger $et\ al.$ (1996), in a factorial design optimisation of the flow injection analysis of 3-aminofluoranthene by the 2NPO/peroxide reaction, calculated the optimum concentration of hydrogen peroxide to be $110 \, \text{mmol dm}^{-3}$ and that of the imidazole catalyst to be $10.7 \, \text{mmol dm}^{-3}$. Copper(II) ions enhanced the chemiluminescence, the optimum concentration being $40 \, \mu\text{mol dm}^{-3}$; other metal ions have lesser effects. These optimum concentrations are dependent on the geometry of the manifold and on the flow rate used, which was $0.8 \, \text{ml min}^{-1}$ total flow.

Aryl oxalates differ in the effect of pH on the intensity and decay of the chemiluminescence. They also differ in their solubilities, which affects their usefulness as detection reagents for HPLC; there are wide variations in their stabilities in the presence of hydrogen peroxide so some are more suitable than others for the delivery of both chemiluminescence reagents by a single pump. Taking all these matters into account, Honda $et\ al.$ (1985a) propose that, in general, the preferred oxalate is bis(pentafluorophenyl) oxalate (PFPO) at pH < 2; 2NPO at pH 2–4; bis(2,4-dinitrophenyl) oxalate (DNPO) at pH 4–6; TCPO at pH 6–8; and

bis(2,4,5-trichloro-6-pentyloxycarbonylphenyl) oxalate (TCPCO) at pH
> 8. The half-life of DNPO is much shorter than that of TCPO,
producing less broadening of the signal but a very brief time window for
its detection; this is an advantage for use with larger capacity flow cells
(several ml), but a disadvantage with smaller ones. Honda suggested the
ratio of maximum chemiluminescence intensity to the decay constant of
the signal as a criterion for evaluating reagents or reaction conditions for
peroxy–oxalate detection in HPLC.

3.6.2   Mechanism

Peroxy–oxalate chemiluminescence (PO-CL) is thought to follow a
chemically initiated electron exchange luminescence (CIEEL) mechanism
as proposed by Koo and Schuster (1977). An electron is transferred from
the fluorophore to an intermediate, which as it decomposes transfers it
back again; as a result the fluorophore is raised to an excited state, and
subsequently radiates. Supporting evidence for this was provided by
Honda et al. (1985b); having calculated relative excitation yields of
different fluorescers, they were able to show that this parameter had a
significant negative correlation with the oxidation potential of the
fluorescer and was at least as important a predictor of chemiluminescence
intensity as the singlet excitation energy. Thus, fluorescers having a low
oxidation potential and a low singlet excitation energy can be sensitively
detected by the peroxy–oxalate system.

Rauhut et al. (1967) had proposed an intermediate dioxetane
(endoperoxide) structure, which decomposed into carbon dioxide with
transfer of energy to an added fluorophore (F), producing its excited
singlet state leading to fluorescence emission (equation 3.5). However, the
chemiluminescence depends on the electronegativity of the aryl group, so
a common 1,2-dioxetanedione intermediate is unlikely to be the sole
source of the emission (Kwakman and Brinkman, 1992).

$$ArO-\underset{\underset{O}{\|}}{C}-\underset{\underset{O}{\|}}{C}-OAr \xrightarrow{H_2O_2} \underset{\underset{O}{\|}}{\overset{O-O}{\underset{|}{C}}}\!-\!\underset{\underset{O}{\|}}{\overset{}{\underset{|}{C}}} \xrightarrow{F} \underset{\underset{O}{\|}}{\overset{O-O}{\underset{|}{C}}}\!-\!\underset{\underset{O}{\|}}{\overset{}{\underset{|}{C}}}F^{\cdot+} \longrightarrow F^* + 2CO_2 \quad (3.5)$$

$$+$$

2ArOH

McCapra (1987) explains the energy transfer by the formation of a
linear peroxide intermediate, ArO·CO·CO·OOH, which decomposes to
radical ion-pairs comprising the fluorophore and a carbon dioxide
molecule. The chemically related luminescence of dioxetanes on warming

(equation 3.6) involves excited (triplet state) carbonyl compounds and these could also have a role in PO-CL.

$$H_3C-\overset{\overset{\displaystyle O-O}{|\quad|}}{\underset{\underset{\displaystyle H_3C\quad CH_3}{|\quad|}}{C-C}}-CH_3 \longrightarrow \overset{H_3C}{\underset{H_3C}{}}C\overset{*}{=}O \; + \; \overset{H_3C}{\underset{H_3C}{}}C=O \qquad (3.6)$$

A more complex mechanism is also suggested by Mann and Grayeski (1990), who have studied the background emission that occurs in the absence of a fluorophore; maximum emission occurs at 450 nm (which could be emitted by excited carbon dioxide molecules) and at 550 nm (which varies with the aryl group and could be an excited carbonyl intermediate containing the phenolic residue). The analytical implication of these spectroscopic findings is that the signal-to-noise ratio could be improved by using a longer wavelength emitter and a cutoff filter to remove the shorter wavelength background; this, however, would probably reduce the fluorescence quantum yield and would require a red-sensitive photomultiplier. Improved chemiluminescent quantum yield (Φ_{CL}) would require new derivatisation reagents or the suppression of dark reactions (Kwakman and Brinkman, 1992).

3.6.3 Dansyl derivatives

Derivatisation needs to produce electron-rich molecules (containing electron-donating substituents) with a high fluorescence quantum yield (Φ_{FL}). Dansyl (5-dimethylaminonaphthalene-1-sulfonyl) chloride is the most widely used derivatisation agent in this field. Derivatives of amines with dansyl chloride give PO-CL. They can be separated by HPLC and function as the fluorescer (F) in equation (3.5) when post-column arrangements are made for the chemiluminescent reaction. Such assays are applicable to lipids containing derivatisable amine functions; fluorescent substances could be detected without derivatisation. Liquid chromatography methods also exist for oestradiol, oxo-bile acids and other steroids in serum, plasma or urine; detectivity is 5–10 fmol. There are also modified dansylating agents, e.g. dansylhydrazine for carbonyl compounds, for PO-CL, though these have so far been mainly used in fluorescence spectroscopy.

3.6.4 Fatty acids

Determinations of —COOH (e.g. in fatty acids or prostaglandins) by derivatisation with luminarine 4 (involving carbodiimide coupling to an aminated coumarin label) are of interest to lipid chemists. The detection limit is 32 fmol but the derivatisation limit is 60 pmol (Kwakman and Brinkman, 1992). Carbodiimide coupling to 3-aminoperylene (Honda et al., 1985b), or to 7-diethylamino-3-[4-(iodoacetylamino)phenyl]-4-methylcoumarin (DCIA) (Grayeski and DeVasto, 1987), or the introduction of a dansyl group (dansylbromoacetylpiperazide) (Kwakman et al., 1991) are other possible derivatisation routes for carboxylic acids.

Emission from perylene in the peroxy–oxalate reaction is 8.5 times as intense as from dansylalanine. Most of this intensity is conserved when the hydrocarbon is aminated (first nitrated and then reduced) and linked to a carboxylic acid group by means of 1,3-dicyclohexylcarbodiimide. Short-chain labelled carboxylic acids were separated by microbore HPLC and detected by post-column peroxy–oxalate chemiluminescence with a limit (at $S/N = 3$) of 0.1 fmol. Derivatisation, however, requires incubation for at least 2 h at 60°C of the carboxylic acid in benzene with perylene-3-ammonium chloride in pyridine.

When 7-diethylamino-3-[4-(iodoacetylamino)phenyl]-4-methylcoumarin is derivatised to carboxylic acids in the C_6 to C_{10} range, the derivative is an efficient chemiluminescence energy acceptor, whereas 4-bromomethyl-7-methoxycoumarin gives no chemiluminescence signal at all. This suggests that, in the latter case, the singlet excitation energy exceeds the enthalpy change ($440\,kJ\,mol^{-1}$) of the peroxy–oxalate reaction. In the former case, a strongly electron-donating diethylamino replaces a methoxy substituent in a coumarin ring. As a result, both the oxidation potential and the singlet excitation energy are lowered—this facilitates CIEEL (see section 3.6.2) — while the fluorescence efficiency is increased. Detectivities of the carboxylic acids ranged from 52 to 74 fmol, with good linearity. This is a 5-fold improvement on fluorescence detection.

3.6.5   Lipid peroxidation

Carbonyl group determinations by dansylhydrazine derivatisation followed by PO-CL are applicable to the analysis of lipid peroxidation products and are covered in section 3.6.6. Mann and Grayeski (1990) have used another label for carbonyls, 3-aminofluoranthene, which is very suitable for chemiluminescence detection but suffers from unfavourable derivatisation conditions and a relatively high detection limit.

3.6.6   Steroids

Nozaki et al. (1988) have reported a method for the extraction of oestradiol from serum using an ODS minicolumn. The derivative is formed at room temperature by 80 min incubation with dansyl chloride and purified using the ODS minicolumn; it is then separated by HPLC using a silica gel column from which it is eluted with hexane–chloroform–ethanol (70:30:0.1) and detected by chemiluminescence in a post-column reaction with bis-2,4,6-trichlorophenyl oxalate (8 mmol dm^{-3} in chloroform containing 100 mmol dm^{-3} triethylamine) and hydrogen peroxide (600 mmol dm^{-3} in methanol containing 6% v/v 0.1 mol dm^{-3} acetate buffer). The overall recovery of oestradiol from serum is ∼90% and the detection limit is 50 pg. The α- and β-isomers elute separately and α-oestradiol can be used as an internal standard, the concentration of the β-isomer being determined from the ratio of peak areas.

Bile acids can also be dansylated for detection by peroxy–oxalate chemiluminescence (Imai et al., 1989; Higashidate et al., 1990). Oxo-bile acids, as their ethyl esters, were derivatised overnight with dansylhydrazine (0.01% in benzene containing 0.077% trifluoracetic acid). Excess reagent was removed by gel permeation chromatography; they were then separated on an ODS column with an eluent of 50 mmol dm^{-3} imidazole buffer/tetrahydrofuran (105:95 v/v) and detected by chemiluminescence from a post-column reaction between 12.5 mmol dm^{-3} hydrogen peroxide in ethyl acetate–acetonitrile (1:1 v/v) and 0.25 mmol dm^{-3} bis[4-nitro-2-(3,6,9-trioxadecoxycarbonyl)phenyl] oxalate (TDPO). Four oxo-bile acids of clinical interest gave well-separated peaks with detectivities of ∼1 fmol. Other oxo-steroids were derivatised overnight with dansylhydrazine (20 mg dm^{-3} in 0.022 mol dm^{-3} ethanolic hydrochloric acid). Purification, separation and detection were as for the oxo-bile acids. Corticosterone, testosterone and progesterone gave well-separated peaks with detectivities of 3, 2 and 4 fmol, respectively.

Oxo-bile acids, as their ethyl esters, were extracted from urine and derivatised overnight with dansylhydrazine (0.01% in benzene containing 0.077% trifluoracetic acid). After purification by high-performance gel-permeation chromatography, they were separated on an ODS column with an eluent containing 50 mmol dm^{-3} imidazole buffer–acetonitrile–tetrahydrofuran (2:1:2 v/v/v) and detected by chemiluminescence as already indicated.

Dansylhydrazine has been used to derivatise the 3α-ketocorticosteroid fluocortin butyl (Koziol et al., 1984); automated dansylhydrazone determination has been used in really routine analysis of ketosteroids in serum with reagent supplies lasting a week (Kwakman and Brinkman, 1992). Dansylhydrazine derivatisation has also been used with photo-initiated PO-CL detection (Nondek et al., 1991). Irradiation of TCPO in the presence of dissolved oxygen leads to the formation of reactive intermediates that transfer energy to dansylhydrazones; although the process has a higher detectivity than hydrogen peroxide-initiated chemiluminescence, it does have the advantage of being easily automated.

PO-CL can detect hydrogen peroxide produced in enzyme reactions at pH~7 and hence can be used for highly specific assays for the substrate of the enzyme, e.g. cholesterol using cholesterol oxidase (Rigin, 1978); 9,10-diphenylanthracene is used as the fluorophore. As with luminol chemiluminescence, enzymic reactions coupled with PO-CL offer a wide range of possibilities; the fluorophore or the enzyme can be immobilised.

3.7   Luciferase-catalysed reactions

Luciferases are enzymes that catalyse the light-emitting reactions in living organisms—bioluminescence. They occur in several species of firefly such as *Luciola mingrelica* and *Photinus pyralis* and in many species of bacterium, e.g. *Photobacterium phosphoreum* and *Vibrio harveyi*. They are extracted from firefly lanterns by differential centrifugation and purified by gel filtration (which diminishes associated luciferin concentration by a factor of 10^4), and a stabiliser is added (Dukhovich *et al.*, 1990). The product is lyophilised and keeps for a few months at $-4°C$.

A luciferin is a substrate for a luciferase. Firefly luciferin emits at 562 nm on reaction with oxygen (equation 3.7), catalysed by luciferase in the presence of adenosine triphosphate and magnesium ions; there is also a weak emission on uncatalysed oxidation by strong oxidants (Gundermann, 1965). A typical luciferin concentration range for the linear relationship with luminescent intensity is $0.01\,nmol\,dm^{-3}$ to $1\,\mu mol\,dm^{-3}$. A very large number of papers have been published reporting applications of ATP determinations, which can include studies of any enzymic reaction that utilises or produces ATP (McCapra, 1987). Kinases and substances involved in reactions catalysed by them can be assayed by ATP-dependent firefly luciferase.

$$ (3.7) $$

Dukhovich *et al.* (1990) have reported a bioluminescent assay for choline-containing phospholipids based on the lipid dependence of the firefly luciferin–luciferase system. Luciferase is incubated with ATP, magnesium ions and detergent; the sample and luciferin are then added and the chemiluminescent intensity is measured; results are expressed net of the blank intensity obtained exactly as stated but without sample. The luciferase is activated only by choline-containing phospholipids, i.e. phosphatidylcholine, lysophosphatidylcholine and sphingomyelin. Phosphatidylinositol, phosphatidylglycerol, phosphatidylethanolamine, lysophosphatidylethanolamine and cardiolipin do not activate the enzyme. The calibration for phosphatidylcholine shows a linear dependence of chemiluminescent emission on phospholipid concentration over the range $0.1–70\,\mu g\,ml^{-1}$; the detection limit is $0.1\,\mu g\,ml^{-1}$. The sensitivity is 50 times greater then in enzymic assays and allows sample dilution to avoid interferences; the assay also involves fewer steps and hence is quicker.

The crustacean *Cypridina hilgendorfii* has a luciferin of very different chemical structure; the mechanism of the bioluminescence is the same as that of the firefly, but without the presence of a co-factor (McCapra, 1987). Nakano and Takahashi (1990) report the use of analogues of *Cypridina* luciferin to detect superoxide generated by ischaemia and reperfusion of a rat's stomach.

The antioxidant activities of tea-leaf catechins can conveniently be measured by the quenching of *Cypridina* chemiluminescence (Mashiko et al., 1993). Ten microlitres of *Cypridina* luciferin analogue (4.4×10^{-8} mol dm^{-3} final concentration), $(2200-y)\,\mu l$ phosphate buffer (pH 7.1), 0.5 ml albumin/buffer ($1\,mg\,ml^{-1}$), $50\,\mu g$ xanthine oxidase ($200\,\mu g$ in 1.8 ml albumin/buffer) and y ml catechin ($1 \times 10^{-6}\,mol\,dm^{-3}$) were mixed in a quartz cell at 25°C in the dark; 0.2 ml hypoxanthine (3 mmol dm^{-3}) was injected; chemiluminescence was measured through the tube bottom by a single photon counter. Tea-leaf catechins attack superoxide radicals. The rate constants of the reaction (where Q is the catechin and P the oxidation product)

$$Q + O_2^{\cdot -} \to O_2^{2-} + P$$

range from 10^4 to $10^6\,dm^3\,mol^{-1}s^{-1}$, as big as the corresponding values for ascorbic acid and cytochrome c. The values were determined in aqueous solution, so differ from membrane antioxidant activities that relate to cancer prevention.

Luminous bacteria are found widely in marine environments. Bacterial luciferase, which acts in accordance with the outline mechanism shown in Figure 3.8, does not have a luciferin substrate as such. Instead, the light emission comes from a complex of luciferase, flavine mononucleotide and a long-chain fatty aldehyde (McCapra, 1987). Thus, bacterial bioluminescence is associated with a pyridine nucleotide rather than with the adenine nucleotide involved in firefly bioluminescence.

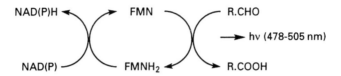

Figure 3.8   Principle of bacterial bioluminescence, in which light is emitted by the oxidation of a long-chain fatty aldehyde by flavine mononucleotide, which is regenerated in a coupled reaction. NAD(P), nicotinamide–adenine dinucleotide (phosphate); FMN, flavine mononucleotide.

Worsfold (1990) reports the use of co-immobilised luciferase and oxidoreductase within the detector of a flow injection manifold to determine NAD(P)H, incorporating the use of merging zones of NADH and decanal to minimise reagent and sample consumption. Dehydrogenases and substances involved in reactions catalysed by them can also be assayed by the NAD(P)H-dependent marine bacterial luciferase. Coupled enzyme assays of bile acids in hyperlipaemic sera have been reported, which are less affected by turbidity than spectrophotometric assays (Lekhakula *et al.*, 1991). Wieland *et al.* (1990a) were able to measure aldehydic lipid peroxidation products at a detectivity of $1 \times 10^{-6} \, mol \, dm^{-3}$ using an oxidoreductase enzyme (to produce flavine mononucleotide) in conjunction with bacterial luciferase in accordance with the principle shown in Figure 3.8. Müller *et al.* (1991) have reviewed *inter alia* the determination of both aldehydes and fatty acids in assays employing pyridinenucleotide-linked bacterial bioluminescence.

3.8   Lucigenin oxidation

Lucigenin and related 9,9′-diacridinium salts give an intense blue-green emission when oxidised by alkaline hydrogen peroxide. Gundermann (1965) and McCapra (1966) postulate the major chemiluminescent emitter to be *N*-methylacridone (blue light), produced via a peroxide, with other excited molecules involved. The reaction is catalysed by pyridine, piperidine, ammonia and osmium tetroxide. McCapra (1987) has proposed the formation of excited peroxide intermediates to explain the chemiluminescence of oxidised acridinium salts (equation 3.8). Lucigenin reacts with various reductants, including those present in normal human blood serum (Veazey and Nieman, 1979), and is affected by metal ions (Montano and Ingle, 1979). Analytical implications of the chemiluminescence of acridans and acridinium salts are discussed by McCapra (1987), Murphy and Sies (1J990) and Worsfold (1990).

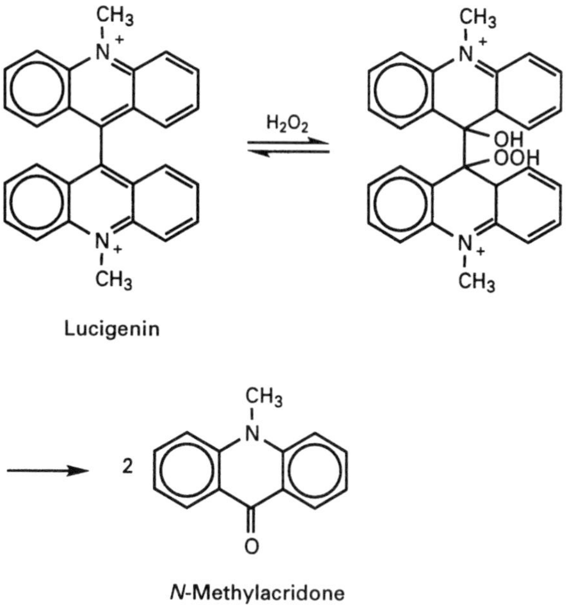

Lucigenin

(3.8)

N-Methylacridone

A chemiluminescent enzymatic assay for cholesterol in serum has been devised using lucigenin (Sasamoto *et al.*, 1995). The sample is mixed with an enzyme solution containing cholesterol esterase and cholesterol oxidase and incubated for 10 minutes. Cholesterol oxidase produces as a by-product hydrogen peroxide, selectivity for which is achieved by pH (11.75–11.9) and by the addition of $10 \, \mu mol \, dm^{-3}$ copper(II) ions, $0.1 \, mol \, dm^{-3}$ guanidine hydrochloride, $1 \, mg \, ml^{-1}$ sodium azide and $40 \, mg \, ml^{-1}$ Triton X-100 in an alkaline solution immediately after the addition of lucigenin. Chemiluminescence is recorded over a 10 s period after a delay of only 5 s (because interference from serum constituents increases with time). Cu^{2+} suppresses by 95% chemiluminescence due to ascorbic acid, glucose and distilled water blank without affecting the signal from hydrogen peroxide or serum. Guanidine hydrochloride diminished the chemiluminescence from hydrogen peroxide, but much more so from serum, thus increasing selectivity. Sodium azide increases the chemiluminescence, possibly by inhibiting peroxidase in serum. Triton X-100 doubles the chemiluminescence from hydrogen peroxide or from serum.

The limit of detection (at two standard deviations above the blank) is $1 \, mg \, l^{-1}$ (26 pmol/assay) using undiluted serum. Chemiluminescence from other sources in serum has to be measured as an enzyme blank and subtracted from the signal for it is equivalent to $3.6 \pm 2.6 \, mg \, l^{-1}$ cholesterol; otherwise the detection limit is $10 \, mg \, ml^{-1}$. The relative standard deviation is 2.3–4.2%; results correlate well ($r = 0.963$) with those from the Wako cholesterol E test across the cholesterol concentration range $0.1–2.7\,g\,l^{-1}$. Detectivity is low enough to measure cholesterol in fractionated lipoproteins.

_N_–Functionalised acridinium ester has been used for a competitive chemiluminescent immunoassay (CLIA) for oestradiol (Sato _et al._, 1996). Antigen was immobilised onto tubes by incubation for 30 min at 56°C with $0.5\,\mu g\,ml^{-1}$ oestradiol-3-thyroglobulin in phosphate buffer (pH 6.4) followed by washing and blocking with phosphate buffer (pH 6.4) containing 0.1% Tween 80. Oestradiol in normal rabbit serum is incubated for 1 hour with three volumes of labelled antibody in assay buffer (10% normal rabbit serum, 0.1% bovine serum albumin and $0.05\,\mu g\,ml^{-1}$ danazol in phosphate buffer, pH 6.4). This is then added to the tubes from which the blocking solution has been aspirated. After incubation for a further hour and washing, chemiluminescence is measured.

3.9 Oxidation of phenylhydrazines

This chapter so far, reflecting most chemiluminescence work to date, has been concerned mainly with applications of well-known reactions such as luminol oxidation and the peroxy–oxalate reaction (Worsfold, 1990; Robards and Worsfold, 1992), in which the usual oxidant is hydrogen peroxide. However, Townshend (1990), Worsfold (1990) and Andrews (1991) have reported on a wide range of usable oxidising agents. Potassium permanganate has proved to be a particularly versatile agent of chemiluminescent oxidation (Townshend, 1990). The range of possible reductants is even wider and consideration will now be given to two 'nontraditional' chemiluminescence systems that have useful applicability to lipid analysis.

The first of these has arisen from an interesting sequence of discoveries made by Professor Townshend's analytical chemistry team at the University of Hull, UK. Chemiluminescence has long been known to be associated with the oxidation of compounds containing the hydrazine functional group. The earliest example to be recorded was luminol (Albrecht, 1928) and further examples (White _et al._, 1967) were discovered among other hydrazides (see section 3.4). Hydrazine itself has also been reported to be chemiluminescent on oxidation by sodium hypochlorite in sodium hydrogencarbonate buffer (pH 11.6) (Wheatley, 1983). The chemiluminescence is affected by metal ions (Faizullah and Townshend, 1985). Detectivity was $5 \times 10^{-10}\,mol\,dm^{-3}$ in the presence of Al^{3+} or Ni^{2+}. Relative standard deviations were about 2%. Hydrazine

sulfate solution injected into polyphosphoric acid carrier is chemiluminescent when oxidised by permanganate (Faizullah, 1985) in nitric acid at
pH ≤ 1.6. Detectivity was $3 \times 10^{-7} \text{mol dm}^{-3}$. Relative standard deviations ranged from 0.6% to 2.5%. There was a lower susceptibility to metal
ion interference, which could be due to the complexing properties of
polyphosphate, but Mn^{2+} did interfere.

Phenylhydrazine and a range of nitro-substituted phenylhydrazines in
propan-2-ol solutions have now been found to be chemiluminescent on
oxidation with acidified potassium permanganate (Townshend and
Wheatley, 1998a). The chemiluminescence was deduced to originate
from the oxidation of the hydrazine functional groups; dinitrogen is a
possible emitting molecule. This finding makes possible a new approach
to chemiluminescence determinations of carbonyl compounds in a lipid
context. The phenylhydrazines have a long history of application in the
analysis of carbonyl compounds (e.g. Brady, 1931) and are of interest as
derivatisation agents for the ultimate products of lipid peroxidation
(Esterbauer and Zollner, 1989; Diplock, 1990).

Townshend and Wheatley (1998b) have developed an assay of 2,4-
dinitrophenylhydrazine based on its chemiluminescence when it is
oxidised by permanganate; $10 \,\mu l$ 2,4-dinitrophenylhydrazine dissolved
in 22.0% propan-2-ol was injected into 0.75mol dm^{-3} aqueous formic
acid carrier, which merged with the oxidant, $5.0 \times 10^{-5} \text{mol dm}^{-3}$
permanganate in 0.041mol dm^{-3} sulfuric acid. Total flow rate was
3.25ml min^{-1}. All solutions were maintained at 40°C. The log–log
calibration for 2,4-dinitrophenylhydrazine in optimum conditions was
linear ($r = 0.9972$, $n = 10$) from $1 \times 10^{-7} \text{mol dm}^{-3}$ to $2 \times 10^{-5} \text{mol dm}^{-3}$;
the detectivity calculated at 3 standard deviations above blank was
$1.1 \times 10^{-7} \text{mol dm}^{-3}$ (1.1 pmol analyte). Relative standard deviations
ranged from 2.3% to 8.0%, except at the detection limit. 2,4-
Dinitrophenylhydrazine, its solvent and the carrier had been selected
for the assay in a factorially designed evaluation of options using
analyte:blank signal ratio as the criterion. Rhodamine-B sensitiser
enhanced all signals but, by increasing blank signals, reduced the signal
to blank ratio. The optimum conditions can be explained in terms of a
combination of flow rate and factors known to increase the rate of
oxidation of 2,4-dinitrophenylhydrazine.

There was evidence of enhancement of the signal by formic acid.
Excited carbon dioxide molecules from the oxidation of formate ions
possibly produce a feeble emission and transfer energy to the emitting
products of analyte oxidation. Other molecules must also be involved in
energy transfer as the propan-2-ol solvent also acts as an enhancer. There
is other recent evidence (Deftereos et al., 1993; Ahmed and Townshend,
1994) that chemiluminescence can be increased by performing

permanganate oxidations in the presence of an ancillary reductant. The enhancing effects of formic acid and propan-2-ol on the signal are offset by diminishing effects due to competition for permanganate. The method was found to be subject to interference from common metal ions from V to Zn in the periodic table, especially Fe^{2+}. Iron salts are known to catalyse the permanganate–formic acid reaction; EDTA largely corrects this latter effect, presumably by the formation of a coordination compound that is less reactive. The effects of formic acid and of iron salts can be explained by the findings of kinetic studies of its reaction with permanganate (Wiberg and Stewart, 1956; Taylor and Halpern, 1959; Monzo *et al.*, 1987).

The oxidative chemiluminescence of phenylhydrazines can be applied to the determination of carbonyl compounds by utilising the attenuation of the signal that occurs on conversion to the corresponding phenylhydrazone (Townshend and Wheatley, 1998c). The optimised assay for 2,4-dinitrophenylhydrazine was carried out on 10 µl sample, which had been derivatised off-line by incubation with 5×10^{-6} mol dm^{-3} 2,4-dinitrophenylhydrazine in 0.05 mol dm^{-3} sulfuric acid for 2 h at 30°C. The log–log calibration for the combined derivatisation/assay of hexanal in optimised conditions is linear ($r = 0.9931$, $n = 5$) from 1×10^{-6} mol dm^{-3} to 2×10^{-5} mol dm^{-3}. The limit of detection calculated as above was 1.7×10^{-7} mol dm^{-3} (1.7 pmol of hexanal). Relative standard deviations ranged from 2.9% to 6.1%. The optimum conditions for 2,4-dinitrophenylhydrazone formation can be understood in terms of a stepwise mechanism of carbinolamine formation via a zwitterion intermediate.

The method was applied to the determination of the aldehyde products of autoxidation, both in fresh linoleic acid (0.33%) and after 2 h at 40°C (58%); these results are an indication of the wide dynamic range of the assay. The aldehyde products of lipoxygenase action on linoleic acid were also determined by this procedure; the assay had to be redesigned for this determination to provide comparable matrices for samples, blanks and the hexanal standards used for recalibration.

3.10 Oxidation of sulfites

Sulfite is a well known reductant and Stauff and Jaeschke (1975) discovered that oxidation of aqueous sulfur dioxide by acidified permanganate, cerium(IV) or hydrogen peroxide is feebly chemiluminescent; the exploitation of the weak chemiluminescence improved the detectivity of atmospheric sulfur dioxide by a factor of 50. A mechanism for the chemiluminescence of sulfite has been proposed by Meixner and Jaeschke (1984) and is shown in Scheme 3.3. Sulfites undergo an addition

(1) $HSO_3^- + MnO_4^- \longrightarrow HSO_3 + MnO_4^{2-}$

(2) $2HSO_3 \longrightarrow S_2O_6^{2-} + 2H^+$

(3) $S_2O_6^{2-} \longrightarrow SO_4^{2-} + SO_2^*$

(4) $SO_2^* \longrightarrow SO_2 + h\nu$ (>300 nm)

Scheme 3.3 Proposed mechanism for the chemiluminescence of acidified sulfite.

reaction with carbonyl compounds and the addition of cyclohexanone to protect sulfite solutions against atmospheric oxidation led to the observation (Al-Tamrah *et al.*, 1987) that this, at appropriate concentrations, enhanced oxidative chemiluminescence. Light emission is also sensitised by other cyclohexyl compounds (e.g. 3-cyclohexylaminopropanesulfonic acid). Paulls and Townshend (1995, 1996) have suggested that the enhancement depends on β-sultine formation and have shown that the phenomenon also occurs with higher cycloalkyl compounds, the optimum ring size being nine.

Work on the exploitation of the oxidative chemiluminescence of sulfites for lipid analysis has been centred on the University of Athens, Greece. Koukli and Calokerinos (1990) have described an air-segmented continuous-flow assay for a range of corticosteroid drugs that enhance the chemiluminescence of sulfite oxidised by cerium(IV). All the corticosteroid molecules contain cycloalkane rings, albeit fused ones, and this forms the rationale for their enhancement effect; in addition, cortisone and hydrocortisone are fluorescers. Because these latter two fluoresce in $0.10 \, mol \, dm^{-3}$ sulfuric acid solution, all the drugs were analysed in such solutions, which lead to maximum emission intensity even in the absence of an enhancer. Some of the drugs require the presence of 1% v/v acetone to make them soluble in sulfuric acid; this diminishes the chemiluminescence but not so severely as would methanol or ethanol. Results from the developed method were not subject to any serious interference from the usual pharmaceutical excipients and agreed well with those obtained by the method recommended in the US Pharmacopeia. Recoveries of added corticosteroid ranged from 91.4% to 106.5% (mean 98.2%). The linear range of the assay was typically 0.100 to $1.00 \, \mu g \, ml^{-1}$ and detection limits ranged from 0.016 to $0.30 \, \mu g \, ml^{-1}$; relative standard deviations covered the range 0.1% to 6.2%.

A similar flow injection assay has been devised (Syropoulos *et al.*, 1990) using bromate as the oxidant. The optimum conditions were found to be $2.0 \times 10^{-3} \, mol \, dm^{-3}$ bromate reacting with $5.0 \times 10^{-4} \, mol \, dm^{-3}$ sulfite (compared with $5.0 \times 10^{-4} \, mol \, dm^{-3}$ cerium(IV) reacting with

$0.010\,\mathrm{mol\,dm^{-3}}$ sulfite); the steroid hormones, all of which enhanced the chemiluminescence, were optimally dissolved in $0.020\,\mathrm{mol\,dm^{-3}}$ sulfuric acid. Acetone 4.0% v/v was added to improve the solubility of testosterone, progesterone and corticosterone. The effects of this were slight; hydrocortisone was assayed both with and without the added organic solvent, which diminished the analytical sensitivity (calibration slope) but also improved the detectivity and the linearity, while the linear range was unaffected. Results once again were not subject to any detectable interference from the usual pharmaceutical excipients and recoveries of added hormone ranged from 95.8% to 104.0%. The typical linear range of the assay was 0.50–$5.00\,\mathrm{\mu g\,ml^{-1}}$ (but up to $20.0\,\mathrm{\mu g\,ml^{-1}}$ for cortisone). Detectivities ranged from $0.010\,\mathrm{\mu g\,ml^{-1}}$ to $0.40\,\mathrm{\mu g\,ml^{-1}}$; typical relative standard deviations were 0.5% at $4.00\,\mathrm{\mu g\,ml^{-1}}$ and 2.3% at $0.50\,\mathrm{\mu g\,ml^{-1}}$. The use of bromate did not prove to be superior to using cerium(IV) even though its colourless solutions would avoid reabsorption of emitted light.

Deftereos and Calokerinos (1994) further developed the sulfite–cerium(IV) flow injection assays for corticosteroids and sex hormones, improving sensitivity, precision and accuracy and making the assays more suitable for post-column detection in liquid chromatography. Aqueous acetonitrile (20%) was found to give the most intense chemiluminescence of all the solvents capable of dissolving water-insoluble steroids; all organic solvents diminished light emission, which reduces the scope of sulfite chemiluminescence in lipid analysis. Sodium dodecyl sulfate micellar system was not a satisfactory solvent. A 500 µl sample in 20% acetonitrile solution was injected into $1.00\times10^{-3}\,\mathrm{mol\,dm^{-3}}$ sulfite in $1.0\,\mathrm{mol\,dm^{-3}}$ sulfuric acid and merged with $1.00\times10^{-3}\,\mathrm{mol\,dm^{-3}}$ cerium(IV) sulfate oxidant in $3.0\,\mathrm{mol\,dm^{-3}}$ sulfuric acid before detection. The optimum sulfite concentration for steroids in aqueous solution was $5.00\times10^{-4}\,\mathrm{mol\,dm^{-3}}$. The increased chemiluminescent intensity obtained by using increased sulfuric acid concentrations is consistent with the view that the emitting molecule is sulfur dioxide. The developed assay had a typical linear range for 20% acetonitrile solutions of 1.00–$20.0\,\mathrm{\mu g\,ml^{-1}}$ and for aqueous solutions of 0.02–$1.00\,\mathrm{\mu g\,ml^{-1}}$. Detection of low steroid concentrations was subject to detectable interference from sorbitol, galactose, fructose and sodium dodecyl sulfate, but not from other pharmaceutical excipients. Recoveries of added hormone ranged from 94.2% to 109.0%. Detection limits ranged from 0.013 to $2.00\,\mathrm{\mu g\,ml^{-1}}$; typical relative standard deviations (cortisone in 20% acetonitrile) were 0.74% at $5.00\,\mathrm{\mu g\,ml^{-1}}$ and 0.84% at $0.80\,\mathrm{\mu g\,ml^{-1}}$.

A flow injection chemiluminometric determination of some bile acids has been reported (Psarellis et al., 1994) that makes use of their sensitisation of the light emission accompanying the oxidation of sulfites by a variety of oxidising agents. The sensitising effect of the bile acids, as of other steroids, presumably arises from the system of fused cycloalkane rings that make up the steroid 'nucleus' in their structure. A 500 μl sample (cholic, deoxycholic, chenodeoxycholic or ursodeoxycholic acids or their sodium salts) was injected into sulfite flowing at $6\,ml\,min^{-1}$, which merged with the oxidant (Ce^{4+}, MnO_4^-, BrO_3^-, or $Cr_2O_7^{2-}$) also flowing at $6\,ml\,min^{-1}$. The reaction occurred in a flat glass coil adjacent to a photomultiplier tube. The preferred oxidant is $1.50 \times 10^{-3}\,mol\,dm^{-3}$ cerium (IV), acidified with $0.50\,mol\,dm^{-3}$ sulfuric acid. Permanganate, $4.00 \times 10^{-4}\,mol\,dm^{-3}$ acidified with $0.10\,mol\,dm^{-3}$ sulfuric acid, gives chemiluminescence of almost equal intensity. Light emission increases up to flow rates of $6\,ml\,min^{-1}$ for sample and oxidant. The optimum sulfite concentration is $7.00 \times 10^{-4}\,mol\,dm^{-3}$. Typical results are a linear range of 2.00–$30.0\,\mu g\,ml^{-1}$ ($r = 0.9995$, $n = 7$) and a detection limit of $1.05\,\mu g\,ml^{-1}$, which is not low enough to be suitable for biological samples.

3.11 Ultraweak chemiluminescence

Spontaneous emission of low-intensity light by living tissues has been reported widely. Ultraweak chemiluminescence from isolated rat liver nuclei is attributed to lipid peroxidation in the nuclear membrane. Lipid peroxidation in membranes is spatially constrained in a quasi-ordered environment; this leads to possible kinetic effects. The effect of temperature on reaction rates is modelled by the Arrhenius equation (3.9):

$$\text{rate} \propto \exp[-E_{act}/RT] \tag{3.9}$$

E_{act} is the activation energy of the reaction, R is the gas constant, and T is the temperature in kelvins. Activation energies can be determined from the slope of the Arrhenius plot of the rate of chemiluminescent emission versus the reciprocal of the temperature; such determinations are independent of sample size, chemiluminescence quantum yield or detector efficiency.

Phase transitions have been observed (Scott, 1994) as abrupt changes in the slope of the Arrhenius plot; the first observation of a phase transition in a nuclear membrane was accomplished in this way. Activation energies of the lipid peroxidation chain in different phases and catalytic effects upon them have been determined. The kinetics of excited state production depends on the composition, conformation and dynamics of the lipid membrane. Investigation of these properties by kinetic observations does not alter membrane dynamics as spin labels or fluorescent probes may.

The thermodynamic response of ultraweak chemiluminescence is a sensitive probe of membrane lipid dynamics. It is being used to study developmental and pathological changes in lipid composition and investigations have been extended to include intact leaves.

Direct chemiluminescence is a rapid and sensitive method for assessing phosphatidylcholine oxidation. Soya phosphatidylcholine was dissolved in a medium chain-length monoacylglycerol and the effects on oxidative status of storage at 40°C in air and under nitrogen were compared. Chemiluminescence correlated well with the thiobarbituric acid test, iodometry, ultraviolet spectrophotometry and isothermal microcalorimetry (Singh et al., 1996a). There was an induction time of 21 days using a 50:50 w/w solution of phosphatidylcholine in monoacylglycerol. The kinetics of the decomposition of hydroperoxides was also studied (Singh et al., 1996b) by analysis of chemiluminescence data obtained in an inert atmosphere at various temperatures. The apparent activation energies of the two steps in the chemiluminescent reaction were 74 and 62 kJ mol^{-1} and the order of reaction changes from 2 to 1 as the hydroperoxides are consumed.

Chemiluminescence can be induced on addition to human placental villous tissue of arachidonic, linoleic and linolenic acids (Das et al., 1997). The maximum emission is reached after 90 s. The chemiluminescence is inhibited by pre-incubation of the placental villi with aspirin, with garlic or with perchloric acid-treated garlic that is allicin-negative; it is proposed that the inhibition of chemiluminescence results from the inhibition of cyclooxygenase.

3.12   Concluding remark

There are very few chemical reactions in which lipids actually emit light. However, there are many methods for the determination of lipid analytes by the use of chemiluminescence reactions. They have arisen from chemiluminescence systems that have been discovered and developed independently of lipid chemistry. However, knowledge of the lipid molecules and their properties determines how these systems can best be exploited.

References

Ahmed, T.E.A. and Townshend, A. (1994) Flow injection chemiluminescence determination of the hydrazones of aromatic ketones. *Anal. Chim. Acta*, **292** 169.

Al-Tamrah, S.A., Townshend, A. and Wheatley, A.R. (1987) Flow injection chemiluminescence of a sulfite. *Analyst*, **112** 883.

Albrecht, H.O. (1928) On aminophthalhydrazide chemiluminescence. *Z. Phys. Chem.*, **136** 321 (in German).

Anderson, D.J., Guo, B., Xu, Y., *et al.* (1997) Clinical chemistry. *Anal. Chem.*, **69** (12) 165R.

Andrews, A.R.J. (1991) Applications of chemiluminescence to clinical, forensic and pharmaceutical analysis. *Anal. Proc.*, **28** (2) 38.

Arora, A., de Mello, A.J. and Manz, A. (1997) Sub-microliter electrochemiluminescence detector —a model for small volume analysis systems. *Anal. Commun.*, **34** (12) 393.

Atwater, J.E., Akse, J.R., Dehart, J. and Wheeler, R.R. (1997) "Reagentless" flow analysis determination of hydrogen peroxide by electrocatalysed electrochemiluminescence. *Anal. Lett.*, **30** 21.

Bissé, E., Gissler J., Wieland, E. and Wieland, H. (1990) Chemiluminescent enzyme method for determination of serum choline-containing phospholipids, in *Bioluminescence and Chemiluminescence: Current Status* (eds P.E. Stanley and L.J. Kricka), Wiley, Chichester, p 467.

Bochev, P.G., Bechev, B.G. and Magrisso, M.J. (1992) Six-sample multiplexing computerised analyser for integral and spectral luminescence measurements. *Anal. Chim. Acta*, **256** 29.

Brady, O.L. (1931) The use of 2,4-dinitrophenylhydrazine as a reagent for carbonyl compounds. *J. Chem. Soc.*, 756.

Burguera, J.L. and Townshend, A. (1979) Determination of amines by using a chemiluminescent reaction in solution. *Anal. Proc.*, **16** 263.

Burguera, J.L., Townshend, A. and Greenfield, S. (1980) Flow injection analysis for monitoring chemiluminescent reactions. *Anal. Chim. Acta*, **114** 209.

Chandross, E.A. (1963) A new chemiluminescent system. *Tetrahedron Lett.*, 761.

Chandross, E.A. (1994) Old and new perspectives on chemiluminescent processes, in *Bioluminescence and Chemiluminescence: Fundamentals and Applied Aspects* (eds A.K. Campbell, L.J. Kricka and P.E. Stanley), Wiley, Chichester, p 7.

Das, I., Patel, S. and Sooranna, S.R. (1997) Effects of aspirin and garlic on cyclooxygenase-induced chemiluminescence in human term placenta. *Biochem. Soc. Trans.*, **25** 995.

Deftereos, N.T., Calokerinos, A.C. and Efstathiou, C.E. (1993) Flow injection chemiluminometric determination of epinephrine, norepinephrine, dopamine and L-dopa. *Analyst*, **118** 627.

Deftereos, N.T. and Calokerinos, A.E. (1994) Flow-injection chemiluminometric determination of steroids. *Anal. Chim. Acta*, **290** 190.

Diplock, A.T. (1990) Measurement of antioxidants and degradation products of peroxides. *Anal. Proc.*, **27** 223.

Dukhovich, A.F., Philippova, N.Yu. and Ugarova, N.N. (1990) Bioluminescent assay for choline-containing phospholipids and luciferin, in *Bioluminescence and Chemiluminescence: Current Status* (eds P.E. Stanley and L.J. Kricka), Wiley, Chichester, p 507.

Esterbauer, H. and Zollner, H. (1989) Methods for the determination of aldehydic lipid peroxidation products. *Free Radicals in Biology and Medicine*, **7** 197.

Faizullah, A.T. (1985) *Continuous flow injection analysis.* PhD thesis, University of Hull, p 61.

Faizullah, A.T. and Townshend, A. (1985) Flow injection analysis with chemiluminescence detection: determination of hydrazine. *Anal. Proc.*, **22** 15.

Frei, B., Yamamoto, Y., Niclas, D. and Ames, B. (1988) Evaluation of an isoluminol chemiluminescence assay for the detection of hydroperoxides in blood plasma. *Anal. Biochem.*, **175** 120.

Geng, Z. (1992) Progress on chemiluminescence analysis in China. *Huaxue Tongbao*, 1 (in Chinese) and *Anal. Abstr.*, 1993, 10C104.

Grayeski, M.L. and DeVasto, J.K. (1987) Coumarin derivatizing agents for carboxylic acid detection using peroxyoxalate chemiluminescence with liquid chromatography. *Anal. Chem.*, **59** 1203.

Gumuslu, S., Erkilic, A., Yucel, G., Serteser, M. and Ozben, T. (1997) The effects of detergents on t-butyl hydroperoxide-induced chemiluminescence. *Int. J. Clin. Lab. Res.*, **26** 203.

Gundermann, K.-D. (1965) Chemiluminescence in organic compounds. *Angew. Chem. Int. Ed.*, **4** (7) 566.

Hastings, J.G.M., Wheat, P.F. and Oxley, K.M. (1989) Rapid microbiology by ATP bioluminescence in the clinical laboratory using a microtiter plate luminometer. *Soc. Appl. Bacteriol. Tech. Ser.*, **26** 229.

Hemmi, A., Yagiuda, K., Funazaki, N., Ito, S., Asano, Y., Imato, T., Hayashi, K. and Karube, I. (1995) Development of a chemiluminescence detector with photodiode detection for flow-injection analysis and its application to L-lactate analysis. *Anal. Chim. Acta*, **316** 323.

Higashidate, S., Hibi, K., Seda, M., Kanda, S. and Imai, K. (1990) Sensitive assay system for bile-acids and steroids having hydroxyl groups utilising high-performance liquid chromatography with peroxyoxalate chemiluminescence detection. *J. Chromatogr.*, **515** 577.

Honda, K., Miyaguchi, K. and Imai, K. (1985a) Evaluation of aryl oxalates for chemiluminescence detection in high-performance liquid chromatography. *Anal. Chim. Acta*, **177** 103.

Honda, K., Miyaguchi, K. and Imai, K. (1985b) Evaluation of fluorescent compounds for peroxyoxalate chemiluminescence detection. *Anal. Chim. Acta*, **177** 111.

Imai, K., Higashidate, S., Nishitani, A., Tsukamoto, Y. *et al.* (1989) Sensitive detection of oxo-steroids and oxo-bile acids by liquid chromatography with peroxyoxalate chemiluminescence detection. *Anal. Chim. Acta*, **227** 21.

Ingle, M., Howorth, J.R., Patchett, B.E., Carter, M.K. and Read, P.D. (1994) Photon counting systems for bioluminescence imaging, in *Bioluminescence and Chemiluminescence: Fundamentals and Applied Aspects* (eds A.K. Campbell, L.J. Kricka and P.E. Stanley), Wiley, Chichester, p 637.

Kawasaki, T., Maeda, M. and Tsuji, A. (1985) Chemiluminescence high performance liquid chromatography using N-(4-aminobutyl)-N-ethylisoluminol as a precolumn labelling reagent. *J. Chromatogr.*, **328** 121.

Kessler, A., Schumacher, M. and Wood, W.G. (1994) Immunoluminometric assays for the quantification of apolipoprotein A-I, B, C-II, apolipoprotein (a) and lipoprotein (a). *Eur. J. Clin. Chem. Clin. Biochem.*, **32** 127.

Kim, B.B., Pisarev, V.V. and Egorov, A.M. (1990) Chemiluminescent reactions of luminol oxidation induced by different peroxides. Stop-flow measurements, in *Bioluminescence and Chemiluminescence: Current Status* (eds P.E. Stanley and L.J. Kricka), Wiley, Chichester, p 393.

Kim, Y.K., Hong, E.K., Lee, C.H. and Lee, W.C. (1993) Measurement of antioxidation activity based on chemiluminescence reaction, in *Bioluminescence and Chemiluminescence: Status Report* (eds A.A. Szalay, P.E. Stanley and L.J. Kricka), Wiley, Chichester, p 244.

Kondo, Y., Kawai, Y., Miyazawa, T., Matsui, H. and Mizutani, J. (1994) Assay for lipoxygenase activity by chemiluminescence. *Biosci. Biotechnol. Biochem.*, **58** 421.

Koo, J.-Y. and Schuster, G.B. (1977) Chemically initiated electron exchange luminescence. A new chemiluminescent reaction path for organic peroxides. *J. Am. Chem. Soc.*, **99** 6107.

Koukli, I.I. and Calokerinos, A.E. (1990) Continuous-flow chemiluminescence determination of some corticosteroids. *Analyst*, **115** 1553.

Koziol, T., Grayeski, M.L. and Weinberger, R.W. (1984) Determination of trace levels of steroids in blood-plasma by liquid chromatography with peroxyoxalate chemiluminescence detection. *J. Chromatogr.*, **317** 355.

Kricka, L.J. (1991) Chemiluminescent and bioluminescent techniques. *Clin. Chem.* (Winston-Salem, NC), **37** 1472.

Kricka, L.J. (1993) Chemiluminescence and bioluminescence. *Anal. Chem.*, **65** 460R.

Kricka, L.J., Aly, A.A. and Xiaoying, J. (1993) Automated enhanced chemiluminescent antioxidant assay using the Dynatech ML 3000 microplate luminometer, in *Bioluminescence and Chemiluminescence: Status Report* (eds A.A. Szalay, P.E. Stanley and L.J. Kricka), Wiley, Chichester, p 239.

Kricka, L.J. and Stanley, P.E. (1996) Bioluminescence and chemiluminescence literature. *J. Biolumin. Chemilumin.*, **11** (1) 39.

Kricka, L.J. and Stanley, P.E. (1992) Bioluminescence and chemiluminescence literature. *J. Biolumin. Chemilumin.*, **7** 47.

Kricka, L.J. and Xiaoying, J. (1995) 4-Phenylboronic acid: a new type of enhancer for the horseradish peroxidase catalysed chemiluminescence of luminol. *J. Biolumin. Chemilumin.*, **10** 49.

Kwakman, P.J.M. and Brinkman, U.A.Th. (1992) Peroxyoxalate chemiluminescence detection in liquid chromatography. *Anal. Chim. Acta*, **266** 175.

Kwakman, P.J.M., van Shaik, H.P., Brinkman, U.A.Th. and de Jong, G.J. (1991) N-(bromoacetyl)-N'-[5-(dimethylamino)-naphthalene-1-sulphonyl]piperidine as a sensitive labelling reagent for the determination of carboxylic acids by liquid chromatography with peroxyoxalate chemiluminescence and fluorescence detection. *Analyst*, **116** 1385.

Lekhakula, S., Boonpisit, S. and Arnornkitticharoen, B.J. (1991) Total bile acids in hyperlipidemic serum determined by bioluminescent and spectrophotometric methods. *J. Biolumin. Chemilumin.*, **6** 259.

Lin, J., An, J., Zheng, X. and Hu, Y. (1991) Combination of the electrogenerated chemiluminescence meter, model ECL-1, with the Apple-II microcomputer. *Fenxi Shiyansi*, **10** 59 (in Chinese) and *Anal. Abstr.*, 1992, 5C81.

Maeda, M., Tsuji, A., Ohshima, N. and Hukuoka, M. (1993) Flow-injection determination of glucose, bile acid and ATP using an immobilised enzyme reactor and a chemiluminescent assay of NAD(P)H. *J. Biolumin. Chemolumin.*, **8** 241.

Mann, B. and Grayeski, M.L. (1990) Background emission from the peroxyoxalate chemiluminescence reaction in the absence of fluorophors. *Anal. Chem.*, **62** 1532.

Mashiko, S., Iwanaga, S., Hatate, H., Suzuki, N., Seto, R., Hara, Y., Oguni, I., Nomoto, T. and Yoda, B. (1993) Antioxidative activity of bioactive compounds: measurement by *cypridina* chemiluminescence method, in *Bioluminescence and Chemiluminescence: Status Report* (eds A.A. Szalay, P.E. Stanley and L.J. Kricka), Wiley, Chichester, p 247.

Matthäus, B., Wiezorek, C. and Eichner, K. (1994) Fast chemiluminescence method for detection of oxidised lipids. *Fat. Sci. Technol.*, **96** 95.

Maxwell, S.J.R., Whitehead, T.P. and Thorpe, G.H. (1993) Detection of oxidative stress in human disease using an enhanced chemiluminescent assay for antioxidant capacity, in *Bioluminescence and Chemiluminescence: Status Report* (eds A.A. Szalay, P.E. Stanley and L.J. Kricka), Wiley, Chichester, p 252.

McCapra, F. (1966) Chemiluminescence in organic compounds. *Q. Rev.*, **20** 485.

McCapra, F. (1987) Potential applications of bioluminescence and chemiluminescence in biosensors, in *Biosensors: Fundamentals and Applications* (eds A.P.F. Turner, I. Karube and G.S. Wilson), Oxford, Oxford University Press, p 617.

Meixner, F. and Jaeschke, W. (1984) Chemiluminescence technique for detecting sulfur dioxide in the ppt-range. *Fresenius Z. Anal. Chem.*, **317** 343.

Metsa-Ketala, T. (1990) Luminescent assay for total peroxyl radical-trapping capability of plasma, in *Bioluminescence and Chemiluminescence: Current Status* (eds P.E. Stanley and L.J. Kricka), Wiley, Chichester, p 389.

Mitchell, P. (1991) Detecting the single cell. *Lab. Equip. Dig.*, **29** 37.

Montano, L.A. and Ingle, J.D., Jr (1979) Investigation of the lucigenin chemiluminescence reaction. *Anal. Chem.*, **51** 919.

Monzo, I.S., Palou, J., Penalver, J. and Valero, R. (1987) Kinetics of the permanganate oxidation of formic acid in aqueous solution. *Int. J. Chem. Kinet.*, **19** 741.

Mueller-Klieser, W. and Walenta, S. (1991) Method, enzyme cocktail and apparatus for determination of the distribution of metabolites in a tissue sample. German Patent 3 935 974, 1991, *Chem. Abstr.* 1991, 115, 68001z.

Mueller-Klieser, W., Kroeger, M., Walenta, S. and Rofstad, E.K. (1991) Comparative imaging of structure and metabolites in tumors. *Int. J. Radiat. Biol.*, **60** 147.

Müller, M.M., Griesmacher, A. and Grabenwoger, M. (1991) Chemiluminometric techniques in clinical chemistry. *Mikrochim. Acta*, **11** 157.

Mullin, J.L. and Seitz, W.R. (1984) Two-phase flow cell for chemiluminescence and bioluminescence measurements. *Anal. Chem.*, **56** 1046.

Murphy, M.E. and Sies, H. (1990) Chemiluminescence: approaches to detecting and determining reactive oxygen species. *Anal. Proc.*, **27** 217.

Nakano, M. (1996) Photon emission and disease. *Photomed. Photobiol.*, **18** 1.

Nakano, M. and Takahashi, A. (1990) A novel assay of superoxide radical generation in *in situ* stomach of experimental rat: ischaemia-reperfusion, in *Bioluminescence and Chemiluminescence: Current Status* (eds P.E. Stanley and L.J. Kricka), Wiley, Chichester, p 397.

Naslund, B., Arner, P., Bernstrom, K., Bolinder, J., Hallander, L. and Lundlin, A. (1990) Assays of free fatty acids and glucose by the horseradish peroxidase catalysed luminol reaction, in *Bioluminescence and Chemiluminescence: Current Status* (eds P.E. Stanley and L.J. Kricka), Wiley, Chichester, p 385.

Neuport, W., Oelkers, R., Brune, K. and Geisslinger, G. (1996) A new reliable chemiluminescence immunoassay (CLIA) for prostaglandin-E_2 using enhanced luminol as substrate. *Prostaglandins*, **52** 385.

Nondek, L., Milofsky, R.E. and Birks, J.W. (1991) Determination of carbonyl compounds in air by high performance liquid chromatography using on-line analysed microcartridges, fluorescence and chemiluminescence detection. *Chromatographia*, **32** 33.

Nozaki, O., Ohba, Y. and Imai, K. (1988) Determination of serum estradiol by normal-phase high-performance liquid chromatography with peroxyoxalate chemiluminescence detection. *Anal. Chim. Acta*, **205** 255.

O'Kane, M.J., Wisdom, G.B., McEneny, J., Mcferran, N.V. and Trimble, E.R. (1992) Pre-β high density lipoprotein determined by immunoblotting with chemiluminescent detection. *Clin. Chem.* (Winston-Salem NC), **38** 2273.

Paulls, D.A. and Townshend, A. (1995) Sensitive determination of sulfite using flow injection with chemiluminescent detection. *Analyst*, **120** 467.

Paulls, D.A. and Townshend, A. (1996) Enhancement by cycloalkanes of the chemiluminescent oxidation of sulfite. *Analyst*, **121** 831.

Pettersen, J. (1994) Chemiluminescence of fish oils and its flavour quality. *J. Sci. Food Agric.*, **65** 307.

Psarellis, I.M., Deftereos, N.T., Sarantoni, E.G. and Calokerinos, A.C. (1994) Flow-injection chemiluminometric determination of some bile acids. *Anal. Chim. Acta*, **294** 27.

Rauhut, M.M., Bollyky, L.J., Roberts, B.G., Loy, M., Whitman, R.H., Ianotta, A.V., Semsel, A.M. and Clarke, R.A. (1967) Chemiluminescence from reactions of electronegatively substituted aryl oxalates with hydrogen peroxide and fluorescent compounds. *J. Am. Chem. Soc.*, **89** 6515.

Rigin, V.I. (1978) Chemiluminescence determination of microamounts of cholesterol by means of immobilised enzymes. *J. Anal. Chem. USSR*, **33** 1265.

Robards, K. and Worsfold, P.J. (1992) Analytical applications of liquid phase chemiluminescence. *Anal. Chim. Acta*, **266** 147.

Roswell, D.F. and White, E.H. (1978) Chemiluminescence of luminol and related hydrazides, in *Methods in Enzymology*, LVII (ed. M.A. DeLuca), Academic Press, New York, p 409.

Sakura, S. (1992) Electrochemiluminescence of hydrogen peroxide–luminol at a carbon electrode. *Anal. Chim. Acta*, **262** 49.

Sakura, S. and Imai, H. (1988) Determination of sub-nmol hydrogen peroxide by electrochemiluminescence of luminol in aqueous solution. *Anal. Sci.*, **4** 9.

Sakura, S. and Terao, J. (1992a) Determination of lipid hydroperoxides by electrochemiluminescence. *Anal. Chim. Acta*, **262** 59.

Sakura, S. and Terao, J. (1992b) Comparison of electrochemiluminescence and amperometric detection of lipid hydroperoxides. *Anal. Chim. Acta*, **262** 217.

Sasamoto, H., Maeda, M. and Tsuji, A. (1995) Chemiluminescent enzymatic assay for cholesterol in serum using lucigenin. *Anal. Chim. Acta*, **310** 347.

Sato, H., Mochizuki, H., Tomita, Y., Izako, T., Sako, N. and Kanamori, T. (1996) Competitive chemiluminescence immunoassay for oestradiol using an *N*-functionalised acridinium ester. *J. Biolumin. Chemilumin.*, **11** 23.

Scott, R.Q. (1994) Thermodynamic response of bioluminescence as a sensitive probe of membrane lipid dynamics, in *Bioluminescence and Chemiluminescence: Fundamentals and Applied Aspects* (eds A.K. Campbell, L.J. Kricka and P.E. Stanley), Wiley, Chichester, p 36.

Singh, S.K., Suurkuusk, M., Eldsaeter, C., Karlsson, S. and Albertsson, A.-C. (1996a) Chemiluminescence is a rapid and sensitive method to assess phosphatidylcholine oxidation. *Int. J. Pharm.*, **142** 199.

Singh, S.K., Suurkuusk, M. and Eldsaeter, C. (1996b) Kinetics of decomposition of hydroperoxides formed during the oxidation of soya phosphatidylcholine by analysis of the chemiluminescence generated. *Int. J. Pharm.*, **142** 215.

Smith, D., Proctor, S.D. and Mamo, J.C.L. (1997) A highly sensitive assay for quantitation of apolipoprotein B_{48} using an antibody to human apolipoprotein B and enhanced chemiluminescence. *Ann. Clin. Biochem.*, **34** 185.

Stanley, P.E. (1992a) A survey of more than 90 commercially available luminometers and imaging devices for low-light measurements of chemiluminescence and bioluminescence, including instruments for manual, automatic and specialised operation, for HPLC, LC, GLC and microtitre plates. I. Descriptions. *J. Biolumin. Chemilumin.*, **7** 77.

Stanley, P.E. (1992b) A survey of more than 90 commercially available luminometers and imaging devices for low-light measurements of chemiluminescence and bioluminescence, including instruments for manual, automatic and specialised operation, for HPLC, LC, GLC and microtitre plates. II. Photographs. *J. Biolumin. Chemilumin.*, **7** 157.

Stanley, P.E. (1993) Survey of some commercially available kits and reagents which include bioluminescence or chemiluminescence in their operation: includes immunoassays, hybridisation, labels, probes, blots and ATP-based rapid microbiology products from more than 40 companies. *J. Biolumin. Chemilumin.*, **8** 51.

Stanley, P.E. (1997) Commercially available luminometers and imaging devices for low-light measurements of chemiluminescence and bioluminescence, and kits and reagents which include bioluminescence or chemiluminescence in their operation. *J. Biolumin. Chemilumin.*, **12** 61.

Stauff, J. and Jaeschke, W. (1975) Chemiluminescence technique for measuring atmospheric trace concentrations of sulfur dioxide. *Atmos. Environ.*, **9** 1038.

Steijger, O.M., den Nieuwenboer, H.C.M., Lingeman, H., Brinkman, U.A.Th., Holthuis, J.J.M. and Smilde, A.K. (1996) Enhancement of peroxyoxalate chemiluminescence by copper(II) in flow injection analysis; optimisation by factorial design analysis. *Anal. Chim. Acta*, **320** 99.

Styrov, V.V., Tyurin, Yu.I. and Shigalugov, S.Kh. (1991) Analytical luminescence method. Principles of luminescence analysis. *Zavod. Lab.*, **57** 1 (in Russian) and *Anal. Abst.*, 1992, 12C32.

Suzaki, E. (1995) Development of video microimage analysis and its applications to the dynamics of a single neutrophil response. *Bunseki Kagaku*, **44** 163 (in Japanese) and *Anal. Abstr.*, 1995, 6F15.

Syropoulos, A.B., Sarantonis, E.G. and Calokerinos, A.E. (1990) Flow-injection chemilumino-metric analysis of some steroids by their sensitizing effect on the bromate–sulfite reaction. *Anal. Chim. Acta*, **239** 195.

Tanaka, K. and Ishikawa, E. (1986) A highly sensitive bioluminescent assay of β-D-galactosidase from *Escherichia coli* using 2-nitrophenyl-β-D-galactopyranoside as a substrate. *Anal. Lett.*, **19** 433.

Taylor, S.M. and Halpern, J. (1959) Kinetics of the permanganate oxidation of formic acid and formate ion in aqueous solution. *J. Am. Chem. Soc.*, **81** 2933.

Townshend, A. (1990) Solution chemiluminescence—some recent analytical developments. *Analyst*, **115** 495.

Townshend, A. and Wheatley, R.A. (1998a) Oxidative chemiluminescence of some nitrogen nucleophiles in the presence of formic acid as an ancillary reductant, *Analyst*, **123** 267.

Townshend, A. and Wheatley, R.A. (1998b) Oxidative chemiluminescence assay of 2,4-dinitrophenylhydrazine. *Analyst*, **123** 1041.

Townshend, A. and Wheatley, R.A. (1998c) The determination of carbonyl compounds by the oxidative chemiluminescence of 2,4-dinitrophenylhydrazine. *Analyst*, **123** 1047.

Ueno, K. (1991) Analytical applications of chemiluminescence—photography. *Senryo to Yakuhin*, **36** 100 (in Japanese) and *Anal. Abst.*, 1992, 10C83.

Ugarova, N.N. (1991) *J. Biolumin. Chemilumin.*, **6** 139.

Veazey, R.L. and Nieman, T.A. (1979) Chemiluminescent determination of clinically important organic reductants. *Anal. Chem.*, **51** 2092.

Wadano, A., Ikeda, T., Matumoto, M. and Himeno, M. (1991) Immobilised catalyst for detecting chemiluminescence in lipid hydroperoxide. *Agric. Biol. Chem.*, **55** 1217.

Wang, L., Yan, F. and Wang, X. (1991) Electrogenerated chemiluminescence and its application in analytical chemistry. *Huaxue Tongbao*, 8 (in Chinese) and *Anal. Abst.*, 1992, 10C46.

Wheatley, A.R. (1983) *Flow analysis using chemiluminescent detection*. PhD thesis, University of Hull, p 62.

White, E.H., Bursey, M.M., Roswell, D.F. and Hill, J.H.M. (1967) The chemiluminescence of some monoacylhydrazides. *J. Org. Chem.*, **32** 1198.

Whitehead, T.P., Thorpe, G.H. and Maxwell, S.J.R. (1992) Enhanced chemiluminescent assay for antioxidant capacity in biological fluids. *Anal. Chim. Acta*, **266** 265.

Wiberg, K.B. and Stewart, R. (1956) The mechanisms of permanganate oxidation. II. The oxidation of formate ions. *J. Am. Chem. Soc.*, **78** 1214.

Wieland, E., David, A., Kather, H. and Armstrong, V.W. (1990a) Bioluminometric determination of aldehydic lipid peroxidation products during the oxidation of low density lipoproteins, in *Bioluminescence and Chemiluminescence: Current Status* (eds P.E. Stanley and L.J. Kricka), Wiley, Chichester, p 455.

Wieland, E., Diedrich, F., Niedmann, P.D. and Seide, D. (1990b) Measurement of free fatty acid, phospholipid, triglyceride and cholesterol ester hydroperoxides by a CL-HPLC method, in *Bioluminescence and Chemiluminescence: Current Status* (eds P.E. Stanley and L.J. Kricka), Wiley, Chichester, p 235.

Worsfold, P.J. (1990) Analytical applications of luminescence detection in liquid chromatography and flow injection analysis, in *Bioluminescence and Chemiluminescence: Current Status* (eds P.E. Stanley and L.J. Kricka), Wiley, Chichester, p 371.

Yamamoto, Y., Brodsky, M.H., Baker, J.C. and Ames, B. (1987) Detection and characterisation of lipid hydroperoxides at picomole levels by high-performance liquid chromatography. *Anal. Biochem.*, **160** 7.

Yoshimura, Y., Ohsawa, K. and Imaeda, K. (1992) Measurement of active oxygen species. *Bunseki*, 207 (in Japanese) and *Anal. Abst.*, 1993, 3D117.

Zhang, J., Shen, X., Tang, L. and Li, X. (1996) Antioxidant activities of baicalin, alizarin, green tea polyphenols and trolox assayed by enhanced chemiluminescence. *Wuli Xuebao*, **12** 350 (in Chinese) and *Chem. Abstr.*, **126** (1997), 5, 56616t.

Zimmermann, T., Albrecht, S., Herrmann, U., Trausch, M., Menschikowski, M., Freidt, T., Müller, R. and Gebhardt, N. (1993) Chemiluminometric detection of the lipid peroxidation state in patients with gastrointestinal cancer, in *Bioluminescence and Chemiluminescence: Status Report* (eds A.A. Szalay, P.E. Stanley and L.J. Kricka), Wiley, Chichester, p 536.

4 NMR in conjunction with GC-MS and UV methods: a case study on marine lipids

Jennifer C. MacPherson, Debra L. Bemis,

Robert S. Jacobs, William H. Gerwick and James Todd

4.1 Overview

In the face of emerging and drug-resistant disease states, there is a pressing need to discover new drugs and/or alternative therapies to existing drugs. Oxylipins, oxygenated fatty acid metabolites of variable chain length, have been reported to be produced in hundreds of marine plants and invertebrates, which implies a fairly high level of conservation of these metabolic pathways (Stanley-Samuelson, 1991; Gerwick and Bernart, 1993; DePetrocellis and DiMarzo, 1994). De Petrocellis and Di Marzo (1994) emphasized how marine models of fatty acid biosynthesis, particularly the twenty-carbon chain eicosanoids, have produced invaluable information concerning previously unknown enzymatic pathways, have revealed the production of unique metabolites, and have demonstrated that eicosanoids can possess unique biological activities that may later be defined in vertebrates. So studies of marine models can serve to provide novel compounds, as well as to provide insights into the evolution of biosynthetic pathways that may contribute to the regulation of human pathological disease.

With the diversity of oxylipins that are formed through a number of enzymatic pathways, it is necessary to use not one, but a series of spectroscopic methods to isolate and identify these metabolites. In this chapter we will discuss the methodology used to identify oxylipins produced by primitive marine organisms: a cyanobacterium, *Lyngbya majuscula*; a red alga, *Agardhiella subulata*; a green alga, *Anadyomene stellata*; and the sole circulating hemocyte from the invertebrate *Limulus polyphemus*.

4.2 Isolation and identification of novel oxylipins from a macrophytic chlorophyte

Studies of polyunsaturated fatty acid (PUFA) metabolism in a variety of macrophytic rhodophytes, phaeophytes, and chlorophytes have led to the discovery of both unique enzymatic pathways and resultant metabolites

(Gerwick and Bernart, 1993; Gerwick, 1994). Our work has focused on a primitive member of the last division, *Anadyomene stellata* (Figure 4.1). We have observed this alga to be capable of enzymatically oxidizing PUFAs, resulting in novel conjugated tetraene-containing metabolites (Mikhailova *et al.*, 1995). Similar enzymatic activity has been observed in the coralline red alga *Bossiella orbigniana*, and the red alga *Lithothamnion corallioides* (Burgess *et al.*, 1991; Hamberg, 1992). Interestingly, this activity in *A. stellata* has been determined to be most prevalent in the chloroplasts when compared to whole algal homogenates and mitochon-

Figure 4.1   *Anadyomene stellata* (×100) collected in the Florida Keys. The large interstitial cells stacked in a tight parallel fashion between the uniserate veins distinguish it from other species, and are easily accessible for experimentation both *in vivo* and *in vitro*. (Photograph courtesy of D.S. Littler and M.M. Littler, Smithsonian Institution.)

drial preparations. The purpose of this section is to illustrate how a combination of spectroscopic methods enabled us to identify and elucidate a series of unique conjugated tetraene-containing PUFAs in *A. stellata* homogenates, and to observe the increased production of these compounds in isolated chloroplasts supplied with a series of PUFA substrates.

The identification and structural elucidation of conjugated tetraene PUFAs pose special difficulties owing to their instability, relatively low levels, and repetitive methine and methylene groups. The molecules were methylated using etheral CH_2N_2 to increase long-term stability. UV spectrophotometry allowed identification of these compounds since the conjugated tetraene system is strongly UV absorbent, a property true of conjugated diene and triene systems as well. These conjugated double bond configurations are known to have characteristic absorbance patterns and λ_{max} rendering them easily identifiable (Figure 4.2). GC-MS analysis of the methyl esters resulted in important information regarding the molecular masses and chemical composition based on observed molecular ions and fragmentation patterns. However, the structural elucidation of these straight-chain PUFA derivatives presented specific challenges since their formation involves only the addition of double bonds rather than varying functionalities, such as cyclization and epoxidation reactions. NMR methodology enabled us to determine the exact location and orientation of the double bonds within these molecules. A more in-depth discussion of NMR methodology will be addressed in the second section of this chapter; however, a brief description of the methodology used in this study is presented here.

4.2.1   Detection in algal homogenates

The whole algal homogenates of *A. stellata* were prepared by grinding in liquid nitrogen followed by homogenization in $CHCl_3$–MeOH (2:1) at $0°C$. The organic phase was dried with Na_2SO_4, filtered, and evaporated under vacuum. After resuspending the sample in 1 ml of hexane, it was methylated with diazomethane gas (Lombardi, 1990). Methyl esters were chromatographed over TLC-grade silica gel under vacuum with increasing percentages of ethyl acetate in hexane. UV spectrophotometry was utilized to determine the presence of the conjugated tetraene moiety, $\lambda_{max}=292, 306, 322$ nm. Compounds containing this functionality were identified in the fraction eluting from silica gel in EtOAc–hexane (1:99, v/v).

Fractions that contained conjugated tetraenes were further purified by normal-phase HPLC (tandem 10 μm 4 mm×30 cm Alltech Si columns, $\lambda=328$ nm). An isocratic separation (EtOAc–hexane (1:99, v/v)) with a

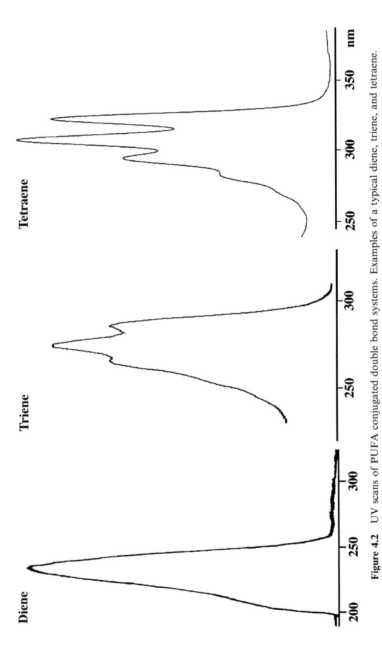

Figure 4.2   UV scans of PUFA conjugated double bond systems. Examples of a typical diene, triene, and tetraene.

flow rate of $3\,ml\,min^{-1}$ was used. Three fractions with significant absorbance were collected and further analyzed by GC-MS and NMR.

Gas chromatography–mass spectrometry (GC-MS) was performed on a Hewlett-Packard Model 5890 Series II GC and an HP5971A quadrupole mass-selective detector interfaced with a Hewlett-Packard Chemstation. A Hewlett Packard Ultra 1 column (11.5 m) with splitless injection was used. The injector temperature was set at 250°C and the initial oven temperature was 100°C with a $10°C\,min^{-1}$ ramp to a final temperature of 240°C. GC-MS analysis resulted in the detection of five PUFAs containing the conjugated tetraene functionality. These PUFAs include 16:5Δ4,7,9,11,13; 18:4Δ6,8,10,12; 20:5Δ5,8,10,12,14; 20:6Δ5,8, 10,12,14,17; and 22:7Δ4,7,9,11,13,16,19 (Scheme 4.1).

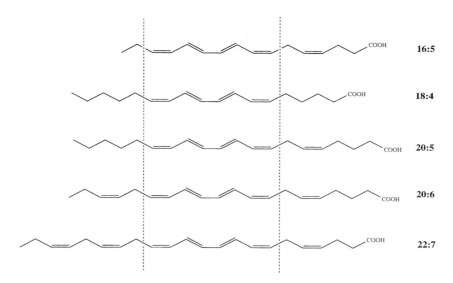

Scheme 4.1 Conjugated tetraene structures isolated from *Anadyomene stellata*. (Reproduced from Mikhailova *et al.* (1995) *Lipids* **30** (7) 583-589 with the permission of AOCS Press.)

The 22:7 compound was quite interesting since it is previously undescribed. The mass spectrum of the methylated analog had an $M^{+} = 340$ which corresponds to the molecular formula of $C_{23}H_{32}O_{2}$ (LR EIMS m/z (rel. intensity) obs.$[M^{+}]$ m/z 340 (6), 260 (18), 232(30), 171 (10), 157 (21), 145 (27), 131 (83), 117 (83), 105 (63), 91 (100), 79 (57), 67 (30)) (Figure 4.3). NMR technology was used to elucidate the structure of this compound. The NMR experiments were conducted on either Brucker ACP 300 MHz or AM 400 MHz instruments. As this novel 22:7 metabolite was isolated in only small quantity, the most sensitive

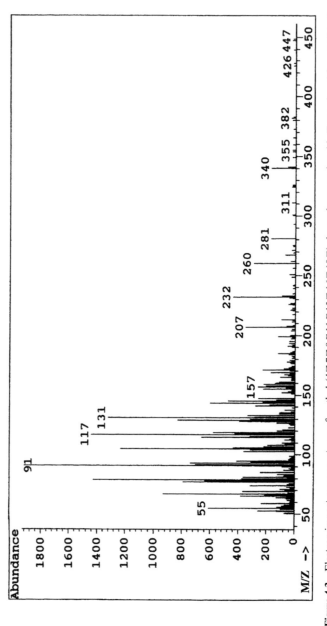

Figure 4.3 Electron impact mass spectrum of methyl-(4Z,7Z,9E,11E,13Z,16Z,19Z)-docosaheptaenoic acid. (Reproduced from Mikhailova *et al.* (1995) *Lipids* **30** (7) 583-589 with the permission of AOCS Press.)

technique, distortionless enhancement by polarization transfer (DEPT), was chosen for observation of its ^{13}C NMR resonance bands (approximately 4-fold better detection over that of the conventional ^{13}C NMR spectrum). The shortcoming of this technique, that non-protonated carbon resonances are not visualized, is of only modest concern in the oxylipin structure class. For example, only the carboxyl carbon is not observed in the DEPT spectrum of this new 22:7 metabolite. COSY analysis eliminated the presence of two of the four possible regioisomers for the 22:7 compound. With the use of two-dimensional nuclear Overhauser enhancement spectroscopy (NOESY), we were able to determine that the conjugated tetraene system begins at C-7 of the hydrophobic chain (Figure 4.4). The combination of these methods allowed us to rigorously determine the structure to be (4Z,7Z,9E,11E, 13Z,16Z,19Z)-docosaheptaenoic acid.

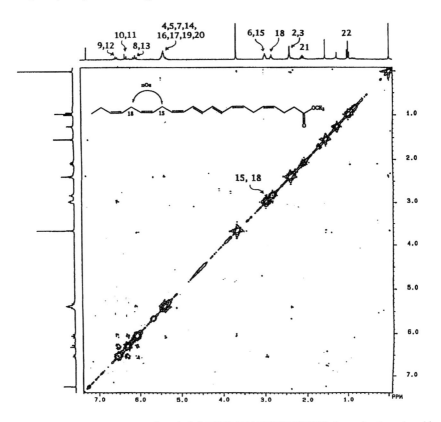

Figure 4.4 NOESY spectrum of methyl-(4Z,7Z,9E,11E,13Z,16Z,19Z)-docosaheptaenoic acid in CDCl$_3$. (Reproduced from Mikhailova *et al.* (1995) *Lipids* **30** (7) 583-589 with the permission of AOCS Press.)

4.2.2 Detection of PUFA biosynthesis; experiments in isolated chloroplasts

The main site of biosynthesis of these conjugated tetraene PUFAs is believed to be the chloroplasts, on the basis of the high level of enzymatic PUFA oxidation observed in this fraction compared to the mitochondrial preparations and the whole algal homogenates. *A. stellata* chloroplasts were isolated by grinding the algae in an appropriate buffer system to keep the chloroplasts intact and viable (Mikhailova *et al.*, 1995). Following filtration of the preparation, the chloroplasts were separated from cellular debris using differential centrifugation techniques. Intact chloroplasts were isolated from broken chloroplasts and other organelles using a continuous density Percoll (Sigma) gradient. After the protein concentration of the isolated chloroplasts was determined (Bradford, 1976), the chloroplasts were used for incubations with fatty acid substrates.

Biosynthesis experiments were conducted with the following non-conjugated PUFA substrates: palmitoleic acid, (6Z,9Z,12Z,15Z)-octadecatetraenoic acid, arachidonic acid (AA), eicosapentaenoic acid (EPA), (7Z,10Z,13Z,16Z)-docosatetraenoic acid, and (4Z,7Z,10Z,13Z,16Z, 19Z)-docosahexaenoic acid. Incubations with substrate (200 μmol dm^{-3} final concentration) were mixed continuously for 3 hours at room temperature prior to extraction in hexane. Ultraviolet spectrophotometric analysis was then conducted on a scanning wavelength spectrophotometer to identify the presence of a conjugated tetraene system.

Compounds containing the conjugated tetraene moiety were found in all preparations, including controls in which no exogenous PUFAs were added. However, marked increases in the absorbance maxima were found in samples incubated with the fatty acid substrates. The largest increases were found in the samples incubated with 6,9,12,15-octadecatetraenoic acid. Comparing peak ratios, this corresponds to an 18-fold increase in concentration of compounds containing the conjugated tetraene moiety. All other substrates resulted in significant increases in absorbance at 306 nm above control levels, ranging between 2.6 (palmitoleic) and 10.8 (EPA) times greater absorbance than controls.

The samples believed to contain the conjugated tetraenes were then treated with ethereal diazomethane, and analyzed by GC-MS. The GC-MS system used was a modified HP 5985B instrument with a solvent-free GC injector (Ray Allen Associates, Boulder CO, USA) connected to a bonded-phase fused-silica capillary column (DB-1, 15 m, 0.26 mm i.d., film thickness 0.25 μm; J&W Scientific, Folsom, CA, USA). Injector temperature was 250°C and initial oven temperature was 80°C for 1 min with a 7°C min^{-1} ramp up to a final temperature of 300°C.

GC-MS analysis resulted in identification of the five conjugated tetraene PUFAs found in the whole algal homogenates. However, all five were not present following incubation of every substrate. In the samples incubated with 6,9,12,15-octadecatetraenoic acid, we detected 16:5, 18:4, 20:5 and 20:6. After the addition of AA, only the conjugated tetraene containing fatty acids with 20-carbon chains were found. The PUFAs 16:5, 18:4, 20:5 and 20:6 were identified in samples following EPA addition, while all five conjugated tetraene containing fatty acids were found in samples incubated with both of the 22-carbon substrates. Interestingly, only 20:6 was identified in the samples prepared with palmitoleic acid.

These data suggest to us the presence of multiple oxidative enzymatic pathways in *A. stellata* chloroplasts that are capable of producing novel conjugated tetraene-containing PUFAs. The involvement of either a variety of substrate-specific enzymes or one nonspecific enzyme in the production of these metabolites has yet to be determined.

4.3   Structure elucidation of cyanobacterial and red algal oxylipins

It is the intent of this section to illustrate something of the power and strategy for using modern NMR techniques in the structure elucidation of new oxylipins. Marine algae have been an extremely rich group of marine life as a source of structurally novel oxylipins (Gerwick, 1994) whose structure elucidation has been chiefly through the application of diverse NMR experiments (Lopez and Gerwick, 1987).

Following the isolation of a new oxylipin derivative, probably the best initial information to gather is that of molecular mass and mass spectral fragmentation data. It is of course desirable to 'de-replicate' known compounds as early in the structure elucidation process as possible; this is most efficiently accomplished with gas chromatography–mass spectrometry (GC-MS) data. Normally free carboxylic acids present in oxylipins are derivatized with CH_2N_2 to fatty acid methyl esters; naturally occurring glycerol esters, lactones, or other derivatives of the carboxylic acid (e.g. ethanolamides as in anandamide) will be unreactive to these reaction conditions (Devane *et al.*, 1992). Oxylipins with free hydroxyl groups are most informatively derivatized to the corresponding trimethylsilyl ethers. While infrared data are of only modest value for structure elucidation of most oxylipins, ultraviolet data can provide pivotal information and should be recorded early in the process if a UV chromophore is detected (e.g. by TLC).

To establish that a new oxylipin is of probable novel structure, a combination of 1- and 2-dimensional NMR techniques is by far the most effective structural probe. Methyl ester derivatives of unsaturated fatty acid precursors to oxylipins possess a moderately well defined peak for the olefinic hydrogens at $\delta 5.4$–5.6, a singlet methoxymethyl group at $\delta 3.3$, and a number of bis-allylic hydrogens in quantity two less than the number of olefin hydrogens. Although allylic hydrogens and the C-2 methylene protons occur in the same $\delta 2.0$–2.4 region, it is usually possible to distinguish the C-2 protons because of their sharp triplet character. In $\Delta 5$ fatty acids, the C-3 hydrogens characteristically resonate at $\delta 1.65$. An envelope of protons at $\delta 1.25$ is characteristic of methylene groups distant from olefin or ketone functionalities, as is found in the more saturated fatty acids. Finally, a $\delta 0.9$ methyl triplet of poor definition is characteristic of ω-6 (or higher) fatty acids, whereas a sharp methyl triplet at $\delta 1.0$ is indicative of an ω-3 fatty acid.

Deviations from these shifts are reflective of additional metabolic functionalization of the fatty acid. For example, additional protons in the $\delta 5.6$–7.5 region indicate conjugated olefin systems, possibly involving a carbonyl. Shifts in the $\delta 3.5$–5.0 region are typical for secondary alcohols, ethers or esters. Shifts in the region of $\delta 2.8$–3.5 reflect the possible presence of an epoxide. Each of these functional groups has a characteristic shift for its associated carbon atoms as well. Rigorous correlation between proton shifts of a suspected fuctionality and their associated carbon atoms is given with greatest sensitivity and reliability by a heteronuclear multiple quantum coherence (HMQC) experiment (Bax *et al.*, 1983) or its more recent counterpart a heteronuclear single quantum correlation (HSQC) experiment (Kay *et al.*, 1992). Spin connectivity is powerfully visualized by the ^{1}H–^{1}H COSY experiment (Aue *et al.*, 1976), run in the double quantum filtered mode for highest resolution (Rance *et al.*, 1983). Given the essentially linear nature of most oxylipins, this is perhaps the single most useful experiment for deducing new oxylipin structures. However, interruptions to this nicely extended spin system are interposed by quaternary centers, such as ketones or bridgeheads containing hydroxyl groups. In these cases, connection of different spin systems is conveniently provided by long-range ^{1}H–^{13}C coupling interactions, nicely visualized by the heteronuclear multiple bond coherence (HMBC) experiment (Summers *et al.*, 1986). Determination of the relative stereochemistry of groups and substituents usually requires a complex interplay of ^{1}H–^{1}H coupling information and through-space nuclear Overhauser effects provided by the NOESY experiment (Meier and Ernst, 1979). There is a wide variety of techniques and approaches for determining absolute stereochemistry in natural products, some of which utilize NMR methods (Ohtani *et al.*, 1989), but these are beyond the scope of this short review.

Example 1

In recent years, our research group has come to focus on the diverse natural products chemistry of the tropical marine cyanobacterium *Lyngbya majuscula*. On a comparative biochemical basis, we have found that the oxylipin metabolism in blue-green algae appears to resemble most closely that seen in green algae and higher plants. For example, a collection from Compass Point, St Thomas, USVI in 1993 was extracted and sequentially fractionated by vacuum liquid chromatography and centrifugal TLC (silica gel and ethyl acetate–hexanes gradients) to give semipure mixtures that were profiled by ^{1}H NMR. In addition to the known metabolites malyngolide (Cardellina *et al.*, 1979), malyngamide F acetate (Gerwick *et al.*, 1987) and malyngamide H (Orjala *et al.*, 1995), a new 'oxylipin-like' compound of a relatively non-polar nature, lyngbyalactone (1), was isolated by HPLC (Phenomenex Maxsil, 10 μm, 500 mm × 10 mm; 10% EtOAc–hexanes, 4.0 ml min^{-1}, t_r=12 min) (Gerwick, 1996). By high-resolution chemical ionization mass spectrometry (HR CIMS), $[M+H]^+$ at 293.212 was observed, consistent with a molecular formula of $C_{18}H_{28}O_3$ (Δ0.3 mmu). Inherent in this molecular formula are 5 degrees of unsaturation. By ^{13}C NMR, two double bonds and one ester carbonyl were in evidence; hence, compound (1) was bicyclic. ^{1}H NMR spectra in both CDCl$_3$ and benzene-d$_6$ showed this new metabolite to possess five vinyl and/or oxymethine protons, two protons typical of epoxides at δ2.87 (H-13) and δ3.15 (H-12), a complex series of overlapping bands between δ1.2 and 2.6, and a methyl triplet signal at δ0.98 (H$_3$-18). As slightly better spectral dispersion was obtained in CDCl$_3$, additional NMR data were obtained in this solvent.

(1) **Lyngbyalactone**

Several starting locations for deducing the spin systems of this new metabolite were evident from ^{1}H–^{1}H COSY(Figure 4.5). That the methyl triplet (H$_3$-18) was characteristically sharp and at δ0.98 suggested ω-3 unsaturation. COSY peaks between the methyl triplet and the complex allylic methylene band at δ2.05 (H$_2$-17) confirmed this assignment. However, owing to spectral overlap in the congested δ2.0–2.2 region,

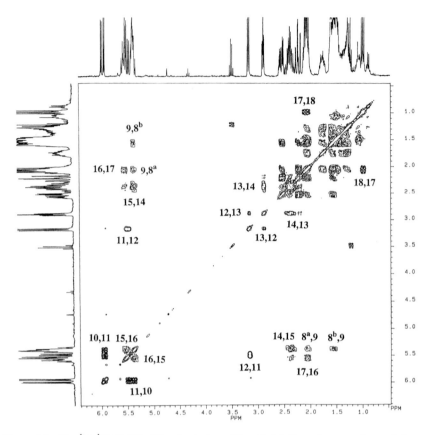

Figure 4.5 The ^{1}H–^{1}H COSY of lyngbyalactone (**1**), with labeling of crosspeak correlations, was used to define its linear spin system.

continuity of the spin system was partially obscured, although a strong correlation was observed from δ2.05 (H$_2$-17) to δ5.53 (H-16). From ^{1}H–^{13}C COSY data (equivalent to HMQC) and DEPT data, it was reasoned that this δ5.53 (H-16) signal must be part of a disubstituted double bond, since all olefinic carbons in (**1**) were protonated (Figure 4.6). The olefinic 'partner' for the δ5.53 proton was located at δ5.37 (H-15) by ^{1}H–^{1}H COSY; comparison of ^{13}C NMR shifts for the ω-2 CH$_2$ (δ20.7, C-17) with literature values showed this to be a *cis*-olefin (Rakoff and Emken, 1983). Therefore, the nature of the first four carbons beginning from the ω-terminus could be deduced from even these limited data (partial structure (**a**)).

A second starting location was given by a clearly defined oxymethine signal in the ^{1}H NMR at δ3.15 (H-12). From ^{1}H–^{13}C COSY data the

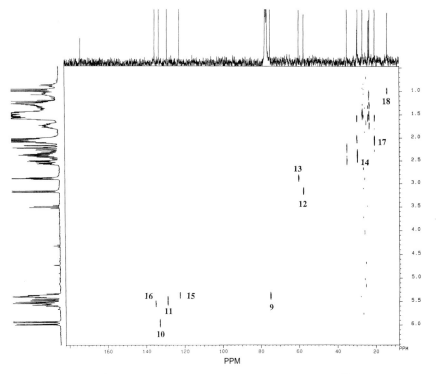

Figure 4.6 The ^{1}H–^{13}C COSY of lyngbyalactone (**1**) clearly identifies the ^{13}C NMR shifts of carbons and their associated protons.

associated carbon atom was located at δ57.4 (C-12), indicating an epoxide ring. By ^{1}H–^{1}H COSY, this proton was coupled to another epoxide proton at δ2.87 (H-13, associated carbon at δ59.9). The geometry of the epoxide ring was deduced to be *trans* by observing a 2.1 Hz coupling between these protons. The δ3.15 band was also coupled to an olefin proton at δ5.45 (H-11), which was in turn coupled to another olefinic proton at δ5.95 (H-10). A 15.7 Hz coupling between these two vinyl protons indicated that the double bond was *trans*. The δ5.95 proton was in turn coupled to a proton in the 2-proton band at δ5.37 (H-9), which could either be an olefinic proton or a deshielded oxymethine. That lyngbyalactone does not have appreciable UV-absorbance in the 220–270 nm region is only consistent with the second possibility. The chemical shift of this oxymethine is consistent with its being both an ester and allylic, supporting the idea of a macrolactone. The δ2.87 epoxide proton (H-13) was coupled in the ^{1}H–^{1}H COSY to an allylic methylene centered at δ2.37 (H$_2$-14); this latter band was further coupled to the olefinic proton at δ5.37 (H-15). Hence, partial structures (**a**) and (**b**) could be joined at this point.

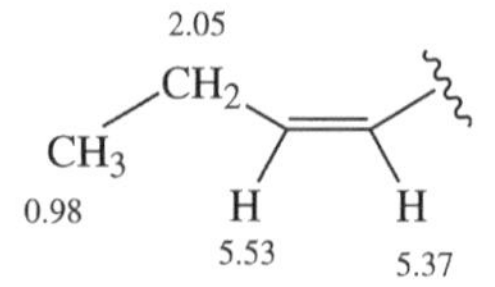

Partial Structure a

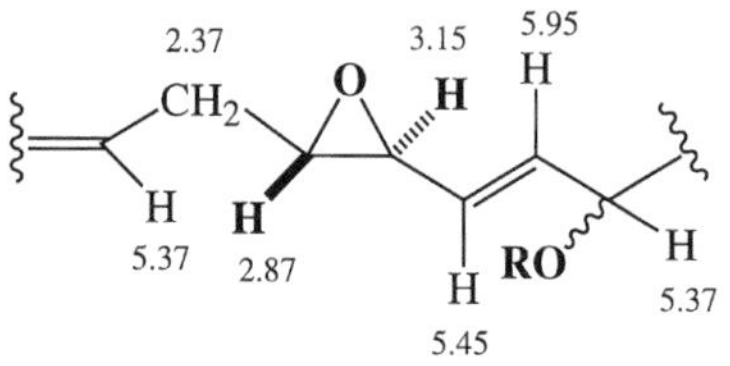

Partial Structure b

The remaining atoms in lyngbyalactone (**1**) consist entirely of —CH₂— groups, and hence can only form a straight chain between the ester carboxyl functionality and the deshielded oxymethine (allylic ester oxymethine) at δ5.37 (H-9). By consideration of the molecular formula, this places seven methylene groups in between these functionalities, defining the position of lactonization at C-9. Hence, a completed planar structure with relative stereochemistry for the two double bonds and epoxide could be formulated; the stereochemistry at C-9 remains unknown. Interestingly, this same epoxylactone metabolite has been isolated twice since from freshwater cyanobacteria (Pappendorf *et al.*, 1997; Stierle *et al.*, 1998); however, in neither case was its former isolation from *Lyngbya majuscula* realized.

Example 2

A second example of the 2D NMR structure elucidation of a marine-derived oxylipin is given by agardhilactone (**2**), a metabolite of the North Atlantic species of red alga *Agardhiella subulata* (Graber *et al.*, 1996). Owing to its taxonomic relationship to a West Coast alga *Sarcodiatheca gaudchaudii*, an alga that we had previously found to contain unusual oxylipin natural products, we targeted this East Coast species. 2D-TLC analysis of the crude lipid extract showed the presence of several minor blue-charring metabolites, a color reaction with 50% aqueous H_2SO_4 and heat that we have come to recognize as typical of oxylipins. Isolation of one of these compounds was accomplished by gradient vacuum silica gel chromatography, derivatization to the acetate derivative, and normal-phase HPLC. A molecular formula of $C_{22}H_{30}O_5$ was established by HR CIMS, indicating eight degrees of unsaturation. Five of these were defined by ^{13}C NMR as two ester carbonyls ($\delta170.1$ and 171.8) and three carbon–carbon double bonds (six olefinic carbons between $\delta127$ and $\delta132$). Hence agardhilactone possessed three rings.

(2) **Agardhilactone**

The backbone structure of agardhilactone consisted of one extended spin system, visible despite extensive spectral overlap in some regions by double quantum filtered COSY (Rance *et al.*, 1983) (Figure 4.7). A methyl triplet at $\delta0.90$ (H$_3$-20) was adjacent to a relatively high-field methylene group at $\delta1.65$ (H$_2$-19), which in turn was next to a deshielded methine at $\delta5.22$ (H-18). The nature of this methine was revealed by the associated carbon shift at $\delta75.9$ by HMQC (C-18), consistent with an oxygen atom substitution. By ^1H-shift reasoning, this was likely an allylic acetate ester. Indeed, continuation of the ^1H spin system showed this oxymethine to be adjacent to a conjugated diene. By coupling-constant analyses, the first of these double bonds (C-16–C-17) was *trans* ($J=15.8\,Hz$) and the second (C-14–C-15) was *cis* ($J=11.0\,Hz$).

The C-14 vinyl proton at $\delta5.47$ was coupled to a bis-allylic methylene, the other olefin (C-11–C-12) of which was *trans* ($J=15.6\,Hz$). The C-11 vinyl proton ($\delta5.49$) was coupled to a higher-field methine at $\delta2.71$ (H-10), which in turn was coupled to two other methines, one at $\delta1.62$ (H-6, ^{13}C at $\delta43.2$) and the other at $\delta3.40$ (H-9, ^{13}C at $\delta60.2$). Clearly, the branching spin system of this oxylipin indicated the presence of a carboxylic ring; the two higher-field methines at shifts consistent with the bridgehead positions and the lower field with an epoxide functionality. This latter hypothesis was confirmed by observing that the $\delta3.40$ oxymethine (H-9) was adjacent to a second oxymethine of similar shifts (H-8, ^1H at $\delta3.46$; ^{13}C at $\delta55.7$). The carbocyclic ring was completed by seeing that both the $\delta3.46$ oxymethine (H-8) and $\delta1.62$ bridgehead

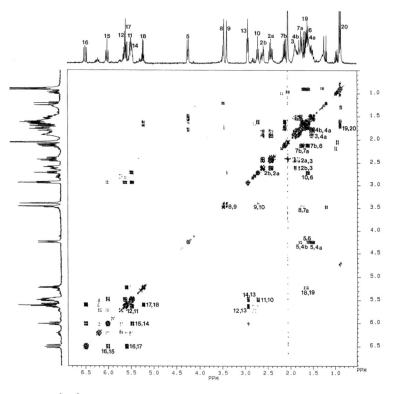

Figure 4.7 The $^1H-^1H$ COSY of agardhilactone (**2**), with labeling of crosspeak correlations, was used to define its branching spin system.

methine (H-6) were coupled to a methylene pair at δ1.74 and δ2.12 (H$_2$-7, ^{13}C at δ27.6).

The latter bridgehead methine (H-6) was also coupled to an oxymethine at δ4.24 (H-5) which had an associated carbon at δ78.5. Several lines of evidence were used to deduce the δ-lactone nature of C-1–C-5, including the diastereotopic nature of the C-2 (δ2.42 and δ2.61) and C-4 methylene (δ1.51 and δ1.78) protons, the latter of which were also coupled to the δ4.24 oxymethine. A combination of NOESY (Meier and Ernst, 1979) and $^1H-^1H$ coupling-constant analyses were used to build a case for a 6S*,8R*,9S*,10R* relative stereochemistry about the cyclopentyl ring. An additional isolation of agardhilactone not derivatized at the C-18 acetate allowed determination of the absolute stereochemistry at C-18 as mainly (80%) of S stereochemistry. We were unable to determine relative or absolute stereochemistry at C-5.

We have proposed that agardhilactone is derived from oxidation of eicosapentaenoic acid (20:5) by an 8-lipoxygenase (Graber and Gerwick, 1996). Lipoxygenases catalyze the formation of a hydroperoxy group which, in this case, reacts with the neighboring 9,10 olefin. There is a consequent loss of a hydroxyl group, leading to the production of an epoxy allylic carbonium ion that induces cyclization ending in the formation of cyclopentyl group and the lactone ring. The hydroxyl group at C-18 could be formed by an ω-3 lipoxygenase-catalyzed hydroperoxidation followed by reduction to a hydroxyl group and an isomerization of the 11,12-olefin from *cis* to *trans*.

As the above examples clearly demonstrate, 2D NMR methods are very powerful in the structure elucidation of oxylipin metabolites, even those of quite complex structure such as agardhilactone (**2**). In certain measure, this derives from the essentially linear carbon framework of these polyunsaturated fatty acid-derived metabolites. This is a feature that both reduces the complexity of these NMR data sets, particularly the ^{1}H–^{1}H COSY, and allows a certain degree of predictability for what might be encountered in unknown features of a new molecule under study.

4.4   Eicosanoid biosynthesis in the amebocyte of *Limulus polyphemus*

This section will demonstrate the use of electrospray ionization mass spectroscopy (ESI-MS) for the detection and identification of low levels of fatty acids and eicosanoid metabolites produced by invertebrate blood cells. Detection of eicosanoids by ESI-MS illustrates the power of this method to identify low levels of cellular metabolites from a sample normally requiring much more rigorous sample preparation techniques. Although reversed-phase HPLC alone can be used to detect and quantify eicosanoids, in many cases this procedure is coupled with a radio-immunoassay or enzyme immunoassay to increase sensitivity or derivatization followed by gas chromatography–mass spectrometry (GC-MS) to positively confirm structural identity of the isomers (Yamamoto *et al.*, 1987; Young and Girard, 1989). Radioactive isotopes and antibodies are expensive and the availability of these materials is usually limited to major eicosanoid metabolites, which makes these methods less amenable to reactive intermediates or novel metabolites. Although derivatization stabilizes reactive compounds, it can interfere with their identification and quantitation owing to (a) decomposition of the compound during the derivatization process or (b) unwanted side reactions resulting in nonquantitative yields (Richmond *et al.*, 1986; Kim and Sawazaki, 1993; Baldwin *et al.*, 1979). Other batch desorption methods, such as fast atom bombardment (FAB) and secondary ion mass spectrometry (SIMS), can be used to analyze non-volatile compounds directly but are relatively insensitive, difficult to quantify, and not compatible with on-line detection. Liquid-inlet ionization methods, like ESI-MS, provide the ability to identify low molecular mass non-volatile substances on-line without prior derivitization. ESI-MS also allows detection of non-UV-active metabolites present in samples, although normally at least one wavelength is monitored in-line to confirm the presence of specific compounds. As fragmentation in ESI-MS occurs via collision-induced dissociation (CID) in the skimmer cone region of the source, the fragmentation ions often correspond to those observed in FAB MS-MS experiments. Identification of molecular ions and fragments associated with the eicosanoid metabolites were therefore based on calculations and published eicosanoid FAB-MS and ESI-MS fragmentation patterns (Kerwin and Torvic, 1996; Duffin et al., 1995; Harrison and Murphy, 1995; Wheelan and Murphy, 1995).

4.4.1   Eicosanoid biosynthesis and identification

Arthropods possess as many as six types of circulating blood cells: plasmocytes, granulocytes, spherulocytes, oenocytes, adipo-hemocytes and prohemocytes. Of these, only two are considered immunocompetent, the plasmocyte and granulocyte (Gupta, 1991). The marine arthropod *Limulus polyphemus* has only a single species of circulating blood cell, a granular amebocyte (Armstrong, 1979). In mammals, the major blood cells believed to be active during the acute initial phase of the inflammatory process are the neutrophil, macrophage and platelet. Gupta (1991) has hypothesized that the *Limulus* amebocyte, as a primitive granulocyte, may function in a manner similar to platelets and macrophages, as well as B- and T-lymphocytes. The acute phase of an inflammatory response is marked by the production and secretion of polyunsaturated fatty acid metabolites, predominately of 18- to 20-carbon chain length. Therefore, it was of interest to determine fatty acid metabolites produced by the amebocyte and their possible role in an immune response. The horseshoe crabs were cooled and blood cells were isolated via cardiac puncture (Armstrong, 1985). The cells were washed, resuspended in sterile 3% NaCl and stimulated with both arachidonic acid ($100\,\mu mol\,dm^{-3}$) and 4-bromo-calcium ionophore, A23187 ($5\,\mu mol\ dm^{-3}$). The metabolites were extracted from the suspension buffer and isolated on octadecylsilyl minicolumns (Powell, 1982; MacPherson et al., 1996). To confirm utilization of the exogenous substrate, parallel experiments were carried out by stimulating the cells with A23187 (5 µmol dm^{-3}) and octadeuterated arachidonic acid (100 µmol dm^{-3}).

The fraction containing the metabolites was analyzed on a Michrom UMA HPLC System (Michrom BioResources Inc., Auburn, CA, USA) using a Zorbax C18 column (1.0 mm i.d. × 150 mm, 5 µm, 80 Å, λ=235 nm) at a flow rate of 50 µl min^{-1} with a linear gradient of MeOH–water 50:50 80:20 over 20 min, and then 20 min isocratically at (80:20). The HPLC effluent was directed into the inlet of an electrospray ionization probe of a Fisons VG Platform II single quadrupole mass spectrometer (Fisons VG Biotech MS, Altrincham, UK). Sample ions were generated in the negative-ion mode using nitrogen nebulization-assisted electrospray with a source temperature maintained at 70°C. The ESI-MS was tuned daily by direct infusion of arachidonic acid (0.2 mmol dm^{-3}). The skimmer cone voltage was set during tuning for low to moderate fragmentation (−35 to −40 V). The mass collection range was 100 to 650 amu and spectra were collected in total scan mode and subjected to background subtraction, and extracted ion profiles were used to locate the molecular ion [M-H]$^{-}$.

Free fatty acids, in some cases eicosanoid precursors, as well as several metabolites were detected in the samples (Figure 4.8). When comparing the two chromatograms, there are at least twice as many peaks evident in the MS chromatogram as in the UV chromatogram. There are also compounds that are present in trace amounts that are not readily observed as peaks even in the total ion chromatogram. But these compounds can be detected by extracting their molecular ion chromatogram from the total ion chromatogram. Saturated and monounsaturated fatty acids were distinguished mainly by the deprotonated molecular ion [M−1], though sometimes the further loss of carbon dioxide [M−45] was observed in the mass spectra of the extracted ion profile. Free fatty acids myristic acid (*m/z* 227), palmitic and palmitoleic acids (*m/z* 255 and 253, respectively), stearic acid (*m/z* 283) were detected, as well as eicosanoid precursor fatty acids: oleic acid (*m/z* 281), linoleic acid (*m/z* 279) and two peaks for linolenic and γ-linolenic acids (*m/z* 277), eicosatetraenoic acid (*m/z* 303), eicosapentaenoic acid (*m/z* 301) and eicosatrienoic acid (*m/z* 305) were detected.

The major eicosanoid metabolite produced by the amebocyte was 8-hydroxyeicosatetraenoic acid (8-HETE). It was possible to identify six fragments in the 8-HETE spectra. The molecular ion species *m/z* 319 (M−1, loss of a proton) was present, as well as *m/z* 301 (M−19, loss of water) and *m/z* 257 (M−63, loss of water and carbon dioxide), *m/z* 127 (cleavage prior to the hydroxyl group) and *m/z* 155 and 163 (cleavage after the hydroxyl group). A unique aspect of ESI-MS analysis is the ability to discriminate quickly whether fragments in a given spectra

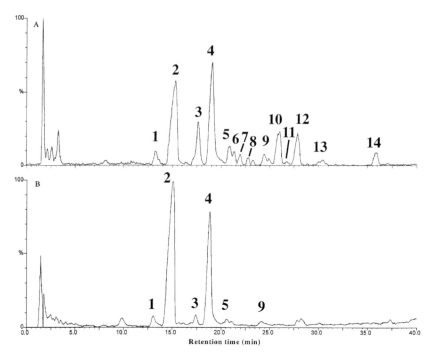

Figure 4.8 Electrospray total ion chromatogram (a) and HPLC chromatogram (b) of metabolites from amebocytes stimulated with both exogenous arachidonic acid and calcium ionophore, A23187. The chromatograms are negative ion profiles generated in total scan mode monitoring from 100 to 650 amu. Peaks: (1) unknown (m/z 297); (2) 8-hydroxyeicosatetraenoic acid; (3) hydroperoxyoctadecenoic acid; (4) calcium ionophore, A23187; (5) unknown (m/z 325); (6) eicosapentaenoic acid (20:5); (7) palmitoleic (16:1); (8) unknown (m/z 241); (9) unknown (m/z 339); (10) eicosatetraenoic acid (20:4); (11) unknown (m/z 363); (12) palmitic acid (16:0); (13) oleic acid (18:1); (14) stearic acid (18:0).

belong to the molecular ion of interest. On altering the skimmer cone voltage, the levels of the fragments in the spectra will change. At low to moderate voltage (-30 to -40 V) the molecular ion predominates, but as the voltage is increased the proportion of the fragments increases as the molecular ion decreases (Figure 4.9). There is a limit to the level of fragmentation that can be produced in small molecules. At higher voltages (-80 V) there is a deterioration in the signal intensity accompanied by rising background noise.

Confirmation that the 8-HETE was produced from the exogenous substrate was provided by incubating the hemocytes with octadeuterated arachidonic acid. The formation of 8-HETE-d_8 was clearly demonstrated. The 8-HETE-d_8 peak had an identical fragmentation pattern to that of

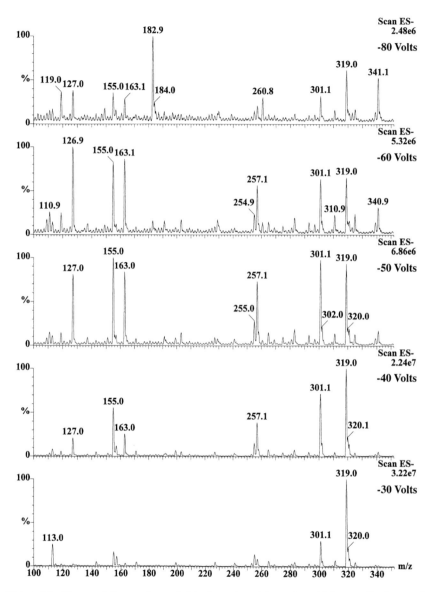

Figure 4.9 Fragmentation pattern of 8-hydroxyeicosatetraenoic acid in negative-ion mode generated by a single quadrupole electrospray ionization mass spectrometer using various levels of cone voltage.

8-HETE except for the additional mass of the deuterium atoms. The molecular ion (M−1, m/z 327), the loss of water (M−19, m/z 309), water plus carbon dioxide (M−63, m/z 265), cleavage prior to the hydroxyl group (m/z 129) and after the hydroxyl group (m/z 158 and M−168) were all present. (Figure 4.10). The retention of all sites of deuteration after oxidation agrees with the results of Hughes and Brash (1991). Other minor metabolites detected in extracted molecular ion profiles were hydroxyoctadecatrienoic acid at m/z 293, hydroxyheptadecatrienoic acid at m/z 279, hydroxyeicosatrienoic acid at m/z 321, hydroperoxyoctadienoic acid at m/z 311, and hydroxyeicosapentaenoic acid at m/z 317. Since multiple precursor fatty acids were seen in the samples, the detection of minor metabolites from sources other than arachidonic acid was not unexpected.

The production of 8-HETE has been observed in both marine and terrestrial organisms. It has been demonstrated that 8-HETE can be produced via both the lipoxygenase and monooxygenase enzymatic systems (Spector *et al.*, 1988; Capdevila *et al.*, 1986) and in general,

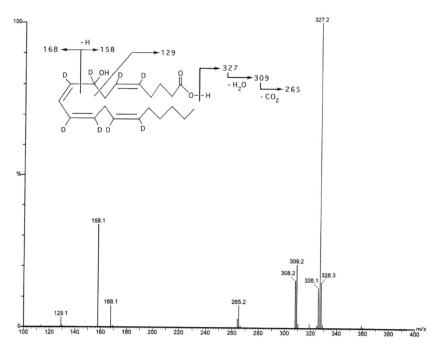

Figure 4.10 The background-subtracted mass spectra of d_8-8-hydroxyeicosatetraenoic acid biosynthesized by *Limulus* amebocytes. In addition to the molecular ion (m/z 327), the other fragments represent loss of water [(M−19), m/z 309], loss of water and carbon dioxide [(M−63), m/z 265], cleavage prior to the hydroxyl group (M−129), cleavage after the hydroxyl group (M−158 and M−168). (Reproduced from MacPherson *et al.* (1996) *Biochimica et Biophysica Acta* **1303** 127-136, ©1996, with the permission of Elsiever Science B.V.)

HETEs are known to influence the inflammatory response. No enzyme that exclusively produces 8-HETE has been isolated from mammalian cells or tissues (Spector *et al.*, 1988), but a stereospecific 8-(*R*)-lipoxygenase has been detected in homogenates and acetone powders prepared from some gorgonians and soft corals (Bundy *et al.*, 1986; Corey *et al.*, 1987; Brash *et al.*, 1987). An increase in HETE formation has been documented in some tissue injury situations and it could possibly act by modulating a number of pathophysiological phenomena (Spector *et al.*, 1988). Hughes and Brash (1991) and Gschwendt *et al.* (1986) speculated that 8-HETE may play a chemotactic role during the host response to immune challenge. In *Limulus polyphemus*, Bursey (1977) detected the migration of amebocytes through connective tissue to form a protective layer around a wound area within 24 hours of exoskeletal perforation. Hemocytes continued to migrate to the site of injury for at least four days, which indicates the possibility that a chemotactic agent, possibly 8-HETE, was being released.

Therefore, analysis of the eicosanoid metabolites was carried out using ESI-MS, demonstrating that it is a sensitive and reliable method for the direct identification of eicosanoid metabolites.

Acknowledgements

The work described here is funded, in part, by a grant from the National Sea Grant College Program, National Oceanographic and Atmospheric Administration, US Department of Commerce, under grant number NA-36RG0537, project number R/MP-71 through the California Sea Grant College System, Smithsonian Marine Station at Fort Pierce Contribution #464, and in part by the California State Resources Agency (R.S. Jacobs) and the Oregon Sea Grant Program, project number R/BT-18 (W.H. Gerwick). The views expressed herein are those of the authors and do not necessarily reflect the views of NOAA or any of its sub-agencies. The US Government is authorized to reproduce and distribute for governmental purposes. Jennifer MacPherson and Debra Bemis' support was funded through the Sea Grant Traineeship Program.

References

Armstrong, P.B. (1979) Motility of the *Limulus* blood cell. *J. Cell Sci.*, **37** 169-180.

Armstrong, P.B. (1985) Amebocytes of the American "horseshoe crab" *Limulus polyphemus*, in *Blood Cells of Marine Invertebrates: Experimental Systems in Cell Biology and Comparative Physiology* (ed. W.D. Cohen), Alan R. Liss, New York, pp 253-260.

Aue, W.P., Bartholdi, E. and Ernst, R.R. (1976) Two-dimensional spectroscopy. Application to nuclear magnetic resonance. *J. Chem. Phys.*, **64** 2229-2264.

Baldwin, J.E., Davies, D.I., Hughes, L. and Gutteridge, N.J.A. (1979) Synthesis from arachidonic acid of potential prostaglandin precursors. *J. Chem. Soc. Perkin Trans. I*, 115-121.

Bax, A., Griftey, R. and Hawkins, B.L. (1983) Sensitivity-enhanced correlation of ^{15}N and ^{1}H chemical shifts in natural-abundance samples via multiple quantum coherence. *J. Am. Chem. Soc.*, **105** 7188-7190.

Bradford, M.M. (1976) A rapid and sensitive method for the quantitation of microgram quantities of protein utilizing the principle of protein-dye binding. *Anal. Biochem.*, **72** 248-254.

Brash, A.R., Baertschi, S.W., Ingram, C.D. and Harris, T.M. (1987) On non-cyclooxygenase prostaglandin synthesis in the sea whip coral, *Plexaura homomalla*: an 8(R)-lipoxygenase pathway leads to formation of an alpha-ketol and a racemic prostanoid. *J. Biol. Chem.*, **262** 15829-15839.

Bundy, G.L., Nidy, E.G., Epps, D.E., Mizasak, S.A. and Wnuk, R.J. (1986) Discovery of an arachidonic acid C-8 lipoxygenase in the gorgonian coral *Pseudoplexaura porosa*. *J. Biol. Chem.*, **26** 747-751.

Burgess, J.R., de la Rosa, R.I., Jacobs, R.S. and Butler, A. (1991) A new eicosapentaenoic acid formed from arachidonic acid in the coralline red algae *Bossiella orbigniana*. *Lipids*, **26** 162-165.

Bursey, C.R. (1977) Histological response to injury in the horseshoe crab, *Limulus polyphemus*. *Can. J. Zool.*, **55** 1158-1165.

Capdevila, J., Yadagiri, P., Manna, S. and Falck, J.R. (1986) Absolute configuration of the hydroxyeicosatetraenoic acids (HETEs) formed during catalytic oxygenation of arachidonic acid by microsomal cytochrome P-450. *Biochim. Biophys. Res. Commun.*, **141** 1007-1011.

Cardellina, J.H. II, Moore, R.E., Arnold, E.V. and Clardy, J. (1979) Structure and absolute configuration of malyngolide, an antibiotic from the marine blue-green alga *Lyngbya majuscula* Gomont. *J. Org. Chem.*, **44** 4039-4042.

Corey, E.J., d'Alarcao, M., Matsuda, S.P.T., Lansbury, P.T. and Yamada, Y. (1987) Intermediacy of 8-(R)-HPETE in the conversion of arachidonic acid to pre- clavulone A by *Clavularia virdis*. Implications for the biosynthesis of marine prostanoids. *J. Am. Chem. Soc.*, **109** 289-290.

DePetrocellis, L. and DiMarzo, V. (1994) Aquatic invertebrates open up new perspectives in eicosanoid research: biosynthesis and bioactivity. *Prostaglandius Leukot. Essent. Fatty Acids*, **51** 215-229.

Devane, W.A., Hanus, L., Breuer, A., Pertwee, R.G., Stevenson, L.A., Griffin, G., Gibson, D., Mandelbaum, A., Etinger, A. and Mechoulam, R. (1992) Isolation and structure of a brain constituent that binds to the cannabinoid receptor. *Science*, **258** 1946-1949.

Duffin, K., Margalit, A. and Isakson, P. (1995) *Proc. 43rd ASMS Conference on MS and Applied Topics*, Atlanta, Georgia, p 659.

Gerwick, W.H. (1994) Structure and biosynthesis of marine algal oxylipins. *Biochim. Biophys. Acta*, **1211** 243-255.

Gerwick, W.H. (1996) Epoxy allylic carbocations as conceptional intermediates in the biogenesis of diverse marine oxylipins. *Lipids*, **31** 1215-1231.

Gerwick, W.H. and Bernart, M.W. (1993) Eicosanoids and related compounds from marine algae, in *Marine Biotechnology, Volume 1: Pharmaceutical and Bioactive Natural Products* (eds D.H. Attaway and O.R. Zaborsky), Plenum Press, New York, pp 101-152.

Gerwick, W.H., Reyes, S. and Alvarado, B. (1987) Two new malyngamides from the Caribbean cyanobacterium *Lynbya majuscula*. *Phytochemistry*, **26** 1701-1704.

Graber, M.A., Gerwick, W.H. and Cheney, D.P. (1996) Isolation and characterization of agardhilactone, a novel oxylipin from *Agardhiella subulata*. *Tetrahedron Lett.*, **37** 4635-4638.

Gschwendt, M., Fürstenberger, G., Kittstein, W., Besemfelder, E., Hull, W.E., Hagedorn, H., Opferkuch, H.J. and Marks, F. (1986) Generation of the arachidonic acid metabolite 8-HETE by extracts of mouse skin treated with phorbol ester *in vivo*; identification by ¹H-n.m.r. and GC-MS spectroscopy. *Carcinogenesis*, **7** 449-455.

Gupta, A.P. (1991) Insect immunity and other hemocytes: roles in cellular and humoral immunity, in *Immunology of Insects and Higher Arthropods* (ed. A.P. Gupta), CRC Press, London, pp 19-105.

Hamberg, M. (1992) Metabolism of 6,9,12-octadecatrienoic acid in the red alga *Lithothamnion corallioides*: mechanism of formation of a conjugated tetraene fatty acid. *Biochem. Biophys. Res. Commun.*, **188** 1220-1227.

Harrison, K.A. and Murphy, R.C. (1995) Isoleukotrienes are biologically active free radical products of lipid peroxidation. *J. Biol. Chem.*, **270** 17273-17278.

Hughes, M.A. and Brash, A.R. (1991) Investigation of the mechanism of biosynthesis of 8-hydroxyeicosatetraenoic acid in mouse skin. *Biochim. Biophys. Acta*, **1081** 347-354.

Kay, L.E., Keifer, P. and Saarinen, T. (1992) Pure absorption gradient enhanced heteronuclear single quantum correlation spectroscopy with improved sensitivity. *J. Am. Chem. Soc.*, **114** 10663-10665.

Kerwin, J.L. and Torvik, J.J. (1996) Identification of monohydroxy fatty acids by electrospray mass spectrometry and tandem mass spectrometry. *Anal Biochem.*, **237** 56-64.

Kim, H.Y. and Sawazaki, S. (1993) Structural analysis of hydroxy fatty acids by thermospray liquid chromatography/tandem mass spectrometry. *Biol. Mass Spectrom.*, **22** 302-310.

Lombardi, P. (1990) A rapid, safe, and convenient procedure for the preparation and use of diazomethane. *Chem. Ind.*, **21** 708.

Lopez, A. and Gerwick, W.H. (1987) Two new icosapentaenoic acids from the Oregon red seaweed *Ptilota filicina*. *Lipids*, **22** 190-194.

MacPherson, J.C., Pavlovich, J.G. and Jacobs. R.S. (1996) Biosynthesis of arachidonic acid metabolites in *Limulus polyphemus* amebocytes: analysis by liquid chromatography-electrospray ionization mass spectrometry. *Biochim. Biophys. Acta*, **1303** 127-136.

Meier, B.H. and Ernst, R.R. (1979) Elucidation of chemical exchange networks by two-dimensional NMR spectroscopy: the heptamethylbenzonium ion. *J. Am. Chem. Soc.*, **101** 6441-6442.

Mikhailova, M.V., Bemis, D.L., Wise, M.L., Gerwick, W.H., Norris, J.N. and Jacobs, R.S. (1995) Structure and biosynthesis of novel conjugated polyene fatty acids from the marine green alga *Anadyomene stellata*. *Lipids*, **30** (7) 583-589.

Ohtani, I., Kusumi, T., Ishitsuka, M.O. and Kakisawa, H. (1989) Absolute configurations of marine diterpenes possessing a xenicane skeleton — an application of an advanced Mosher method. *Tetrahedron Lett.*, **30** 3147-315.

Orjala, J., Nagle, D. and Gerwick, W.H. (1995) Malyngamide H, an ichthyotoxic amide possessing a new carbon skeleton from the Caribbean cyanobacterum *Lyngbya majuscula*. *J. Nat. Prod.*, **58** 764-768.

Pappendorf, O., Konig, G.M., Wright, A.D., Chorus, I. and Oberemm, A. (1997) Mueggelone, a novel inhibitor of fish development from the fresh water cyanobacterium *Aphanizomenon flosaquae*. *J. Nat. Prod.*, **60** 1298-1300.

Powell, W.S. (1982) Rapid extraction of arachidonic acid metabolites from biological samples using octadecylsilyl silica. *Methods. Enzymol.*, **86** 467-477.

Rakoff, H. and Emken, E.A. (1983) Synthesis and properties of mono-, di- and trienoic fatty esters containing a 12,13 double bond. *J. Am. Chem. Soc.*, **60** 546-552.

Rance, M., Sorensen, O.W., Bodenhavren, G., Wagner, G., Ernst, R.R. and Wuthrich, K. (1983) Improved spectral resolution in COSY ¹H NMR spectra of proteins via double quantum filtering. *Biochem. Biophys. Res. Commun.*, **117** 479-485.

Richmond, R., Clarke, S.R., Watson, D., Chappell, C.G., Dollery, C.T. and Taylor, G.W. (1986) Generation of hydroxyeicosatetraenoic acids by human inflammatory cells: analysis by thermospray liquid chromatography–mass spectrometry. *Biochim. Biophys. Acta*, **881** 159-166.

Spector, A.A., Gordon, J.A. and Moore, S.A. (1988) Hydroxyeicosatetraenoic acids (HETEs) *Prog. Lipid Res.*, **27** 271-323.

Stanley-Samvelson, D.W. (1991) Comparative physiology of invertebrate animals. *Am. J. Physiol.*, **260** (5 pt2) 849-853.

Stierle, D.B., Stierle, A.A., Bugni, T. and Loewen, G. (1998) Gloeolactone, a new epoxy lactone from a blue-green alga. *J. Nat. Prod.*, **61** 251-252.

Summers, M.F., Marzilli, L.G. and Bax, A. (1986) Complete ^{1}H and ^{13}C assignments of coenzyme B_{12} through the use of new two-dimensional NMR experiments. *J. Am. Chem. Soc.*, **108** 4285-4294.

Wheelan, P. and Murphy, R.C. (1995) *Proc. 43rd ASMS Conference on MS and Applied Topics*, Atlanta, Georgia, p 539.

Yamamoto, S., Yokota, K., Tonai, T., Shono, F. and Hayashi, Y. (1987) Enzyme immunoassay, in *Prostaglandins and Related Substances* (eds Benedetto, C. *et al.*,) IRL Press, Oxford, pp 197-208.

Young, R.N. and Girard, Y. (1989) Assay methods for various lipoxygenase products, in *Leukotrienes and Lipoxygenase: Chemical, Biological and Clinical Aspects* (ed. J. Rokach), Elsevier, New York, pp 209-307.

5 Pulse-NMR in the food science laboratory

Wouter L.J. Meeussen

5.1 Radio pulses and spin reorientation

In order to observe the behaviour of a system of spins, it is necessary to be able to manipulate it in some way. In principle, it would be simplest to change the magnitude or direction of the applied magnetic field. In practice, it is easier to superpose a second magnetic field on the reference field. This can be done by irradiating the sample with a wave of exactly the precession frequency and at right angles to reference field (Kessler *et al.*, 1988). Each rotating spin sees this radio wave as a steady magnetic field. The spins will precess around this second field too, with a precession frequency proportional to its strength. For a field that is 200 times weaker than the reference field, the precession rate will also be 200 times less: 0.1 MHz. This corresponds to a complete precession in 10 µs. If we leave the radio transmitter switched on for a quarter of that time, the spins will have rotated a quarter of a full cycle. Such a well designed and timed pulse is called a 90° pulse. It flips the spins from their original z-direction into the $x–y$ plane, at right angles to it. The result is that the spins are now precessing at 20 MHz around the reference field again, not in their stable up or down states, but rather in an unstable state half way up and down, and with an energy that is half way between that of the up and down state. Quantum theory dictates that they should radiate the excess energy away and revert to the stable (allowed) states. The amount of energy to be radiated, however, is so small that the emission probability is also very small. Other relaxation processes will have acted long before then. It is, however, the small amplitude of this radiation that is measured and amplified to an observable signal: the free induction decay (FID). In ideal circumstances it could take many minutes for the stored energy to be radiated away.

5.2 Signal detectors

For routine use as an instrument to measure solid fat content, the availability of an oscilloscope was not essential in older models. Any aspiration to widen the usefulness of the machine to other applications, however, does depend on a visual feedback of the signal form. Best suited are oscilloscopes with on-board memory, but simpler (and much cheaper) models are practicable too. Without it, only the numerical data at the sample points chosen can be read out, not satisfying the need for overall control by the user.

Modern apparatus has an in-built signal display, by-passing the need for an external oscilloscope.

Signal detection and amplification are the parts that determine sensitivity and signal to noise ratio of most measurement instruments. In pulse-NMR, two types of detector are available: the phase-sensitive detector (PSD) and the absolute (or diode) detector. The PSD is used in applications where the sign (positive or negative) of the signal as a function of time is important. It contains more information than the signal from the diode detector, and is more linear as a function of the quantity of hydrogen atoms present. The diode detector is less linear, but more sensitive. For solid/liquid applications, the diode detector is preferred, but for relaxation measurements, the PSD must be used.

5.3 T_1 relaxation

When a spin system in equilibrium with the lattice is in some way disturbed, it will relax back to equilibrium with a rate called the relaxation time. The slowest of these processes is the 'spin–lattice relaxation' or T_1 relaxation (Farrar and Becker, 1971).

The T_1 relaxation results in the spins being oriented along the field again, just as before the pulse was applied. In order to understand the mechanism, it is important to realise that we are dealing with rotating magnetic moments in the x–y plane at about 20 MHz, and that we need a process that can couple this system to the motions of the atoms around it. It can be shown that the only processes that can do this must have a component of motion at about the same frequency. Candidates are molecules that come to a close distance from the rotating spins and, while oscillating at a similar frequency, have some of the energy of the spins transferred to them in the same way that coupled oscillators transfer energy back and forth. After completion of the transfer, the molecule should move away in order to 'uncouple' the oscillators. The main question now is to find a way of estimating under what conditions such oscillating (parts of) molecules most often occur. The concept of 'molecular motion frequency spectrum' is helpful here. We can think of the molecular motions as being built up by a superposition of vibrations of a whole spectrum of frequencies. Small and mobile molecules will have a flat but broad spectrum with all frequencies occurring up to the highest values of about 10^{12} Hz (far infrared at 30 cm^{-1}). Viscous fluids will have a less broad spectrum, and solids will only show very low frequency lattice vibrations. In all these cases, the total area under the spectrum is of course normalised to unity since it represents a probability density distribution.

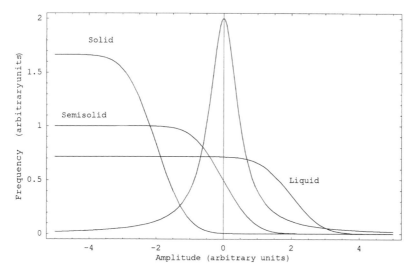

Figure 5.1 Interaction (coupling) of spin Larmor frequency (central Lorentzian band) with a liquid, a semi-solid and a solid material: spread over a wide frequency range, or confined only to low-frequency modes.

The coupling with the spins, rotating in the x–y plane, can only occur at about the precession frequency. We can now check each of the above spectra for their contributions at that frequency: solids will have a very low contribution because there is very little net molecular motion, their spectrum extends upwards towards 20 MHz before dropping steeply (Padua and Schmidt, 1992). Viscous liquids will do nicely if their highest frequencies extend beyond 20 MHz. Very mobile liquids will do worse again : their frequency spectrum is spread over such a broad range that only a low probability density will be found around any given frequency.

Thus, the T_1 relaxation time for liquids is about 10–100 ms, while for solids it is between 1 and 10 s.

5.4 T_2 relaxation

Another relaxation mechanism is the 'spin–spin relaxation' or T_2 relaxation. Immediately after a 90° pulse, the spins are rotating with the same phase: they all point in the x-direction simultaneously, and a quarter turn later, they all point towards y. Therefore, their magnetic moments add up to a macroscopically measurable FID signal.

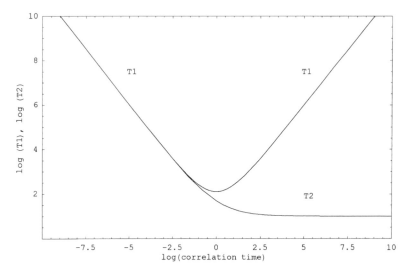

Figure 5.2 Log–log plot of T_1 and T_2 as functions of molecular correlation times: T_1 first decreases, goes through a minimum, and rises again; T_2 decreases and then levels off.

The T_2 relaxation process leaves the spins in the x–y plane, but randomly oriented with no net magnetic moment. Their resultant vector sum becomes zero. In the strictest sense, the T_2 relaxation is not a real relaxation process since it only describes the disappearance of the FID signal, not the return to equilibrium. No net energy is transferred. It is the *correlation* between the motions of the individual spins that gets lost. The relaxation mechanism is similar to the T_1 mechanism, but lower frequencies than the precession frequency are effective. Any slow close encounter with a polarisable part of a molecule will be able to de-phase the spins. In solids, this mechanism is very effective since the low molecular frequencies contribute and the atoms are close together for long periods of time. In liquids the mechanism is less effective because the molecular frequencies are distributed over a broader spectrum, leaving less contribution from the low frequency side.

Thus, solids will T_2 relaxation times of 5–20 µs, while liquids reach 10 ms. It is important to remember that T_1 and T_2 of solids lie wide apart, but for liquids they are only an order of magnitude apart.

5.5 T_2^* relaxation

A third relaxation mechanism is caused by the imperfection of the magnetic field in the instrument. It has exactly the same effect as the T_2

relaxation mechanism : loss of phase coherence. Very small inhomogene-ities in the local field strength cause the precession frequency to be position dependent. A difference in field strength of 1 in 2000 causes a change in precession frequencies with a same amount. At 20 MHz, and thus 0.05 µs per revolution, this causes de-phasing after 100 µs. This is shorter than the T_2 for liquids, but is irrelevant for solids: the T_2 relaxation time of solids is about the same as the T_2^* relaxation time.

The functional form of the net magnetisation versus time is not a simple exponential decay, but rather a Gaussian: the same function that describes the distribution of instantaneous phases around the mean.

5.6 Pulse sequences and spin echos

Rotation of the spins from the 'up' direction to the x–y plane can be performed by a single 90° pulse. A pulse of twice that length is called a 180° pulse. On a system in equilibrium, it causes a transition from up spins into down spins. Its effect on a system of spins rotating in the x–y plane is somewhat more complex: it causes a rotation of the system around an axis in the x–y plane. This brings the spins back into the x–y plane, but with a difference: they are now rotating in the opposite sense. Consider what effect this has on a system of spins undergoing T_2^* relaxation. At places in the sample where the reference field is slightly higher than average, the precession frequency after a 90° pulse will be slightly higher, and those spins 'run ahead' of the others. Similarly, other

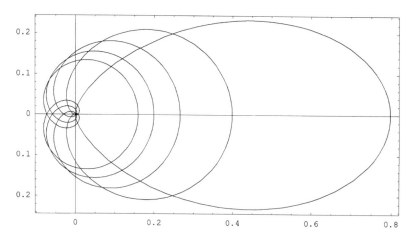

Figure 5.3 Fanning out of the spin vectors during loss of phase coherence in the x–y plane (arbitrary units).

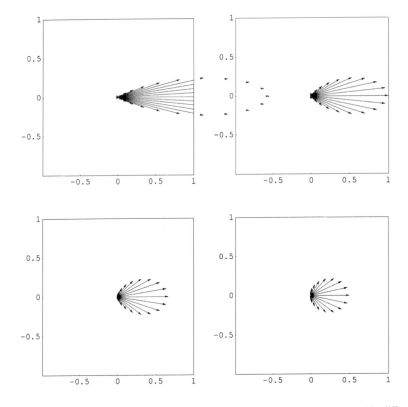

Figure 5.4 Plot in the x–y plane of the spread of the spin amplitude (arbitrary units) in different directions at five time intervals after the pulse. As the phase difference increases, the vectors become more and more evenly distributed over all angles. Their vector sum then approaches zero. The function turns out to be Gaussian.

spins will lag behind. After a certain duration τ, a 180° pulse will change the direction of rotation, and the front-runners find themselves at the back of the pack. Since they run slightly faster than average, they will catch up after again the same duration τ. Similarly for the slower spins: they lag behind, then find themselves at the front of the pack, which catches up with them.

So the pulse sequence '90°, duration τ, 180°, duration τ' causes a rephasing of the spins that will show up on an oscilloscope as a spin-echo. The procedure can be repeated indefinitely, and is called the 'Carr–Purcell–Meiboom–Gill' or CPMG sequence: 90°, duration τ, (180°, duration 2τ), ... The spins will temporarily re-focus each time half way through the 2τ duration to an amplitude that it would have had if there were no T_2^* relaxation present. This CPMG procedure allows us to circumvent T_2^* and measure T_2.

A different pulse sequence can be used for measurement of the T_1 relaxation time. It starts from a well-equilibrated spin system: a duration of 5 times T_1 is sufficient to erase any previous conditions. Such duration of about 5 T_1 is called a 'relaxation delay'. First, a single 90° pulse is applied. This gives a FID governed by T_2^* with a certain initial amplitude $M(0)$. After a duration $\tau(1)$, a new 90° pulse is applied. Depending on the degree of T_1 relaxation during $\tau(1)$, a FID like the first one will be measured, but with a lower initial amplitude $M(1)$. Only the spins that have returned to equilibrium during $\tau(1)$ will be available for rotation into the x–y plane (the others are already in x–y, but are defocused). By choosing longer and longer durations $\tau(t)$, the return to equilibrium of the spins along an exponential decay curve can be monitored.

A more accurate way to perform this experiment uses a 180° pulse instead of the initial 90° pulse. This flips the spins in the $-z$ direction, from where they relax to the $+z$ direction. It is then sufficient to find the delay $\tau(t)$ for which the FID is zero. This method does, however, depend on using a phase-sensitive detector in order to be able to measure the sign of the FID.

These types of measurement, for both T_2 and T_1, can be performed more easily if a purpose-made built-in pulse programming device is available. This can either perform calculations fitting a single decay curve to the data, or provide an output of net magnetisation versus time, letting the user decide how to further analyse the data.

5.7 Data analysis: superposition of multiple decay curves

Measurement of relaxation processes yields data in the form of a set of signal amplitudes versus time. The signal amplitude can in the best case reflect the presence of only a single relaxation mechanism. Analysis is then quite straightforward. The logarithm of the amplitude versus time is a straight line. If two or more relaxation processes are active at the same time, things become mathematically hard to handle. The situation gets especially complex if two exponential decays whose half-lives differ by less than a factor of 3 are superposed. The resulting curve is then very similar to a single decay curve with an about average relaxation time. Moreover, the presence of noise will obscure the local values of the slope of the decay curve, making further analysis futile. Very often, a numerical curve-fitting routine is used. This has the disadvantage that the 'goodness of fit' of such procedure is a property that is hard to visualise.

A more accessible method goes as follows. First, smooth the data to eliminate high-frequency noise, then choose a zero offset to compensate for any offset in the data. Then tabulate the logarithm of the data versus time (or other sampling basis). This provides a data set that should be linear for a single decay mechanism. Now, calculate the local slope along the line using a simple point-by-point difference calculation adapted to the sampling rate. For a single decay mechanism, all these slopes along the curve should be equally spread around the true mean, with a Gaussian distribution caused by the noise spectrum. But if two superposed relaxation processes are present, the first part of the decay will be steeper than the second part. Consequently, the distribution of slopes will tend towards a double distribution. This method leaves open the interpretation of the number of relaxation processes. It casts the problem into a statistical form where relaxation time separation versus noise amplitude plays a role.

5.8   Measurements of solid fat content

The main application of pulse-NMR in the food science laboratory is measurement of solid fat content (SFC). This method has superseded the earlier, cumbersome solid fat index (SFI) measurements that were based on the density difference between liquid and crystalline fat.

In essence, the fat to be measured is melted, put in a sample tube, solidified in a standard way, tempered for a standard time at the measurement temperature, and finally introduced into the NMR apparatus for measurement. This last step requires only about 10 s. The preceding temperature conditioning takes about 2 hours. The measured SFC value obtained should be interpreted as a product characteristic: it is the maximal amount of solids obtainable from such fat at the chosen tempering temperature.

A quite different kind of determination is needed when the question is 'What is the actual amount of solids in this sample?', as for example, when a fat sample is inspected for graininess or oiling-out. The fat should be measured directly in a measurement tube, without changing its temperature.

5.8.1   The direct method

The actual SFC measurement is programmed into the apparatus. It consists of relaxation delay (RD) of 2 s, followed by a single 90° pulse. The FID is sampled after 11 µs and again after 70 µs. These values, M_{11} and M_{70}, are microvolt readings that are to be processed into a percentage solids output according to the formula (5.1).

$$\%\text{Solids} = \frac{100(M_{11} - M_{70})f}{(M_{11} - M_{70})f + M_{70}} \tag{5.1}$$

In this formula, M_{70}, the FID signal amplitude after 70 μs, is taken to be proportional to the amount of liquid oil present. This is so because the liquid material has a T_2^* that is much longer (about 1000 μs), and has not appreciably relaxed during this period. The crystalline material, however, has a T_2^* of about 10 μs, and so no longer contributes to the net magnetisation.

The magnetisation M_{11} contains the full contribution from the liquid phase plus part of the contribution from the solid phase. The difference $(M_{11}-M_{70})$ contains the net contribution from the solid phase, but its value is quite a bit lower than it originally was (at 0 μs, or immediately after the 90° pulse). The factor f compensates for this partial relaxation of the solid signal during the 11 μs delay. The NMR apparatus is built to minimise the delay between the 90° pulse and the first measurement. The technical difficulties arise from having to measure a very weak radio signal from the sample using an antenna (coil) immediately after having given a very strong radio pulse through that same coil. The excitation pulse must have damped down to below the value to be measured, and that takes some time. This 'receiver dead time' of about 11 μs is one of the quality parameters of a good apparatus: it should be a short as possible.

The parameter f, called the 'f-factor' or 'the constant', is the weak spot of this technique. It has values of about 1.35 to 1.45. It is not accurately known, and depends mainly on the quality of the coil and the magnet. So, for practical reasons, it is taken to be independent of the fat to be measured. The NMR manufacturer provides calibration samples (solid polymer surrounded by liquid paraffin oil) that allow fine adjustment of the f-factor. A few other calibration parameters are also present in the system, but need only be used if one wants to fine-tune the system. One is a factor to compensate for the small decay of the liquid signal after 70 μs (value of about 1.005), and a second one is an offset (or bias) of the zero signal voltage of the detector.

When SFC data are to be compared between different laboratories, problems with reproducibility often arise. A measured value of, say, 40% solids in one laboratory could turn out to give 42% solids in another, both laboratories strictly adhering to standard methodology and calibration. In such cases, it is wiser to 'translate' the results numerically in case of dispute than to try and get both laboratories strictly in line. Attempts to compare calibration samples on both machines lead to frustrating results: both sets will not be strictly compatible. It seems that all calibration samples get their nominal values from measurement on a single machine at the manufacturer's laboratory. Different machines have slight differences in coil geometry and field homogeneity and these differences will translate into different measurement outcomes.

5.8.2   The indirect or 'absolute' method

A method exists that avoids the use of calibration samples: the 'absolute' method. It is based on measurement of the liquid signal exclusively, correcting for the change of magnetisation of the liquid with temperature. The liquid signal is measured at 60°C, when the sample is completely molten. After crystallisation, the same sample is measured again, and the ratio of their signal amplitudes (after 70 μs) is equal to their ratio of respective liquid contents. Since the amount of signal per unit mass varies slightly with filling height and this secondary effect becomes worse at extreme filling heights (beyond the coil height), it is necessary to have the same effective mass of sample for both measurements. This can be achieved either by limiting the amount of sample to within the linear portion of the coil (bottom as well as top), or by overfilling and subsequent correction for expansion with temperature of the liquid and contraction of the solid versus the melt.

If a probe head is used with a sufficiently long coil, than underfilling the tubes will guarantee measurement of the same weight of sample for both the molten and the liquid state for each tube. A liquid sample at 60°C needs to be measured, and its amount of signal per gram is used to calculate the liquid signal for all other tubes (of the same sample) on the basis of their weight. No correction for liquid expansion should be used in that case.

For a S/L-probe head, it is easiest to overfill the sample tubes, guaranteeing optimal signal to noise ratio. The amount of signal is then determined by the weight of sample inside the volume 'as seen' by the coil. The liquid signal at 60°C per gram of oil needs to be measured here too, once for one sample type, and the expansion correction calculated for all samples. These expansion correction calculations can be implemented in a spreadsheet program for routine use. Small differences in sample tube cross section will be a source of error with this method.

Of course, the amount of signal per gram of oil depends on the number of protons per gram of sample. This allows measurement of mixtures of oil and water also: simply recalculate H-atom content per gram of sample.

5.8.3   The effect of polymorphism and crystal size and perfection on T_1

In the standard *direct* method, the measurement begins with a relaxation delay (RD) of 2 s. This is based on the estimation of the T_1 relaxation time being less than 0.4 s: the RD is then 5 times T_1.

Samples that have been crystallised by chilling them to 0°C and subsequently tempered for 30 min at the measuring temperature will consist of very small crystals with an enormous surface to volume ratio. The forced crystallisation promotes mixed crystal formation in the β' polymorph. Wide-angle X-ray diffraction of such material shows broad diffraction peaks, typical of material with small crystal sizes and a relatively poor crystal perfection. As mentioned earlier, such material will have a T_1 relaxation time that is more liquid-like (shorter) than that of a normal crystal.

When longer tempering times are used, as is the case for measuring cocoa butter, a more perfect crystalline material is obtained, and a RD of 4 s should be used. The same argument holds for samples that either were never chilled (monitoring of a fractionation experiment, for instance) or samples that have aged (core samples taken from a block of frying fat).

In such cases, it is prudent to perform an independent check on T_1 by repeated measurement at different RD. Given that slow crystallisation into the β polymorph can lead to T_1 values of about 2 s, a RD of 10 s would be necessary. This is only possible if temperature control of the sample holder is available. Lacking this, the sample would slowly warm up and partially melt (the magnet is kept at 40°C for stability reasons).

The dramatic effect of the polymorphic form on the T_1 relaxation time can be demonstrated by measurements on tristearine, tabulated in Table 5.1. The table shows a clear jump in T_1 from about 0.26 s for the α polymorph to about 1 s for the β polymorph. During further tempering, the T_1 relaxation time further increases because of increasing crystal perfection.

The difficulty of attributing a T_1 relaxation time as a material characteristic to a product like tristearine becomes obvious when the experiment is repeated, but with a tempering at 60°C this time (Table 5.2). The α phase melts at this temperature, and a more perfect crystalline β matrix is produced. The T_1 relaxation time increases to 2 s and even tends towards 3 s after a few days of tempering.

Table 5.2 also shows that after 2 and 4 min tempering, a double-peaked distribution is present. At longer times, only a single-peaked distribution is visible. The peak separation corresponds to a factor of about 5 in T_1 relaxation time between the α and β polymorphs. Based on the rule of thumb that a RD (relaxation delay) of 5 times T_1 is necessary, 1 s is enough for the α phase, but for the β phase from 6 to 11 s may be needed. The usual value of 4 s for tempered fats is a compromise to minimise partial melting during measurement.

Table 5.1 The frequency distribution of relaxation times, taken as slopes of the logarithm of the decay curve. The sample is a pure triglyceride that has three polymorphic forms: α(m.p.=54°C), β′(m.p.=65°C) and β (m.p.=72°C)

T_1(s)	\multicolumn{13}{c}{Time sample was kept at 50°C}
	1	2	4	8	16	32	1	2	4	8	16	32	64
	\multicolumn{6}{c}{(min)}	\multicolumn{7}{c}{(h)}
0.20	2	1											
0.22	**35**	**33**	13										
0.25	**21**	**28**	**41**										
0.28	13	15	15	10									
0.32	6	6	7	**35**									
0.35	1	1	3	**25**									
0.40	1	1	3	10									
0.45	1	1	2	2									
0.50				4	3	2							
0.56	1	2	1	2	4		2				1	1	
0.63				1	5	2	2	1	2	1			
0.71				**2**	16	4	3	2	2		1	1	
0.79				1	**42**	8	10	11	4	3	4	4	2
0.89					27	**36**	**58**	22	14	4	6	7	1
1.00					11	**44**	28	**38**	35	15	5	8	1
1.12					6	13	8	**32**	**48**	20	24	13	10
1.26					2	5	1	5	10	**49**	**42**	28	21
1.41					1	3	3	2		19	24	**43**	**42**
1.58						1	1				6	9	23
1.78							1			2	2	1	11
2.00											1	2	1
2.24										2	1	1	2
2.51													
2.82											1		
3.16											1		
3.55													
3.98													
Sum	81	88	89	91	116	116	115	115	115	115	119	118	114
Mean T_1	0.26	0.26	0.29	0.36	0.83	0.98	0.93	1.02	1.04	1.22	1.28	1.28	1.43

A 2 g sample in a NMR tube is melted, and then chilled in a bath at 10°C. This produces the α polymorph. This form of the fat can be stored at low temperatures, but it is slowly converted into the β polymorph at temperatures above 25°C. The experiment consisted in controlling this transition by tempering the sample for different times in a bath at 50°C. After tempering, the sample was quickly cooled again, freezing the obtained system for subsequent T_1 measurement at room temperature.

5.8.4 Discrimination between liquid oil and water using T_2 measurements

Apart from measuring solid to liquid ratios, one of the applications where low-resolution pulse-NMR is invaluable is nondestructive measurement

Table 5.2 Tempering at 60°C

	Time sample was kept at 60°C												
	1	2	4	8	16	32	1	2	4	8	16	32	64
T_1(s)				(min)						(h)			
0.20	4												
0.22	29	14											
0.25	31	31											
0.28	11	18											
0.32	2	10											
0.35	4												
0.40	1	2											
0.45	2	3											
0.50	1	1		1									
0.56	1	1	12		3		2						
0.63		3	10	1	1		1	1	3	1	2		
0.71			7	3	2	1	1		1	3	3	1	
0.79			11	2	3	2		3	5	4		4	3
0.89			14	5	4	2	3	4	3	1	1	5	5
1.00		1	10	13	8	6	4	6	5	2	7	2	5
1.12			18	10	8	14	7	7	5	5	5	4	3
1.26			11	22	11	13	9	10	7	5	5	3	5
1.41			11	17	22	14	20	12	9	14	10	4	2
1.58				24	30	37	20	24	17	12	14	11	5
1.78				21	23	23	38	42	29	26	18	14	2
2.00					3	6	13	5	26	24	34	37	13
2.24						1	1	2	5	11	15	13	17
2.51							1		4	4	2	7	18
2.82							2		1	2	2	6	31
3.16									1		1	1	13
3.55													0
3.98										1		3	1
Sum	86	84	104	119	118	119	119	118	119	117	118	117	123
Mean T_1	0.26	0.30	0.95	1.35	1.41	1.48	1.60	1.51	1.66	1.70	1.75	1.88	2.26

of water content versus oil content for seeds. Imagine a plant breeding testing station having to screen oil seeds for oil content. The simplest procedure would be to dry the seeds to a constant low moisture level, and simply measure solid/liquid ratio. The solid will be mainly cell wall material (cellulose), the liquid will be mainly oil, but with an unknown contribution from moisture. If the premise of a constant low moisture level is correct, then differences in total liquid content will mainly reflect differences in oil content. If the plant breeder is satisfied with a sound classification of large amounts of seed samples at minimum cost, such a solid/liquid ratio screening, supplemented with a calibration through a set of classical (destructive) tests for moisture and oil content, will solve his problems. At about 15 s per tube, a thousand tubes of seeds can be measured in a working day.

If more extensive and precise analysis is desired, then a discrimination between the solid signal and the signals from moisture and from oil is needed. This is possible by decomposition of the T_2 decay curve into its components (Shanbhag et al., 1970; Mousseri et al., 1974; Leung et al., 1976).

The oil phase has a T_2 of 90–150 ms. The T_2 of the water is very much dependent on the physical state of the water molecules. In low to medium moisture oil seeds, two populations of water molecules seem to be present, with T_2 values of 0.5 and 5 ms. The longer of these is conveniently about 10 times shorter than the T_2 of the oil. Graphical decomposition of relaxation signal contribution is easy in such cases, provided that the amounts (signal amplitude contributions) of the respective phases are comparable. Measurements on seeds with between 3 % and 20 % moisture are feasible. (Gambhir and Agarwala, 1985; Roefs et al., 1989).

For pure water at 20°C in a bulk phase, a T_2 of 2.0–2.4 s can be found. (As an interesting aside, it is noteworthy how difficult it is to find literature references for this value; Callaghan et al., 1983). This is about the same as its T_1, as should be expected for a mobile, low-viscosity fluid. The enormous decrease in T_2, from 2 s to 0.5 ms, is caused by the reduced mobility of adsorbed or 'bound' water. The enormous dipole moment of the water molecule causes hydrogen bonding with any polar substrate. This reduces the mobility of a number of layers of water molecules. This reduced mobility will translate into a decreased T_2, according to the same logic by which T_2 for solids or viscous liquids is longer than for normal liquids. As an exercise, the 'viscosity' of this adsorbed water could be calculated from comparison with T_2 relaxation times of water–glycerol mixtures. For apolar substances, a similar logic holds: these molecules accept no hydrogen bonds, causing the surrounding water molecules to 'optimise' their interactions with water farther away. This again reduces their mobility ('cage' or 'clathrate' formation, hydrophobic interactions).

As discussed earlier, a pulse echo sequence is needed to discriminate between T_2 and T_2^* relaxation mechanisms. The CPMG pulse sequence, as defined above, contains an unspecified inter-pulse delay 2τ. By manipulating this parameter, some extra information can be obtained from the T_2 relaxation spectrum. For long τ, an abnormally low T_2 will be measured. This is caused by diffusion of the molecules during the interval towards positions with a different local field value. As a consequence, the reversal in precession rate is no longer an exact 'mirror image' of its prior dephasing : it goes faster or slower in the second leg of the trip than on the first leg. Both the diffusion coefficient and the degree of inhomogeneity worsen the effect. For liquid free water, a interval 2τ between the 180° pulses of no more than 1 ms is needed, because of the fast self-diffusion of water $(2.3 \times 10^{-9}\,\mathrm{m^2\,s^{-1}}$ at 25°C). In 1 ms time, a distance of about 1 μm in either direction is probably travelled; if the magnetic field has a 0.001% lower value at this point, then there will be a 0.001% mismatch between the phase of the spin at the time of echo. This mismatch would amount to 0.001% of (1 ms times 20×10^6 cycles per second) or 0.2 cycles or 72° in phase. So the spin would be practically orthogonal to the direction it should have had, and will not contribute to the echo. A magnet inhomogeneity of 0.001% over 1 μm amounts to 10% per cm, a typical diameter of a NMR tube. This inhomogeneity is local, however, and is not maintained coherently over large distances, but is a varying noise-like function of distance. In practice, the permanent magnets are better than this.

5.8.5   Field gradient techniques for water droplet measurement

Some benchtop models of wide-line pulse-NMR can be equipped with a field gradient module. With this extension, a very precise gradient pulse can be superimposed on the steady field. This gradient field effectively makes the 'local' field strength a function of the z-coordinate. As discussed earlier, the effect of a nonhomogeneous field is to desynchronise the spin phase angles, resulting in a diminished echo amplitude. By using a well-timed gradient pulse, this effect can be applied more selectively than would be possible with an omnipresent and lasting field gradient. With a stronger gradient, a higher sensitivity to proton diffusion is obtained. It results in a discrimination between hindered and unhindered diffusion (Singer, 1978). The technique goes as follows. After a 90° initiator pulse, a (τ–180°–τ–echo) repeat sequence is set up as for the standard T_2 measurement, but this time, each τ delay period also contains a field gradient pulse of a chosen strength and duration. Let us assume that the protons do not change position during the (τ–180°–τ–echo) episode. In that case, the protons will experience the gradient pulse twice at exactly the same magnitude. Its effect during the 'upstroke' (or first τ period) is exactly cancelled during the 'downstroke' (or second τ period). But if the proton moves to a different z-position between both gradient pulses, then the downstroke gradient pulse will not cancel out the upstroke. So this proton will not contribute to the echo signal. Varying the delay between upstroke and downstroke will allow the experimenter to explore diffusion of the proton over larger distances. The net effect of the gradient pulse is to dramatically enlarge the effect of even small diffusion distances on the echo amplitude (Hrovat and Wade, 1981).

Suppose the proton belongs to a molecule that is trapped in a spherical cavity such as an emulsion droplet. In such case, lengthening the time available for diffusion (τ) will first show a diminished echo amplitude with τ, but, when a diffusion distance equal to the droplet diameter is reached, no further decrease will be seen. The echo amplitude will then level off at a value indicative of the cavity (droplet) diameter (Packer and Rees, 1971).

A realistic sample will contain drops of different sizes, and an overall response is measured, generated by contributions from the different size classes. No clear discontinuity between normal decay and plateau value can be expected in such case. Complex numerical routines are the only way to back-track from measured data towards an initial size distribution assumption.

In practice, the pulse programming, data manipulation and results generation are performed by the constructor's software. This frees the user from the need to tune and optimise the hardware, but entails the risk of severe oversimplification. Instead of generating a complete set of gradient pulse intervals, and leaving the burden of data interpretation to the user, such software tries very hard to come up with a simple clear-cut figure for the droplet radius in an 'acceptable' time. The aim is often to obtain a measurement tool that can be handled by an unskilled operator. In order to simplify matters, an assumption on particle size distribution must then be made. Samples that are nicely monodisperse can then be measured in 20 min or less. Samples with a bimodal distribution might be detected as problematical, but could just as well pass the fitting routine and come out at nonsensical values. Similar problems might arise with samples of ill-defined viscosity: concentration of polymeric material or proteins near or at the particle boundary might change the relation between (diffusion) time and physical distance travelled. Small drops could have a relatively depleted inside and a relatively enriched outer shell, while larger drops might have lost relatively less material because of their larger volume to surface ratio. The viscosity of the fluid inside the droplets could be different from that measured on a bulk sample of the internal phase. The viscosity, as experienced by the diffusing protons (as part of a small molecule), could be much lower than that measured by any macroscopic rheological technique. In all such cases, an independent diffusion rate experiment must be performed on the bulk phase, prior to the particle size calculations.

It must be stressed, however, that very few particle-sizing techniques allow measurement of opaque samples without dilution, and most of them are plagued by difficulties equivalent to or worse than those mentioned above. For repeated and well-standardised quality control purposes, the field gradient technique might well be a tool suited to the job. Since the hardware discussed here is a simplified and scaled-down version of larger and more versatile research instruments, a continual improvement in electronics and data-handling software is to be expected. For research purposes in the food laboratory, the current capabilities are, in my opinion, just not inspiring enough to warrant investment.

5.8.6 Different types of probe heads for use on whole seeds, chocolate, SFC and general relaxation measurements

The different applications in the food laboratory suitable for pulse-NMR techniques have led to a specialisation of the hardware. The 'probe head' is the heart of the machine, where the sample is positioned between the pole shoes, and surrounded by the coil for emission and reception of the radio frequency pulse.

As discussed earlier, the solid/liquid ratio measurements require a receiver dead time of less than 10 µs. The shorter this dead time, the less extrapolation is needed to estimate the contribution of the solids signal. This can only be achieved with radiofrequency coils of small impedance, hence small radius and small number of windings (about 20 turns). A first trade-off is apparent: the duration of the 90° pulse will be shorter if the coil has larger impedance (higher pulse amplitude), but the ringing in such a coil takes longer to die down, increasing the receiver dead time.

The radius also determines the sample volume, hence signal to noise ratio and sensitivity. Larger volumes, up to 13 ml, are advantageous for measurement of whole seeds or products where the sample is somehow diluted. These can be accommodated only within a magnet with a larger air gap. Since for permanent magnets this implies a lower field strength, such systems operate at a lower resonance frequency (10 or even 4 MHz). So, the larger volume, proportional to the square of the sample tube radius, is partly offset by the lower field strength (inversely proportional to the air gap).

If solid/liquid ratio measurements are not essential for a given application, then a probe head with longer dead time but higher sensitivity can be chosen. These probe heads are optimal for implementing the 'indirect' or 'absolute' method of solid fat measurement.

A still different parameter to optimise is the magnet diameter. Larger diameters can provide a 0.47 T field over a larger gap, and allow samples not only with larger diameters but also with larger filling height. Such magnets are of course heavier and more expensive.

Finally, it should be stressed that the pulse-NMR technique can in principle achieve any signal to noise ratio desired if the operator is willing to accumulate enough repeat pulse measurements: S/N ratio goes up with the square root of the number of repeats. In order to make use of this advantage, and in order to perform some of the longer T_1 or T_2 measurements, the sample needs to be thermostated during the measurement. Precious little space being available, this requirement will again work to the detriment of sample diameter. A water circulation jacket, held in an insulating foam support (hydrogen free), allows thermal regulation of the sample. The presence of the thin air layer between the glass wall of the sample tube and the outer glass wall of the circulation jacket reduces the heat transfer from the sample. The user should be aware that samples in the course of crystallisation can show a 10°C rise in temperature inside the circulation jacket, compared to a 1°C rise if directly submerged in an external water bath. The temperature control is effective for maintaining the temperature, not as a heat exchanger.

The fluid used as heat transfer medium is most often plain water. An alternative is a fluid like Fluorinert ® (from 3M®), which contains no hydrogen atoms, is not poisonous, and has acceptable heat transfer properties (low viscosity). Water as a heat exchange medium will somewhat deteriorate the signal, but as long as the flow rate is sufficient this effect is only minor. In fact, the high flow rate has the same signal quenching effect as extremely fast diffusion.

References

Callaghan, P.T., Jolley, K.W. and Humphrey, R.S. (1983) Diffusion of fat and water in cheese as studied by pulsed field gradient nuclear magnetic resonance. *J. Colloid Interface Sci.*, **93** (2) 521-529.

Farrar, T.C. and Becker, E.D. (1971) *Pulse and Fourier Transform NMR (Introduction to Theory and Methods)*, Academic Press, New York.

Gambhir, P.N. and Agarwala, A.K. (1985) Simultaneous determination of moisture and oil content in oilseeds by pulsed nuclear magnetic resonance. *J. Am. Oil Chem. Soc.*, **62** (11) 103-108.

Hrovat, M.I. and Wade, C.G. (1981) NMR pulsed-gradient diffusion measurements. I. Spin-echo stability and gradient calibration. *J. Magn. Reson.*, **44** 62-75.

Kessler, H., Gehrke, M. and Griesinger, C. (1988) Two-dimensional NMR spectroscopy:background and overview of the experiments. *Angew. Chem. Int. Ed. Engl.*, **27** 490-536.

Leung, H.K., Steinberg, M.P., Wei, L.S. and Nelson, A.I. (1976) Water binding of macromolecules determined by pulsed NMR. *J. Food Sci.*, **41** 297-300.

Mousseri, J., Steinberg, M.P., Nelson, A.I. and Wei, L.S. (1974) Bound water cavity of corn starch and its derivatives by NMR. *J. Food Sci.*, **39** 114-116.

Packer, K.J. and Rees, C. (1971) Pulsed NMR studies on restricted diffusion. I. Droplet size distributions in emulsions. *J. Colloid Interface Sci.*, **40** (2) 206-218.

Padua, G.W. and Schmidt, S.J. (1992) Proton nuclear magnetic resonance measurements on various sugar solutions. *J. Agric. Food Chem.*, **40** 1524-1527.

Roefs, S.P.F.M., Van As, H. and Van Vliet, T. (1989) Pulse-NMR of casein dispersions. *J. Food Sci.*, **54** (3) 704-708.

Shanbhag, S., Steinberg, M.P. and Nelson, A.I. (1970) Bound water defined and determined at constant temperature by wide-line NMR. *J. Food Sci.*, **35** 612-615.

Singer, J.R. (1978) NMR diffusion and flow measurement and introduction to spin phase graphing. *J. Phys. E: Sci. Instrum.*, **11** 281-291.

6   Mass spectrometric techniques in the analysis of triacylglycerols

Päivi Laakso and Pekka Manninen

6.1   Introduction

Mass spectrometry (MS) is without doubt the most powerful technique for structure elucidation of triacylglycerols especially in combination with chromatographic separation. Electron ionisation and chemical ionisation are the basic ionisation techniques that have been coupled with gas chromatography (GC). Instead of GC-MS, most effort has recently been expended on developing and applying methods based on high-performance liquid chromatography (HPLC) and supercritical fluid chromatography (SFC) combined with MS. The general aim of mass spectrometric analysis of triacylglycerols is to generate ions providing information on (1) the molecular weight, which defines the combined number of acyl carbons and double bonds in the acyl chains of a molecule; (2) the fatty acyl residues and their molecular associations; and (3) the distribution of fatty acyl residues between the primary (sn-1/3) and the secondary (sn-2) glycerol positions. Furthermore, quantitative determination based on MS data is of great importance. In this chapter we do not present a thorough review on the early MS studies of triacylglycerols; instead, we focus on those techniques that have most practical value today. In addition, some basic information on the ionisation techniques will be provided. For those who want more information on MS in general, books by Johnstone and Rose (1996), McLafferty and Turecek (1993), and Niessen and van der Greef (1992) are highly recommended. In addition, the book by Murphy (1993) on mass spectrometry of lipids is a valuable reference for lipid chemists.

6.2   Ionisation techniques

6.2.1   Electron ionisation

Electron ionisation (EI) is the most widely used ionisation technique in mass spectrometry. Ions are formed by the exchange of energy during the interaction of the electrons emitted from a glowing filament and vaporised sample molecules. Unimolecular reactions require a high vacuum, typically below 10^{-5} torr. Standard EI mass spectra are obtained with electron energy of 70 eV, which results in excess ionisation energy imparted to the molecule and, thus, in most cases leads to extensive fragmentation of odd-electron molecular ions ($M^{+\cdot}$). The only possibility for the user to affect the ionisation process in EI is to change the ionisation energy in those instruments where it is possible. In general, EI is a hard ionisation technique, typically providing information on fragment ions but not always on the molecular ion. The abundances of major fragment ions of several molecules produced by EI at 70 eV are quite reproducible regardless of the mass spectrometer used. Spectral libraries may therefore be of great value when interpreting an EI spectrum of an unknown compound.

EI mass spectra of triacylglycerols may exhibit low-abundance $M^{+\cdot}$ or $[M-18]^{+}$ ions valuable for molecular weight determination and several fragment ions useful for structure elucidation, such as $[M-RCOO]^{+}$, $[M-RCOOH]^{+}$, $[M-RCOOCH_2]^{+}$, $[RCO+128+14n]^{+}$, $[RCO+74]^{+}$ and RCO^{+}, where R = aliphatic hydrocarbon chain and $n = 0, 1, 2, \ldots$ (Ryhage and Stenhagen, 1960; Barber et al., 1964; Aasen et al., 1970; Lauer et al., 1970; Murphy, 1993). Generally, the most abundant peaks in the EI spectra correspond to $[M-RCOO]^{+}$ and RCO^{+} ions, which can be used for fatty acid identification. Furthermore, information on the regiospecific distribution of fatty acids in triacylglycerol molecules may be achieved from the $[M-RCOOCH_2]^{+}$ ions, which are only formed by the cleavage of an acyl chain from the sn-1 and sn-3 positions (Ryhage and Stenhagen, 1960; Barber et al., 1964).

6.2.2   Chemical ionisation

Chemical ionisation (CI) is based on gas phase ion–molecule reactions involving a small amount of sample and a large amount of reactant gas at fairly high pressure (~1 torr). The reactant gas (most often methane, ammonia, or isobutane) is ionised with EI to produce primary reactant gas ions (radical cations), which will further react with other reactant gas molecules to produce stable secondary reactant gas ions (e.g. $CH_5^+, NH_4^+, C_4H_9^+$). These secondary reactant gas ions interact with analytes typically by proton exchange, hydride abstraction or charge exchange reactions or to give adduct ions. Utilising CI, the degree of fragmentation of molecules can be controlled by selecting the reactant gas and adjusting the ionisation conditions, i.e. reactant gas pressure and temperature in the ion source. The extent of fragmentation depends on the proton affinity difference between the reactant gas ion and the analyte: the greater the difference, the more fragmentation is expected to occur. The proton affinities of the most commonly used reactant gases in positive-ion mode decrease in the order $NH_4^+ > C_4H_9^+ > CH_5^+$; thus, methane is a hard CI gas causing severe fragmentation, whereas ammonia is a softer choice causing much less fragmentation. As compared with EI, CI is a gentle ionisation technique resulting in less fragmentation and greater abundance of ions containing the intact molecule. In addition to positive ions, CI can be carried out in negative-ion mode. The proton affinities of reactant ions decrease in the order of $NH_2^- > OH^- > Cl^-$: the higher the proton affinity of the reactant ion, the more likely is proton abstraction from the analyte.

Triacylglycerols have been ionised by CI using ammonia (e.g. Murata and Takahashi, 1977; Kallio and Currie, 1993a; Evershed, 1996; Huang *et al.*, 1997), methane (e.g. Rezanka *et al.*, 1986; Myher *et al.*, 1988; Anderson *et al.*, 1993), isobutane (Fales *et al.*, 1975; Spanos *et al.*, 1995), and a mixture of methane–nitrous oxide (75:25, v/v) (Stroobant *et al.*, 1995) as reactant gases. It is also possible to use HPLC eluents, such as a mixture of propionitrile–acetonitrile with or without dichloromethane, as a source of reactant ions (Marai *et al.*, 1983; Kuksis *et al.*, 1991a,b). Molecular weight information on triacylglycerols is achieved from the m/z values of the $[M+NH_4]^+$ ions formed by ammonia as a reactant gas in positive-ion CI or from the m/z values of the $[M+H]^+$ ions formed by methane, isobutane, or a mixture of propionitrile–acetonitrile in positive-ion mode. Negative-ion CI with ammonia or methane–nitrous oxide mixture results in the formation of abundant $[M-H]^-$ ions. Triacylglycerols may also yield abundant $[M+Cl]^-$ ions utilising chloride-attachment negative-ion CI (Kuksis *et al.*, 1991a,b). In addition to ions containing the intact molecule, CI mass spectra of triacylglycerols typically exhibit informative fragments for structure elucidation of fatty acyl residues, such as $[M-RCOO]^+$, $[MH-RCOOH]^+$, $[RCOO-H_2O]^-$, $[RCOO-H_2O-H]^-$, and $RCOO^-$ ions. In literature, both of the forms $[M-RCOO]^+$ and $[MH-RCOOH]^+$, having equal m/z values, are used to describe the loss of one acyl chain from a triacylglycerol. However, the fragmentation pathways of triacylglycerols have not been thoroughly studied and, therefore, in this chapter we have used the same fragment expressions as presented in the original publications.

6.2.3 Atmospheric pressure ionisation

Instead of ionisation techniques requiring vacuum, ions can also be generated in ion sources operating at atmospheric pressure. Atmospheric pressure ionisation (API) is performed in either liquid or gas phase as reviewed by Bruins (1991). Problems related to API technique concern transfer of ions from the atmospheric pressure region into a vacuum system, and prevention of the strong cooling effect of a mixture of gas and ions when it expands into a vacuum, resulting in the formation of cluster ions. Attempts to minimise cluster ion formation have been made, by passing the formed ions via a heated capillary to the analyser or by introducing a drying gas countercurrent to the ion stream. API techniques are especially valuable in coupling HPLC or SFC to MS.

6.2.3.1 *Electrospray ionisation*

Electrospray ionisation (ESI) is a technique based on liquid phase ionisation. The general processes occurring in (ESI)MS are: (1) formation of a fine mist of electrically charged droplets from electrolytes dissolved in a solvent under the influence of an electric field; (2) shrinkage of charged droplets by solvent evaporation and repeated droplet disintegration; (3) ejection of sample ions from the highly charged, very small droplets into the gas phase; and (4) entrance of the ions to the mass spectrometer and ion analysis (Bruins, 1991; Kebarle and Tang, 1993). When the nebulising action of the electric field is assisted by a high-velocity gas flow, the technique is called either ion spray or nebuliser-assisted ESI. As an example, a schematic of the Finnigan (ESI)MS system is presented in Figure 6.1a. The sample solution is introduced through the ESI needle assembly, which is typically held at a potential of ± 3 to $\pm 8\,kV$ relative to the ion source walls. The electric field results in increasing charge density

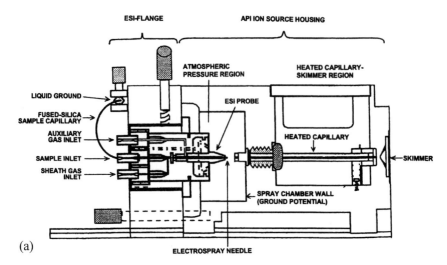

(a)

Figure 6.1   A schematic drawing of (a) the Finnigan MAT electrospray ionisation (ESI) probe attached to the atmospheric pressure ionisation (API) ion source housing, and (b) the Finnigan MAT atmospheric pressure chemical ionisation (APCI) probe assembly, to be attached to the API ion source housing (Finnigan MAT, 1993). (Reprinted with permission from Finnigan MAT.)

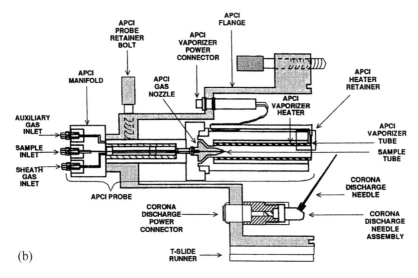

Figure 6.1 (Continued).

at the solution surface, which leads to formation of a spray of highly charged droplets. ESI is not a very useful technique for combination with conventional HPLC separations: ESI can only be carried out with flow rates up to 20 μl min^{-1}. Nebuliser-assisted ESI allows higher flow rates to be used (e.g. up to 200 μl min^{-1}), but splitting of the normal HPLC effluent flow is still required. The most recent constructions of mass spectrometers equipped with API sources can handle flow rates up to 1 ml min^{-1} in nebuliser-assisted ESI mode. The sensitivity of (ESI)MS depends on both the flow rate and sample concentration: the best sensitivity is achieved with flow rates less than 200 μl min^{-1} and with sample concentrations less than 10^{-5}–10^{-4} mol l^{-1} (Ikonomou et al., 1991; Kostiainen and Bruins, 1994).

The use of (ESI)MS in the analysis of lipids has been reviewed by Myher and Kuksis (1995a) and Kuksis (1997). Most lipid applications concern determination of phospholipids, which are readily ionisable in solution, thus being ideal components for ESI. Triacylglycerols are not typically in an ionised form in solution, which explains the small number of publications so far concerning (ESI)MS analyses of triacylglycerols (Duffin et al., 1991; Kuksis and Myher, 1995; Schuyl et al., 1995; Myher et al., 1997). The (ESI)MS spectra of triacylglycerols exhibit typically only adduct ions, such as $[M + NH_4]^+$ and $[M + Na]^+$ ions, with little or no fragmentation. The m/z values of adduct ions characterise the molecular weights of triacylglycerols. Additional structure information on fatty acyl residues can be achieved by collision-induced dissociation

(CID) of e.g. $[M + NH_4]^+$ ions on triple-quadrupole instruments (MS/ MS determination) (Duffin *et al.*, 1991), or by inducing fragmentation by introducing additional energy to the ions formed on single-quadrupole instruments (HPLC-(ESI)-CID-MS determination) (Kuksis and Myher, 1995; Myher *et al.*, 1997).

6.2.3.2 *Atmospheric pressure chemical ionisation*

Atmospheric pressure chemical ionisation (APCI) is a gas-phase ionisa- tion technique occurring via identical reactions to CI. As an example, a schematic of the Finnigan (APCI)MS is presented in Figure 6.1b. In this system, the HPLC effluent is introduced into the APCI source via a fused- silica sample capillary (sample tube). A sheath gas (= nebulising gas) flow is introduced to the sample capillary exit region to assist in aerosol formation. The small droplets formed will be evaporated while passing through a vaporiser ($T_{max} \approx 600°C$), and the gas stream is focused by auxiliary gas flow before entering into the ionisation region. The solvent molecules are ionised by corona discharge to form reactant ions, which will further ionise the analyte molecules. APCI is a solvent-mediated ionisation process, i.e. the HPLC eluents serve as a source of CI gases. In addition to HPLC, components separated by SFC can also be efficiently ionised by APCI. In general, ion–molecule reactions in APCI are mostly dependent on acid–base chemistry (Bruins, 1991). In positive-ion APCI, reactant ions are acids, such as protonated methanol, acetonitrile and water. Ammonium ion is a weak acid which is selective towards sample molecules, whereas protonated methane is a strong nonselective acid. The weakest possible reactant ion is formed when mixed reactant gases are used. In negative-ion APCI, reactant ions are bases. There are no demands for ionic species to be present in solution, and therefore the APCI technique is also suitable for the ionisation of non-polar molecules.

Most triacylglycerols yield simple (APCI)MS spectra consisting of abundant ions containing the intact molecule, such as $[M + H]^+$ or $[M + NH_4]^+$ ions, and fragment ions formed by the loss of one acyl chain from a molecule (Byrdwell and Emken, 1995; Laakso and Voutilainen, 1996; Laakso and Manninen, 1997; Manninen and Laakso, 1997a). The $[M + H]^+$ and $[M + NH_4]^+$ ions provide the molecular weight informa- tion, which defines the combined number of carbon atoms and double bonds in the acyl chains of a triacylglycerol, whereas the $[M - RCOO]^+$ ions characterise the fatty acid moieties and their molecular association in a triacylglycerol. Furthermore, the $[M - RCOO]^+$ ion abundances may provide information on the distribution of fatty acyl residues between the *sn*-2 and the *sn*-1/3 positions (Laakso and Voutilainen, 1996; Manninen and Laakso, 1997a). The use of (APCI)MS in lipid analysis has been reviewed by Byrdwell (1998).

6.2.4   Other ionisation techniques

Ionisation of triacylglycerols has also been examined by several other soft ionisation techniques, such as field ionisation (FI), field desorption (FD), fast atom bombardment (FAB) and thermospray (TSP) (reviewed by Games, 1978; Jensen and Gross, 1988; Matsubara and Hayashi, 1991; Kim and Salem, 1993). Most of these techniques produce mass spectra containing abundant ions consisting of the intact molecule and a few characteristic fragment ions. Molecular weight information is achieved, for example, by FD, which produces $M^{+\cdot}$ and $[M+H]^{+}$ ions in addition to fragments, but interfering adduct ions (e.g. $[M+Li]^{+}$, $[M+Na]^{+}$, $[M+K]^{+}$) may also be formed if metal salts are present in the sample (Fales et al., 1975; Schulten et al., 1987). Formation of cationised triacylglycerol ions $[M+Na]^{+}$ has been utilised in FAB ionisation and applied to the characterisation of various edible oils (Evans et al., 1991; Lamberto and Saitta, 1995). Also TSP mass spectra of triacylglycerols provide molecular weight information ($[M+NH_4]^{+}$ and/or $[M+H]^{+}$ ions) in addition to fragment ions formed by the loss of one or two acyl chains from the molecule (Kim and Salem, 1987). Abundant $[M+NH_4]^{+}$ ions in addition to minor fragment ions have also been achieved by combining HPLC separation with a particle-beam interface to MS followed by ammonia positive-ion CI (Huang et al., 1994). FAB ionisation of neat samples of triacylglycerols results in extensive fragmentation of triacylglycerols both in positive-ion ($[M-RCOO]^{+}$, $[RCO+74]^{+}$, RCO^{+} ions) and negative-ion ($RCOO^{-}$ ions) mode practically without the formation of molecular ions (Evans et al., 1991; Hori et al., 1994). Similarly, discharge-assisted TSP (plasma spray) ionisation of triacylglycerols yields solely fragment ions $[M-RCOO]^{+}$ (Valeur et al., 1993).

6.2.5   Collision-induced dissociation

Parent (precursor) ion–daughter (product) ion relationships can be studied by tandem mass spectrometry (MS/MS) by decomposing an ion utilising collisional activation. For this purpose, a triple-quadrupole, a multisector magnetic or a hybrid instrument is needed. In this context, we concentrate on describing briefly collision-induced dissociation using triple-quadrupole instruments. Triple-quadrupole instruments consist of three quadrupoles (Q1, Q2, Q3) situated in series between the ion source and the ion detection system. Quadrupoles Q1 and Q3 are scanning quadrupoles, whereas Q2, the collision chamber, is an ion transmission-only (RF-only) device. A CID process is most often performed by introducing inert gas, such as helium, argon or xenon, into the collision chamber to collide with ions, followed by dissociation of ions into smaller fragments. The process can be controlled by changing the collision gas and its pressure and collision energy. The translational kinetic energy of ions can be affected by collision energy, which is the potential difference between the ion source and the collision chamber. Triple-quadrupole instruments offer three scanning possibilities: in (1) daughter ion scanning, Q1 is set to pass a selected parent ion and the daughter ions formed by collisional activation of the parent ion in Q2 are scanned with Q3; (2) in parent ion scanning, Q1 is set to scan all those ions that decompose in Q2 to form the daughter ion selected by Q3; or (3) in neutral loss scanning: both Q1 and Q3 are set to scan for a neutral loss, i.e. Q3 is set to scan a selected number of mass units lower than Q1.

Several studies of triacylglycerols using MS/MS techniques have been published (e.g. Duffin et $al.$, 1991; Evans et $al.$, 1991; Hogge et $al.$, 1991; Anderson et $al.$, 1993; Kallio and Currie, 1993a; Spanos et $al.$, 1995; Stroobant et $al.$, 1995; Taylor et $al.$, 1995). In general, triacylglycerols are ionised followed by CID of the ions consisting of the intact molecule, such as $[M+H]^+$, $[M+NH_4]^+$ or $[M-H]^-$ ions, into characteristic fragment ions, such as $[M-RCOO]^+$, $[RCO+74]^+$, $RCOO^-$, and $[M-H-RCOOH-100]^-$ ions. Most often the fragment ions provide information concerning the fatty acyl residues and their molecular associations in each molecular weight species of triacylglycerols without chromatographic separations. In addition, it may be possible to achieve information on the distribution of fatty acyl residues between the sn-2 and the sn-1/3 positions of a triacylglycerol (Kallio and Currie, 1993a; Stroobant et $al.$, 1995). Chromatographic separation becomes a necessity if the research is focused on the analysis of triacylglycerols consisting of isomeric fatty acyl residues, because their differentiation as such (equal m/z values) is not possible by MS.

In addition to MS/MS experiments, single-quadrupole instruments equipped with an API source usually offer an CID possibility, i.e. fragmentation is induced by introducing additional energy to the ions formed before the first skimmer. This technique does not allow the parent ion of interest to be chosen, thus, all ions will get the extra energy and are susceptible to fragmentation. (ESI)MS combined with HPLC enables often more efficient separation of compounds compared with the mass separation provided with a quadrupole analyser. Furthermore, HPLC-(ESI)-CID-MS results often in greater sensitivity than MS/MS determinations owing to enhanced ion transfer properties (reviewed by Kuksis and Myher, 1995).

6.3 Analysis of triacylglycerols

6.3.1 Sample introduction techniques without chromatographic separation

6.3.1.1 Direct probe introduction

The first mass spectrometric studies of triacylglycerols were carried out by direct introduction of a sample with a probe into the ion source of a mass spectrometer without chromatographic separation (Ryhage and Stenhagen, 1960; Barber et al., 1964; Sprecher et al., 1965; Aasen et al., 1970; Lauer et al., 1970; Hites, 1970, 1975). At present, direct sample introduction is a valuable technique for fast analysis of small sample amounts. In general, two types of direct probes are available for sample introduction: direct insertion probes (DIP, also called solid probes) and direct exposure probes (DEP, also called desorption CI probes). A DIP contains a sample crucible (e.g. quartz or platinum) whereas a DEP contains a thin filament loop (e.g. rhenium or tungsten) for sample application at the tip of the probe. After a sample is loaded, solvent, if present, is evaporated before introduction of the probe into the MS. Under vacuum, the probe heating is started to evaporate analytes followed by ionisation using either EI or CI. The DIP is suitable for analyses of both solid and liquid samples; however, thermally labile components may be decomposed before evaporation owing to the relatively slow heating of the probe. The DEP allows rapid evaporation of a sample from the heated filament exposed directly into the electron beam during EI analyses, or in the reactant gas plasma during CI analyses. Thus, DEP is suitable also for analyses of thermally labile components.

The direct probe sample introduction technique has been applied to the analysis of natural mixtures of triacylglycerols as well as for chromatographically separated triacylglycerol fractions. Both EI (Hites, 1970; Hogge et al., 1991; Taylor et al., 1991; Demirbüker et al., 1992) and positive-ion CI (Anderson et al., 1993; Spanos et al., 1995) have been used to achieve qualitative information concerning the molecular weight of triacylglycerols and the molecular association of fatty acyl residues: EI spectra exhibit several fragment ions characterising the fatty acid moieties in addition to low-abundance $M^{+\cdot}$ and $[M-18]^+$ ions, whereas positive-ion CI spectra typically exhibit abundant $[M+H]^+$ or $[M+NH_4]^+$ ions and fragment ions corresponding to the loss of one fatty acyl residue from the molecule. If the sample contains more than one molecular weight species of triacylglycerols, most often the CID technique is required for structural assignment of triacylglycerols. For example,

Hogge *et al.* (1991) applied EI to ionise trimethylsilyl derivatives of the triacylglycerols of castor beans, rich in ricinoleic acid (12-hydroxy-9-octadecenoic acid), resulting in the formation of $[M-CH_3]^+$ ions. CID of the $[M-CH_3]^+$ ions yielded diagnostic fragment ions such as $[RCO+74]^+$, $[M-RCOO]^+$, $[M-RCOOH]^+$, $[M-CH_3-RCOOH]^+$ and $[M-RCOOH-16]^+$, which were used for fatty acid identification. Similarly, collisional activation of $[M+H]^+$ ions, produced by isobutane CI of milk fat triacylglycerol fractions (Spanos *et al.*, 1995) or by methane CI of the triacylglycerols of *Vernonia galamensis* oil rich in vernolic acid (*cis*-12,13-epoxy-*cis*-9-octadecenoic acid) (Anderson *et al.*, 1993), resulted in the formation of abundant $[MH-RCOOH]^+$ ions.

In addition to qualitative identification of triacylglycerols according to the *m/z* values of characteristic ions, the ion abundances of molecular ions or adduct ions, such as $M^{+\cdot}$, $[M-18]^+$, $[M+H]^+$, $[M+NH_4]^+$ or $[M-H]^-$ ions, can be used for defining the proportions of different molecular weight species of triacylglycerols in a sample. In order to use direct probe sample introduction for semiquantitative determination of triacylglycerols, it is worth optimising the mass spectrometric conditions affecting the ionisation process for the instrument used. As an example, ammonia negative-ion CI of triacylglycerols is most efficiently carried out using relatively high reactant gas pressure (8500 mtorr) at relatively high ion source temperature (200°C) to maximise the formation of $[M-H]^-$ ions and minimise the formation of $[M+35]^-$ cluster ions (Laakso and Kallio, 1996). This is essential, since the $[M+35]^-$ ions would disturb the interpretation of the molecular weight area in the mass spectra. Optimisation of the reactant gas pressure and the ion source temperature for the analyte of interest often greatly improves the sensitivity of the determination. Depending on the instrumentation, and especially on the type of direct probe used, severe discrimination of triacylglycerols may occur during the evaporation process. The mass spectrometric response of triacylglycerols has been reported to depend on the molecular size (Schulte *et al.*, 1981), the degree of unsaturation (Rezanka and Mareš, 1991), and on both aforementioned factors (Hites, 1970). However, the discrimination according to the molecular size and degree of unsaturation can be minimised or even avoided by optimising the CI conditions and sample treatment (Laakso and Kallio, 1996). A rapid desorption of triacylglycerols from the sample filament is needed to minimise thermal degradation of unsaturated molecules and the reducing effect of double bonds on the mass spectrometric response of triacylglycerols. In addition, rapid heating improves the sensitivity. The proper heating rate depends, for example, on the scanning rate of the instrument: too high heating rates result in poor peak shapes and unacceptable repeatability because only few data points can be acquired. Moreover, the amount of sample applied to the heated wire has to be small enough. The importance of summing all mass spectra obtained during the evaporation of a sample has been well reported (Hites, 1970; Schulte *et al.*, 1981; Kallio and Currie, 1993a; Laakso and Kallio, 1996). The evaporation of triacylglycerols from the sample crucible or from the heated filament occurs gradually according to the molecular weight rather than being a simultaneous process. Thus, the mass spectra taken from different parts of the evaporation peak may differ substantially (Schulte *et al.*, 1981). In addition, the ion abundances have to be corrected according to the isotopes ^{13}C and ^{18}O (Hites, 1970; Laakso and Kallio, 1996).

In general, DEPs instead of DIPs are recommended for direct sample introduction of triacylglycerols into a mass spectrometer because of their fast heating capabilities. The use of a DEP allows fast analyses (2–3 min) (Laakso and Kallio, 1996). Several studies on determination of the molecular weight distribution of triacylglycerols using direct probe sample introduction MS have been carried out, for example analyses of several vegetable oils (EI, Hites, 1970; ammonia CI, Murata and Takahashi, 1977, Merritt *et al.*, 1982), animal fats (ammonia CI, Schulte *et al.*, 1981), green alga (methane CI, Rezanka *et al.*, 1986), plant oils (ammonia CI, Rezanka and Mareš, 1991) and human plasma triacylglycerols (ammonia CI, Mareš *et al.*, 1991). More recently, ammonia negative-ion CI has been applied to the analysis of various seed oils, such as rapeseed oil (Kallio and Currie, 1993a), seed oils of edible Finnish wild berries (Manninen *et al.*, 1995a,b; Laakso and Kallio, 1996; Johansson *et al.*, 1997a,b), and spruce and pine seed oils (Tillman-Sutela *et al.*, 1995). In addition, it has been used to characterise butterfat triacylglycerols containing *cis*- and *trans*-monoenoic fatty acyl residues (Laakso and Kallio, 1993) as well as postparturition changes in molecular species composition of cow colostrum triacylglycerols (Figure 6.2) (Laakso *et al.*, 1996). The technique may be of limited value in the analysis of highly unsaturated fish oils: one m/z value may represent two different triacylglycerols—for example, triacylglycerols 60:1 and 62:15 would both yield the same $[M-H]^-$ ion (m/z 971.7).

Direct probe sample introduction followed by CID of a selected parent ion may provide qualitative information on the fatty acyl residues of a triacylglycerol. However, it has not been possible to differentiate the regiospecific positions (*sn*-2 vs *sn*-1/3) of fatty acid moieties with positive-ion CI (Anderson *et al.*, 1993; Spanos *et al.*, 1995). Kallio and Currie (1991, 1993a) were the first to apply ammonia negative-ion CI to the ionisation of triacylglycerols followed by collisional activation of the $[M-H]^-$ ions with xenon or argon. CID spectra of $[M-H]^-$ ions exhibit abundant $RCOO^-$ and $[M-H-RCOOH-100]^-$ ions, and weak $[M-H-RCOOH]^-$, $[M-H-RCOOH-56]^-$ and $[M-H-RCOOH-74]^-$

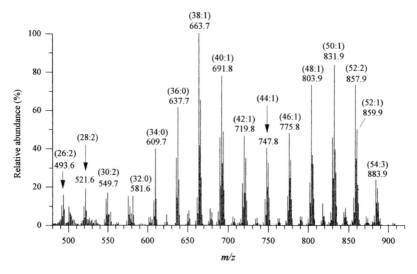

Figure 6.2 The [M−H]⁻ ion region of the mass spectrum of milk fat triacylglycerols (sample collected 2 days after parturition) produced by direct exposure probe sample introduction and ammonia negative-ion CI (reactant gas pressure 8500 mtorr, ion source temperature 200°C, Finnigan MAT TSQ-700). The m/z values of [M−H]⁻ ions characterise the combined number of acyl carbons (ACN) and double bonds (n) in the acyl chains of a triacylglycerol (ACN:n), and the [M−H]⁻ ion abundances define the proportions of different molecular weight species. As an example, some identifications of triacylglycerols are presented in parentheses in the spectrum. The determination of triacylglycerols with an odd number of acyl carbons is also possible with this technique.

ions (Figure 6.3). Similar daughter ion spectra of [M−H]⁻ ions have also been produced by using a mixture of methane–nitrous oxide as a CI gas (Stroobant *et al.*, 1995). The m/z values of the RCOO⁻ ions define the fatty acid moieties by the number of acyl carbons and double bonds without information on the positions of the double bonds or the configuration of fatty acids. The abundances of the RCOO⁻ ions can be used for calculation of the proportions of different combinations of the three fatty acids in triacylglycerols. For this purpose, the molar response factors of fatty acids have to be determined as they may vary according to the chain length, the degree of unsaturation and the regiospecific position of the fatty acid moieties. Furthermore, the abundances of the [M−H−RCOOH−100]⁻ ions provide data on the distribution of fatty acids between primary (*sn*-1/3) and secondary (*sn*-2) positions (Figure 6.3). Stroobant *et al.* (1995) have presented the fragmentation pathway for the formation of [M−H−RCOOH−100]⁻ ions (ketone enolate ions) from the deprotonated molecular ions. The formation of ketone enolate containing the hydrocarbon chain of the fatty acid

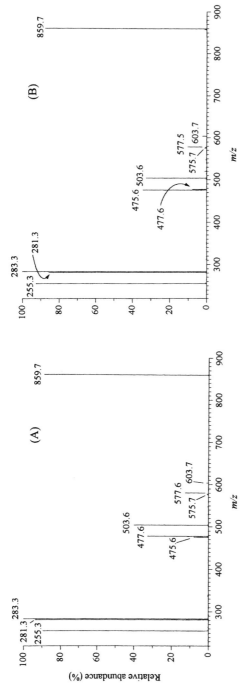

Figure 6.3 Collision-induced dissociation (CID) spectra of (a) *rac*-1-palmitoyl-2-stearoyl-3-oleoyl-*sn*-glycerol, (b) *rac*-1-palmitoyl-2-oleoyl-3-stearoyl-*sn*-glycerol, (c) *rac*-1,2-dioleoyl-3-γ-linolenoyl-*sn*-glycerol, and (d) 1,3-dioleoyl-2-γ-linolenoyl-*sn*-glycerol. Triacylglycerols were ionised by ammonia negative-ion CI followed by CID of selected [M−H]⁻ ions. CID spectra were recorded with a triple-quadrupole instrument (Finnigan MAT TSQ-700) at collision gas (Ar) pressure 1.5 mtorr and collision energy 15 eV. Ion identifications in (a) and (b): RCOO⁻: *m/z* 255.3 (16:0), 281.3 (18:1), 283.3 (18:0); [M−H−RCOOH−100]⁻: *m/z* 475.6 (−18:0), 477.6 (−18:1), 503.6 (−16:0); [M−H−RCOOH−74]⁻: *m/z* 501.7 (−18:0), 503.6 (−18:1), 529.6 (−16:0); [M−H−RCOOH−56]⁻: *m/z* 519.6 (−18:0), 521.6 (−18:1), 547.6 (−16:0); [M−H−RCOOH]⁻: *m/z* 575.7 (−18:0), 577.6 (−18:1), 603.7 (−16:0); [M−H]⁻: *m/z* 859.7. Ion identifications in (c) and (d): RCOO⁻: *m/z* 277.3 (18:3*n*−6), 281.3 (18:1); [M−H−RCOOH−100]⁻: *m/z* 497.5 (−18:1), 501.5 (−18:3*n*−6); [M−H−RCOOH−74]⁻: *m/z* 523.5 (−18:1), 527.5 (−18:3*n*−6); [M−H−RCOOH−56]⁻: *m/z* 541.6 (−18:1), 545.6 (−18:3*n*−6); [M−H−RCOOH]⁻: *m/z* 597.4 (−18:1), 601.4 (−18:3*n*−6); [M−H]⁻: *m/z* 879.7.

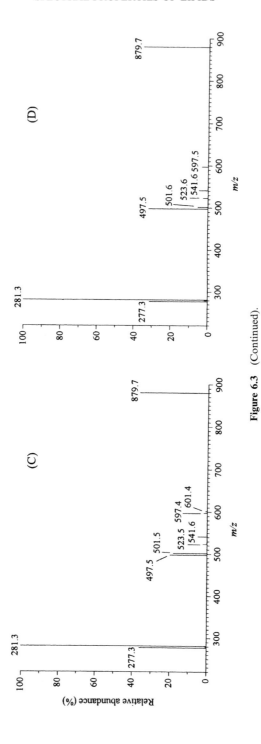

Figure 6.3 (Continued).

esterified to the *sn*-2 position is more favourable than the formation of ketone enolate formed by the loss of the fatty acyl residue from the *sn*-2 position.

Parameters affecting the CID process are collision gas pressure and collision energy. The optimum for the regiospecific analysis of triacylglycerols is a state where (1) the abundances of RCOO⁻ ions are reproducible and independent on the structure and stereochemical positions of fatty acyl residues in triacylglycerols, and (2) the difference in the abundance for the loss of a fatty acyl residue from the *sn*-2 and *sn*-1/3 positions is as great as possible. The ratio of [M−H−RCOOH−100]⁻ ions produced by cleavage of a fatty acyl residue from the *sn*-2 position to that from the *sn*-1/3 positions is not significantly affected by the collision gas pressure and collision energy (Laakso, 1995). On the contrary, energy requirements for the loss of RCOO⁻ ion from the *sn*-2 and *sn*-1/3 positions are different: an increase in the collision energy typically decreases the response of the RCOO⁻ from the *sn*-2 position (Kallio and Currie, 1993a; Laakso, 1995). The effect of collision gas pressure on the quality of daughter ion spectra is less critical than that of collision energy. Therefore, MS/MS measurements should be carried out at constant collision energy, whereas collision gas pressure can be adjusted to produce the desired fragmentation efficiency with high collection efficiency of the daughter ions. In addition, mass spectrometers may have other parameters that should be controlled during MS/MS analyses. For example, the Finnigan TSQ-700 has an MSMSC factor (a mathematical correction factor used to increase the transmission of ions in MS/MS mode) that has a dramatic effect on the quality of the produced CID spectra.

Under optimised MS/MS conditions, the regiospecific distribution of fatty acids can be determined based on the abundances of the [M−H−RCOOH−100]⁻ ions. Figure 6.4 shows the ratio of [M−H−B−100]⁻ and [M−H−A−100]⁻ ions of nine binary mixtures of *rac*-A-A-B and *sn*-A-B-A type of triacylglycerols, where A and B are different fatty acids (the fatty acid in the middle refers to the *sn*-2 position). The ratio is relatively independent of the fatty acid moieties in triacylglycerols. Thus, it is possible to measure the presence of regioisomeric triacylglycerols and their proportions from simple mixtures of triacylglycerols with reasonable accuracy. Kallio and Rua (1994) have introduced a computerised program to calculate the molar proportions of different fatty acid combinations as well as the regiospecific distribution of fatty acids between the *sn*-2 and *sn*-1/3 positions of triacylglycerols having equal molecular weight (same m/z values of [M−H]⁻ ions). The technique has been applied to the regiospecific analysis of triacylglycerols of several seed oils (Kallio and Currie, 1993a,b; Kallio *et al.*, 1997),

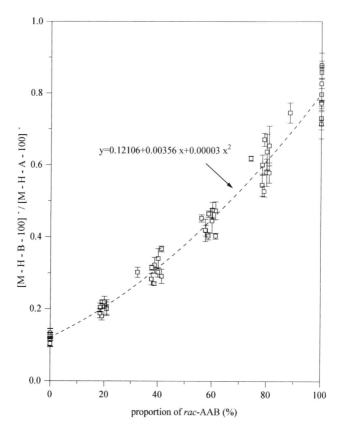

Figure 6.4 The ratio of the abundances of $[M-H-B-100]^-$ and $[M-H-A-100]^-$ ions achieved from the CID spectra (collision gas (Ar) pressure 1.5 mtorr, collision energy 15 eV, Finnigan MAT TSQ-700) of nine binary mixtures of rac-A-A-B and sn-A-B-A type of triacylglycerols (A and B are fatty acid moieties, the fatty acid in the middle refers to the sn-2 position). Each point in the figure is the average of at least four measurements. Components of the binary mixtures were: rac-18:1-18:1-16:0 and sn-18:1-16:0-18:1; rac-18:1-18:1-18:3(n−6) and sn-18:1-18:3(n−6)-18:1; rac-16:0-16:0-18:0 and sn-16:0-18:0-16:0; rac-16:0-16:0-14:0 and sn-16:0-14:0-16:0; rac-18:0-18:0-16:0 and sn-18:0-16:0-18:0; rac-18:1-18:1-18:0 and sn-18:1-18:0-18:1; rac-16:0-16:0-18:1 and sn-16:0-18:1-16:0; rac-14:0-14:0-18:1 and sn-14:0-18:1-14:0; rac-18:1-18:1-14:0 and sn-18:1-14:0-18:1 (Laakso, 1995).

human milk (Currie and Kallio, 1993; Kallio and Currie, 1993b; Kallio and Rua, 1994), butterfat (Laakso and Kallio, 1993), and Baltic herring (Kallio, 1992; Kallio and Currie, 1993b).

6.3.1.2   Flow injection analysis
Continuous introduction of sample liquid with a constant flow rate, typically a few microlitres per minute, into the ionisation chamber is a

technique used with API instruments. Continuous sample introduction is especially useful in API tuning procedures. In addition to tuning purposes, the authors do not find this technique particularly valuable in the analysis of triacylglycerols; only one application has been reported in the literature (Duffin *et al.*, 1991). Instead of continuous introduction, the sample can be injected via a sample loop into the ionisation region without chromatographic separation. This fast technique is also valuable for tuning purposes to optimise single parameters. In addition, loop injection technique could be utilised, for example, to determine purity of standards or products of synthesis, to follow reactions, to determine molecular weight profiles, or in some cases to differentiate the distribution of fatty acyl residues between the *sn*-2 and the *sn*-1/3 positions.

6.3.2 Sample introduction *via chromatographic separation*

6.3.2.1 Gas chromatography–mass spectrometry

Gas chromatography–mass spectrometry (GC-MS) is generally the simplest and most routinely used chromatography–MS combination. A capillary column is connected to the mass spectrometer through a heated interface so that the sample is supplied directly into the ion source of the mass spectrometer. Since molecules to be separated with GC should have sufficient volatility, such molecules are already in the gaseous phase as introduced into the mass spectrometer.

High-temperature gas chromatography (HT-GC) with capillary columns separates triacylglycerols according to the increasing acyl carbon number (ACN) on non-polar stationary phases (e.g. Geeraert *et al.*, 1983; Geeraert and Sandra, 1985). On medium-polarity stationary phases, e.g. polarisable phenylmethylpolysiloxanes, the unsaturation of triacylglycerols also affects the elution (e.g. Geeraert and Sandra, 1987). In some cases, triacylglycerols with equal ACN but differing by one double bond have been separated. In addition, isobaric triacylglycerols containing short-chain fatty acyl residues (C_2–C_{10}) can be separated from those with longer-chain fatty acyl residues by HT-GC (Myher *et al.*, 1988; Kuksis, 1994). Polar stationary phases, such as poly(ethylene glycol) phases, have too low a heat resistance to be used in HT-GC.

On-column injection is the only acceptable sample introduction method for triacylglycerols, because techniques involving vaporisation of the sample lead to thermal decomposition of unsaturated triacylglycerols (Grob, 1981). Discrimination of less volatile compounds also occurs (Grob, 1979). Several authors have reported thermal degradation and polymerisation of unsaturated triacylglycerols at high elution temperatures. It has been recommended that highly unsaturated triacylglycerols, such as fish oils, should not be separated by HT-GC.

Separation of triacylglycerols with HT-GC has been reviewed by several authors (Mareš, 1988; Kuksis, 1994; Myher and Kuksis, 1995b; Ruiz-Gutiérrez and Barron, 1995).

Electron ionisation.   GC-MS is usually applied to the analysis of milk fat and other relatively saturated triacylglycerol mixtures owing to the restrictions of the chromatography. GC-(EI)MS provides characteristic RCO^+, $[RCO + 74]^+$, $[RCO + 128]^+$ and $[M-RCOO]^+$ fragment ions (e.g. Murata and Takahashi, 1973; Fukatsu and Tamura, 1988). In addition, unsaturated triacylglycerols form fragment ions $[RCO-1]^+$ and $[M-RCOO-1]^+$ (Wakeham and Frew, 1982; Rezanka *et al.*, 1986; Ohshima *et al.*, 1989). The serious disadvantage of GC-(EI)MS is low abundance or non-existence of ions providing molecular weight information, i.e. $M^{+\cdot}$ and $[M-18]^+$ ions. The abundance of these ions is typically up to two orders of magnitude lower than that of $[M-RCOO]^+$ fragment ions (Rezanka *et al.*, 1986). Therefore, direct molecular weight information is usually not available on spectra, and the ACN identification of triacylglycerols has most often been executed according to the retention times of reference compounds.

An (EI)MS spectrum produced from a gas chromatographic peak of a natural mixture of triacylglycerols shows several ions, which complicates the interpretation of the spectrum (Figure 6.5). The variation in the abundances of the fragment ions depends not only on the peak size and the proportion of respective molecular species in the peak but also on fragment type and mass range (Kalo and Kemppinen, 1993). Triacylglycerols containing unsaturated fatty acid moieties tend to fragment more willingly compared with saturated triacylglycerols (Wakeham and Frew, 1982). The relative abundance of RCO^+ ion corresponding to an unsaturated fatty acid moiety has been found to decrease in the series of triolein (100%), trilinolein (25–50%) and trilinolenin (10–25%) (Wakeham and Frew, 1982; Rezanka *et al.*, 1986). Unfortunately, data on the behaviour of corresponding molecular ions are not available, mostly because of the mass range limitations of the instruments. Wakeham and Frew (1982) have also reported that RCO^+ and $[M-RCOO]^+$ ions are very weak or non-existent in the spectra of triacylglycerols containing polyunsaturated acids (e.g. 22:6).

Chemical ionisation.   GC-(CI)MS of triacylglycerols has been conducted both in positive-ion and negative-ion mode. Positive-ion CI with methane as a reactant gas yields $[M-RCOOH]^+$ fragment ions, but no molecular ions are detected for milk fat triacylglycerols (Myher *et al.*, 1988). With ammonia as reactant gas, pseudo-molecular $[M+NH_4]^+$ adduct ions together with fragment ions are formed. In the early study of ammonia positive-ion GC-(CI)MS with a CI gas pressure of 0.9 torr, the $[MH-RCOOH]^+$ ion was reported to be the base peak in the mass spectra of milk fat triacylglycerols (Figure 6.6) (Murata, 1977). The abundance of $[M+NH_4]^+$ ions was reported to be 20 times higher than that of the corresponding molecular ions by GC-(EI)MS. A recent study of peanut cup samples containing Salatrim by Huang et al. (1997) showed that $[M+NH_4]^+$ ions can be detected as a base peak when a higher reactant gas pressure (1.8 torr) is applied. However, the abundances of $[MH-RCOOH]^+$ and $[RCOO+74]^+$ ions identifying the fatty acid moieties in triacylglycerols were very low.

GC-(CI)MS has been performed also in negative-ion mode with ammonia as reactant gas (Evershed et al., 1990; Evershed, 1994, 1996). The $RCOO^-$ or $[RCOO-H_2O-H]^-$ ion exists as a base peak in the spectra together with abundant $[RCOO-H_2O]^-$ ions. However, no molecular ions were detected for milk fat triacylglycerols (Figure 6.7). In contrast, in the study of *Heisteria silvanii* seed oil triacylglycerols by Spitzer et al. (1997), $[M-H]^-$ ion was formed and detected as the base peak (Figure 6.8). In addition, abundant $RCOO^-$ ions were formed. In both reports no value of reactant gas pressure was given. Therefore, it is difficult to determine if the reason for the different results was the reactant gas pressure, the molecular weight of triacylglycerols or the degree of unsaturation of the triacylglycerols.

6.3.2.2 *High-performance liquid chromatography–mass spectrometry*

Reversed-phase HPLC is a commonly used chromatographic technique for the analysis of triacylglycerols (Ruitz-Gutiérrez and Baron, 1995; Myher and Kuksis, 1995b; Laakso, 1996; Nikolova-Damyanova, 1997). The separation is based on both the molecular size and on the degree of unsaturation of triacylglycerols. In general, the molecules elute in ascending order of the ACN, each of the double bonds reducing the retention of the molecule to the stationary phase. The retention time-reducing effect of double bonds depends on the efficiency of the chromatographic system applied. In addition, the distribution of double bonds between the acyl chains (Kuksis et al., 1991a; Laakso, 1997), the chain-length asymmetry (Nurmela and Satama, 1988; Maniongui et al., 1991; Myher et al., 1993; Marai et al., 1994) and the configuration of double bonds (Laakso and Kallio, 1993), as well as the positions of double bonds in the hydrocarbon chain (e.g. α- and γ-linolenic acid) (Phillips et al., 1984; Perrin et al., 1987; Aitzetmüller and Grönheim, 1992; Laakso, 1997) affect the separation. Triacylglycerols having the same equivalent carbon number (ECN = ACN–2n, where n is the number of double bonds in the acyl chains of a triacylglycerol) are called critical pairs because they are not always chromatographically separable.

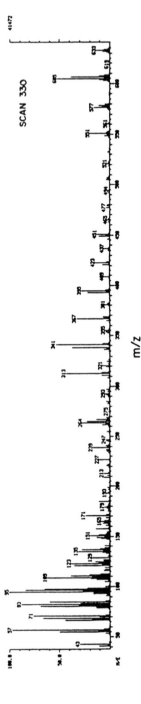

Figure 6.5 A GC-(EI)MS spectrum of zooplankton triacylglycerols with 54 acyl carbons. Triacylglycerols were separated on a capillary column according to the ACN. Some ion identifications: RCO^+ ions: m/z 211 (14:0), 237 (16:1), 239 (16:0), 265 (18:1), 267 (18:0), 293 (20:1), 321 (22:1); $[RCO-1]^+$ ions, e.g. m/z 264 (18:1); $[RCO+74]^+$ ions, e.g. m/z 313 (16:0), 339 (18:1), 341 (18:0), 367 (20:1), 395 (22:1); $[M-RCOO]^+$ or $[M-RCOOH]^+$ ions, e.g. m/z 551 ($-C_{22}$), 577 ($-C_{20}$), 605 ($-C_{18}$), 633 ($-C_{16}$). The probable fatty acid combinations were C_{14}/C_{20}, C_{16}/C_{20}, $C_{16}/C_{16}/C_{22}$, $C_{16}/C_{18}/C_{20}$, and $C_{18}/C_{18}/C_{18}$. (Reprinted with permission from Wakeham and Frew (1982). Copyright (1982) The American Oil Chemists' Society).

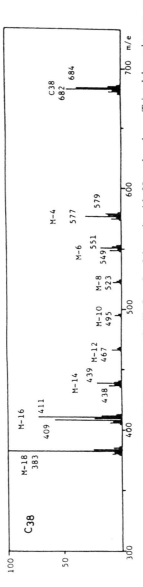

Figure 6.6 An ammonia positive-ion GC-(CI)MS spectrum of milk fat triacylglycerols with 38 acyl carbons. Triacylglycerols were separated on a packed glass column according to the ACN. The abbreviation C38 represents the $[M+NH_4]^+$ ions defining the number of acyl carbons and double bonds in a triacylglycerol. The $[MH−RCOOH]^+$ ions abbreviated as M−18, M−16, M−14, M−12, M−10, M−8, M−6 and M−4 characterise the loss of one fatty acid from a triacylglycerol; the numbers refer to the number of carbon atoms in an acyl chain. (Reprinted with permission from Murata (1977). Copyright (1977) The American Chemical Society.)

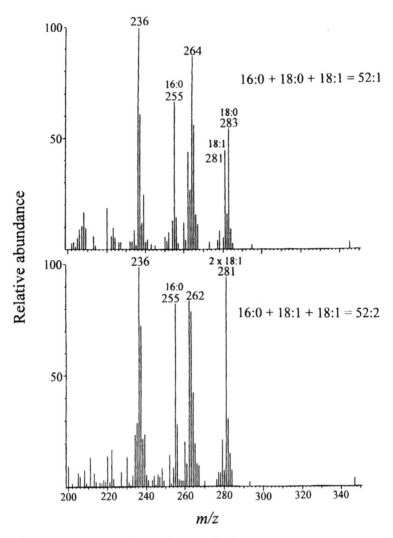

Figure 6.7 An ammonia negative-ion HT-GC-(CI)MS spectrum of milk fat triacylglycerols with 52 acyl carbons. Triacylglycerols were separated on a capillary column mainly according to the ACN. Ion identifications: $RCOO^-$ ions, m/z 255 (16:0), 281 (18:1), 283 (18:0); $[RCOO-H_2O-H]^-$ ions, m/z 236 (16:0), 262 (18:1), 264 (18:0); $[RCOO-H_2O]^-$ ions: m/z 237 (16:0), 263 (18:1), 265 (18:0). (Reprinted with permission from Evershed (1996). Copyright (1996) The American Society for Mass Spectrometry.)

Argentation chromatography by HPLC on a silver-loaded cation-exchange column has been proved to be an efficient technique for separating molecules according to differences in unsaturation (Nikolova-Damyanova, 1992; Christie, 1995; Dobson *et al.*, 1995; Ruitz-Gutiérrez

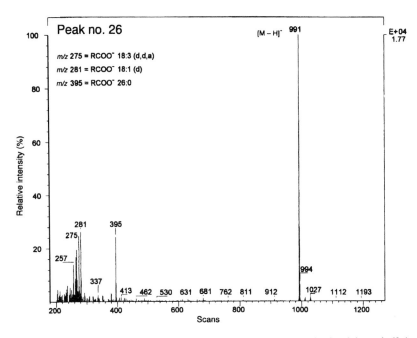

Figure 6.8 An ammonia negative-ion HT-GC-(CI)MS spectrum of triacylglycerol 62:4 of *Heisteria silvanii* seed oil. The triacylglycerol has been identified as 18:1/18:3/26:0 (18:3 = heisteric acid; *cis*-7,*trans*-11-octadecadiene-9-ynoic acid); the distribution of fatty acyl residues between the *sn*-1, *sn*-2, and *sn*-3 positions could not be distinguished. Abbreviations: d, double bond; a, triple bond. (Reprinted with permission from Spitzer *et al.* (1997). Copyright (1997) The American Oil Chemists' Society.)

and Baron, 1995; Laakso, 1996). The separation is based on the weak interaction between the π-electrons of the double and triple bonds and the silver ion. The separation of triacylglycerols depends on the number of double bonds, the distribution of double bonds between the fatty acyl residues within a single molecule, the configuration and position of double bonds within a fatty acyl residue, and the regiospecific distribution (*sn*-2 vs. *sn*-1/3 positions) of acyl chains in a triacylglycerol. For example, fatty acyl residues having a *cis*-configuration form stronger complexes with silver ions than *trans*-fatty acids (Christie, 1988; Brühl *et al.*, 1993; Laakso and Kallio, 1993; Smith *et al.*, 1994). In addition, α- and γ-linolenic acid-containing triacylglycerols have been separated by argentation HPLC (Christie, 1991; Laakso and Voutilainen, 1996).

Direct liquid inlet technique. The combination of HPLC with MS is not as simple as that of GC-MS owing to the large volumes of gas produced from the liquid eluents under vacuum. So far, most effort on developing the analysis of triacylglycerols by HPLC-MS has been done by Kuksis and his colleagues utilising direct liquid inlet (DLI) techniques. This technique requires the splitting of the HPLC effluent; thus, only a small portion, approximately 1% of the total flow (1.5–$2\,\mathrm{ml\,min^{-1}}$) is sprayed into the ion source via a diaphragm pinhole orifice placed adjacent to the ion source. Ionisation of triacylglycerols is achieved by solvent-mediated CI. Marai *et al.* (1983) and Kuksis *et al.* (1983) introduced the formation of prominent $[M+H]^+$ and $[MH-RCOOH]^+$ ions of triacylglycerols by combining reversed-phase HPLC separation with MS detection, using acetonitrile and propionitrile as HPLC eluents. In addition, low-abundance adduct ions of acetonitrile $[M+41]^+$ and propionitrile $[M+55]^+$ as well as weak RCO^+ ions were recorded, but these ions were not used for identification. The ion abundances of $[M+H]^+$ and $[MH-RCOOH]^+$ ions are affected by the molecular weight, degree of unsaturation and positional distribution of fatty acyl residues in a triacylglycerol molecule (Myher *et al.*, 1984). For example, the abundance of $[M+H]^+$ ions increases with increasing degree of unsaturation of a triacylglycerol, whereas saturated triacylglycerols produce negligible amounts or no $[M+H]^+$ ions. Furthermore, the loss of a fatty acid moiety either from the *sn*-1 or the *sn*-3 positions results in about four times more abundant $[MH-RCOOH]^+$ ion than the loss of a fatty acid moiety from the *sn*-2 position. Also, the fatty acid structure has an effect on the abundance of the $[MH-RCOOH]^+$ ion; for example, 18:1 and 18:2 have been reported to be released at about the same rate, but about 20% more slowly than the 16:0 (Myher *et al.*, 1984). In general, the total ion current instead of the abundance of protonated molecular ion has been used for triacylglycerol quantitation, as this varies much less according to the triacylglycerol structure (Myher *et al.*, 1984). However, it is essential to determine response factors of triacylglycerols for accurate quantitation purposes. As an example, the ion current response of corn oil and peanut oil triacylglycerols decreased with increasing degree of unsaturation (Myher *et al.*, 1984). The total ion current response cannot be used for quantitation of chromatographically unresolved triacylglycerols. Instead, the responses of $[M+H]^+$ ions, if molecules differ in molecular weight, or $[MH-RCOOH]^+$ ions, if molecules have identical molecular weights, are useful for quantitation.

Detection sensitivity of triacylglycerols can be greatly improved by applying CI in negative-ion instead of positive-ion mode. Kuksis *et al.* (1991a,b) introduced the formation of negative ions of triacylglycerols by inclusion of 1% dichloromethane in the HPLC eluents (acetonitrile and propionitrile). Chloride-attachment negative-ion CI spectra of triacylgly-cerols exhibit exclusively $[M+Cl]^-$ ions of both chloride isotopes, $^{35}Cl^-$

and $^{37}Cl^-$. By using this ionisation technique, it is possible to achieve about 100-fold increase in sensitivity compared with positive-ion CI (Kuksis *et al.*, 1991a,b). In addition, negative-ion CI with chloride-attachment has been reported to give apparently correct proportions of $[M + Cl]^-$ ions of triacylglycerols regardless of the degree of unsaturation or the chain length of the fatty acyl residues, thus being a better choice for quantitation than positive-ion CI. As no fragmentation during chloride-attachment negative-ion CI occurs, it is often necessary to combine information achieved by positive-ion and negative-ion CI (Figure 6.9). This kind of approach has been utilised, for example, for the analysis of milk fat (Kuksis *et al.*, 1986, 1991a; Myher *et al.*, 1993; Marai *et al.*, 1994), menhaden oil (Kuksis *et al.*, 1991a) and human plasma triacylglycerols (Kuksis *et al.*, 1991b).

Atmospheric pressure ionisation techniques. Mass spectrometric techniques utilising ionisation at atmospheric pressure have been of great interest, especially for interfacing HPLC with MS. So far, only a few applications on electrospray ionisation, more precisely nebuliser-assisted ESI, of triacylglycerols have been published. A prerequisite for ESI is ion formation in solution. Non-polar triacylglycerols are not in ionic form under typical HPLC conditions and therefore ionic species have to be added to the eluent system to facilitate ionisation, e.g. by post-column addition or as sheath liquid addition. Duffin *et al.* (1991) used continuous flow introduction ($2\,\mu l\,min^{-1}$) of a sample solution, without chro- matographic separation, into the ionisation chamber. The samples were dissolved in chloroform–methanol (70:30, v/v), which was modified with different ionic species. Kuksis and co-workers (Kuksis and Myher, 1995; Myher and Kuksis, 1995a; Myher *et al.*, 1997) have presented an approach having more practical value: triacylglycerols were separated on a reversed-phase HPLC column using a gradient of isopropanol and acetonitrile for elution, and ionisation was facilitated by post-column addition of 1% ammonia in isopropanol with a flow rate of $0.15\,ml\,min^{-1}$. One example of ESI of triacylglycerols separated by HPLC on a silver-loaded silica column has been presented by Schuyl *et al.* (1995). Components were eluted with a ternary gradient consisting of toluene, hexane and ethyl acetate and ionisation was done by post-column addition of sodium acetate in methanol. Typically, ESI of triacylglycerols yields abundant $[M + Na]^+$ and $[M + NH_4]^+$ ions in the presence of sodium or ammonium ions in the solvent, respectively, without fragmentation. Additional structural information on triacylglycerols can be achieved on triple quadrupole instruments by CID of $[M + NH_4]^+$ ions with argon. The daughter ion spectra of $[M + NH_4]^+$ ions exhibit $[M - RCOO]^+$ and RCO^+ ions, which define the fatty acid

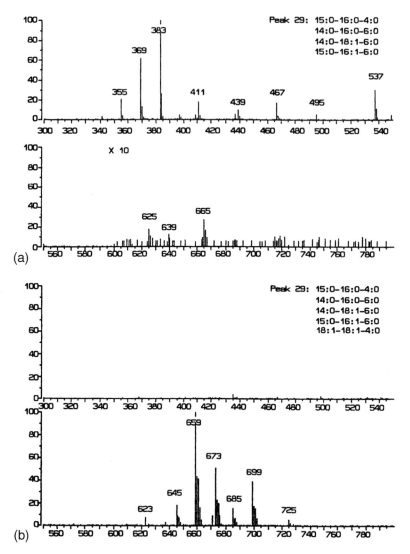

Figure 6.9 Direct liquid inlet HPLC-MS spectra of a chromatographic peak of milk fat triacylglycerols produced by (a) positive-ion CI, and (b) negative-ion CI. Triacylglycerols were separated by reversed-phase HPLC using a gradient of acetonitrile–propionitrile (a) or acetonitrile–propionitrile containing dichloromethane (b). Positive-ion CI of triacylglycerols resulted in the formation of abundant [MH−RCOOH]$^+$ fragment ions (upper trace of (a)) and weak or no [M + H]$^+$ ions (lower trace of (a)); e.g. 15:0–16:0–4:0 formed the ions m/z 625, [M + H]$^+$; m/z 369, [MH−16:0]$^+$; m/z 383, [MH−15:0]$^+$; and m/z 537, [MH−4:0]$^+$. Chloride-attachment negative-ion CI of triacylglycerols yielded abundant [M + Cl]$^-$ ions (lower trace of (b)) without fragmentation (upper trace of (b)); e.g. m/z 659 represents triacylglycerol 35:0, m/z 673 triacylglycerol 36:0, m/z 685 triacylglycerol 37:1, m/z 699 triacylglycerol 38:1, and m/z 725 triacylglycerol 40:2. (Reprinted with permission from Kuksis *et al.* (1991a). Copyright (1991) Elsevier Science Publishers B.V.)

moieties but not their regiospecific distribution in triacylglycerols (Duffin *et al.*, 1991). Kuksis and Myher (1995) and Myher *et al.* (1997) have shown that fragmentation of $[M + NH_4]^+$ ions can also be achieved with single-quadrupole instruments by increasing the capillary exit voltage, which is the voltage between the capillary exit and the first skimmer. At lower capillary exit voltages (170 V), only $[M + NH_4]^+$ and $[M + Na]^+$ ions are formed, with little or no fragmentation. When the voltage is increased to 250 V, abundant $[M - RCOO]^+$ ions are produced and the $[M + NH_4]^+$ ion almost disappears (Figure 6.10). The $[M + Na]^+$ ions are very stable and require extreme CID conditions to yield structurally informative fragment ions, thus they typically remain in the spectra (Duffin *et al.*, 1991; Kuksis and Myher, 1995; Myher *et al.*, 1997). Quantitative determination of triacylglycerols using (ESI)MS needs careful calibration since the ion current response is affected by the analyte polarity: $[M + Na]^+$ and $[M + NH_4]^+$ ions of triacylglycerols containing short-chain or unsaturated fatty acyl residues were observed in greater abundance than those of triacylglycerols containing long-chain or saturated fatty acid moieties (Duffin *et al.*, 1991). So far, nebuliser-assisted (ESI)MS has been utilised for characterisation of a lipid mixture recovered from a mammalian cell culture reactor (Duffin *et al.*, 1991) and a polyunsaturated oil of microalgal origin rich in docosahexaenoic acid (Myher *et al.*, 1997).

Atmospheric pressure chemical ionisation is a gas-phase process suitable also for the ionisation of non-polar molecules. No buffers in solution are needed and therefore most existing HPLC methods for the separation of triacylglycerols can be combined with (APCI)MS as such without modifications. Byrdwell and co-workers (Byrdwell and Emken, 1995; Neff and Byrdwell, 1995) were the first to apply (APCI)MS to the

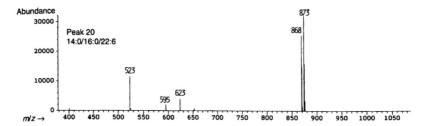

Figure 6.10 Nebuliser-assisted ESI-CID-MS spectrum of a chromatographic peak of polyunsaturated triacylglycerols of microalgal origin (capillary exit voltage 215 V). Triacylglycerols were separated by reversed-phase HPLC using a gradient of isopropanol and acetonitrile for elution. Ionisation was facilitated by post-column addition of 1% ammonia in isopropanol. Ion identifications: m/z 868, $[M + NH_4]^+$; m/z 873, $[M + Na]^+$; $[M - RCOO]^+$ ions: m/z 523, $[M - 22:6]^+$; m/z 595 $[M - 16:0]^+$; m/z 623, $[M - 14:0]^+$. (Reprinted with permission from Myher *et al.* (1997). Copyright (1997) The American Oil Chemists' Society.)

detection and identification of triacylglycerols in combination with HPLC. They separated triacylglycerols on an ODS (octadecylsilyl) column using a gradient of propionitrile and hexane for elution. The total flow of $1.0\,ml\,min^{-1}$ was split between the APCI source ($\sim 400\,\mu l\,min^{-1}$) and an evaporative light-scattering detector ($\sim 600\,\mu l\,min^{-1}$). More recently, silver-ion HPLC of triacylglycerols has been combined with (APCI)MS (Laakso and Voutilainen, 1996). Separations were accomplished on a silver-loaded Nucleosil 5SA cation-exchange column and components were eluted with a ternary solvent gradient consisting of dichloromethane–1,2-dichloroethane (1:1, v/v), acetone, and acetone–acetonitrile (4:1, v/v). The total flow ($0.8\,ml\,min^{-1}$) was introduced into the APCI source without splitting. Most triacylglycerols yield very simple (APCI)MS spectra exhibiting abundant $[M + H]^+$ and $[M - RCOO]^+$ ions, which define the molecular weight and the molecular association of fatty acyl residues, respectively (Figure 6.11). The relative abundances of $[M + H]^+$ and $[M - RCOO]^+$ ions are strongly affected by the degree of unsaturation of a triacylglycerol: e.g., saturated molecules do not produce any $[M + H]^+$ ion, whereas $[M + H]^+$ ion is the base peak in the mass spectra of more unsaturated triacylglycerols, which typically contain four or more double bonds in the acyl chains. In addition, Byrdwell and Emken (1995) have reported the formation of weak adduct ions with water and propionitrile, such as $[M + H + H_2O]^+$, $[M - RCOO + H_2O]^+$ and $[M + H + 55]^+$ ions. The propionitrile adduct ions may interfere with the interpretation of mass spectra (Byrdwell *et al.*, 1996); eluent systems consisting of chlorinated solvents and acetonitrile do not produce any adduct ions. The ionisation process can be substantially affected by adjusting the API parameters. For example, the ionisation efficiency of triacylglycerols decreased with increasing the capillary temperature in a Finnigan APCI source (Laakso and Voutilainen, 1996). On the other hand, the capillary temperature should be high enough to provide a relatively low and stable signal background. The vaporiser temperature has to be high enough to produce good sensitivity for the analytes.

The regiospecific distribution of fatty acids has an effect on the (APCI)MS spectra of triacylglycerols. The abundance of $[M - RCOO]^+$ ion formed by the loss of a fatty acyl residue from the *sn*-2 position is less than that formed by the loss of a fatty acyl residue from the *sn*-1/3 positions (Kusaka *et al.*, 1996; Laakso and Voutilainen, 1996; Mottram and Evershed, 1996). The chain-length and degree of unsaturation of the cleaving fatty acid moiety has an important effect on the $[M - RCOO]^+$ ion abundance (Table 6.1).

Byrdwell *et al.* (1996) have approached the quantitative analysis of triacylglycerols by (APCI)MS based on determining (1) calibration curves

Table 6.1 (APCI)MS of triacylglycerols[a]: effect of fatty acid structure ad regiospecific position on the ion abundances

Triacylglycerol	$[M+H]^+$ ion abundance	$[M-RCOO_{sn-2}]^{+}$[b] ion abundance	$[M-RCOO_{sn-1/3}]^{+}$[c] ion abundance	Ratio[d]
sn-14:0-18:3-14:0	(m/z 773.9) 21.2±1.8[e]	$[M-18:3]^+$ (m/z 495.6) 69.6 ± 6.2	$[M-14:0]^+$ (m/z 545.7) 100	0.696
sn-14:0-18:1-14:0	(m/z 777.9) 2.2±0.2	$[M-18:1]^+$ (m/z 495.6) 18.2 ± 1.0	$[M-14:0]^+$ (m/z 549.6) 100	0.182
rac-14:0-14:0-18:1	(m/z 777.9) 5.3 ± 0.5	$[M-14:0]^{+}$[f] (m/z 549.6) 91.1 ± 7.8	$[M-18:1]^+$ (m/z 495.6) 100	0.911
sn-18:1-14:0-18:1	(m/z 832.0) 10.4±0.6	$[M-14:0]^+$ (m/z 603.7) 13.2 ± 0.8	$[M-18:1]^+$ (m/z 549.7) 100	0.132
rac-18:1-18:1-14:0	(m/z 832.0) 9.4±0.7	$[M-18:1]^{+}$[f] (m/z 549.7) 100	$[M-14:0]^+$ (m/z 603.7) 57.7 ± 2.6	1.733
rac-16:0-18:1-18:2	(m/z 858.0) 40.0 ± 2.7	$[M-18:1]^+$ (m/z 575.6) 18.2 ± 0.9	$[M-18:2]^+$ (m/z 577.6) 84.5 ± 3.0; $[M-16:0]^+$ (m/z 601.9) 100	0.099
sn-18:1-16:0-18:1	(m/z 859.9) 8.2 ± 0.2	$[M-16:0]^+$ (m/z 603.7) 11.8 ± 0.4	$[M-18:1]^+$ (m/z 577.7) 100	0.118
rac-16:0-18:0-18:1	(m/z 862.0) 3.6 ± 0.5	$[M-18:0]^+$ (m/z 577.7) 21.5 ± 2.4	$[M-18:1]^+$ (m/z 579.7) 100; $[M-16:0]^+$ (m/z 605.7) 59.2 ± 6.7	0.135
rac-16:0-18:1-18:0	(m/z 862.0) 2.8±0.3	$[M-18:1]^+$ (m/z 579.7) 65.3 ± 4.9	$[M-18:0]^+$ (m/z 577.7) 100; $[M-16:0]^+$ (m/z 605.7) 91.7 ± 5.2	0.341
rac-18:0-16:0-18:1	(m/z 862.0) 3.5 ± 0.2	$[M-16:0]^+$ (m/z 605.7) 17.4 ± 0.5	$[M-18:0]^+$ (m/z 577.7) 58.0 ± 2.6; $[M-18:1]^+$ (m/z 597.7) 100	0.110
sn-18:0-16:0-18:0	(m/z 863.8) ND[g]	$[M-16:0]^+$ (m/z 607.7) 13.8 ± 0.6	$[M-18:0]^+$ (m/z 597.7) 100	0.138
18:3-18:3-18:3	(m/z 873.9) 100	$[M-18:3]^+$ (m/z 595.6) 43.8 ± 3.2		
sn-18:1-18:3n-6-18:1	(m/z 881.9) 100	$[M-18:3]^+$ (m/z 603.7) 59.2 ± 2.1	$[M-18:1]^+$ (m/z 599.7) 90.8 ± 3.4	0.652
rac-18:0-18:1-18:2	(m/z 886.0) 42.8±1.3	$[M-18:1]^+$ (m/z 603.9) 25.8 ± 1.8	$[M-18:0]^+$ (m/z 601.7) 100; $[M-18:2]^+$ (m/z 605.7) 68.2 ± 1.6	0.153

Data from Laakso, 1997; unpublished results.

[a]Loop-injection analysis with Finnigan MAT TSQ-700 equipped with an API source: eluent flow dichloromethane–acetonitrile (1:1, v/v) $0.8 \, ml \, min^{-1}$, APCI vaporiser temperature 400°C, heated capillary temperature 200°C, sheath gas (N_2) pressure 50 psi, auxiliary gas (N_2) flow $5 \, ml \, min^{-1}$.

[b]Loss of a fatty acyl chain from the sn-2 position.

[c]Loss of a fatty acyl chain from the sn-1/3 positions.

[d]Ratio of the ion abundance of $[M-RCOO_{sn-2}]^+$ and $[M-RCOO_{sn-1/3}]^+$.

[e]Normalised ion abundance ± standard deviation ($n = 5$).

[f]Loss of the same fatty acid also from the sn-1 or the sn-3 position.

[g]ND, not detected.

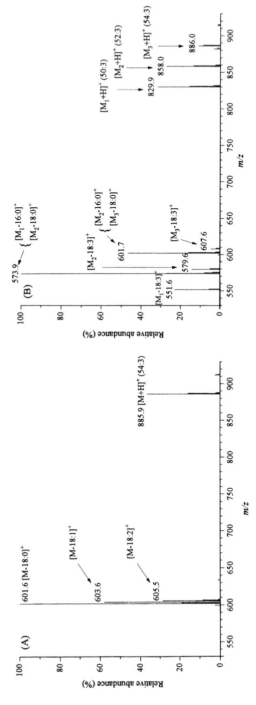

Figure 6.11 (APCI)MS spectra of chromatographic peaks of linseed oil triacylglycerols separated by (a) reversed-phase HPLC (eluents acetonitrile and dichloromethane-1,2-dichloroethane), and (b) argentation HPLC on a silver-loaded Nucleosil 5SA cation-exchange column (eluents dichloromethane–1,2-dichloroethane, acetone, and acetone–acetonitrile). Triacylglycerol identifications: (a) 18:0/18:1/18:2, and (b) M₁ = 16:0/16:0/18:3, M₂ = 16:0/18:0/18:3, M₃ = 18:0/18:0/18:3 (the distribution of fatty acyl residues between the *sn*-1, *sn*-2, and *sn*-3 positions is not differentiated).

for mono-acid triacylglycerol standards, (2) response factors for a synthetic triacylglycerol mixture, (3) triacylglycerol response factors for a randomised sample in comparison to its statistical composition, and (4) triacylglycerol response factors calculated from fatty acid response factors achieved by comparing the fatty acid composition determined by GC-FID with that calculated from (APCI)MS data. The uncorrected relative amounts of triacylglycerols in the (APCI)MS spectra were achieved by summing the areas of $[M + H]^+$ and $[M - RCOO]^+$ peaks in extracted ion chromatograms attributed for each triacylglycerol species. The first two approaches did not result in acceptable results: calibration curves should have been determined for each triacylglycerol, which is not a practical approach in the analysis of complex natural mixtures. Furthermore, the triacylglycerol response factors were found to be concentration dependent, resulting in a concave appearance of the curves. The approaches (3) and (4) resulted in less average relative error than was obtained with HPLC-FID data, and were in good agreement with the predicted compositions of the samples studied. In order to determine triacylglycerol response factors, a randomised sample of the unknown sample is needed. The fatty acid approach means that response factors determined by GC-FID are unique for each sample. In the future, the quantitative determination of triacylglycerols by (APCI)MS needs to be further developed.

Reversed-phase HPLC-(APCI)MS has been applied to the analysis of genetically modified soybean lines (Neff and Byrdwell, 1995), randomised and normal soybean oils and lard samples (Byrdwell et al., 1996), and perilla, corn and olive oils (Kusaka et al., 1996). Recently, cloudberry seed oil (rich in α-linolenic acid, $18:3n-3$), evening primrose and borage oils (rich in γ-linolenic acid, $18:3n-6$) as well as blackcurrant seed oil (rich in α- and γ-linolenic acids) have been analysed using silver-ion HPLC-(APCI)MS (Laakso and Voutilainen, 1996) and reversed-phase HPLC-(APCI)MS (Laakso, 1997).

6.3.2.3 Supercritical fluid chromatography–mass spectrometry
Supercritical fluid chromatography (SFC) can separate triacylglycerols according to the increasing ACN on non-polar stationary phases (e.g. Proot et al., 1986; Richter et al., 1988; Baiochhi et al., 1993; Borch-Jensen et al., 1993; Staby et al., 1994; Manninen et al., 1995a; Laakso et al., 1996), according to the increasing ACN and combined number of double bonds on 25% cyanopropylphenylmethylsiloxane stationary phases (e.g. van Oosten et al., 1991; Manninen et al., 1995a,b), and according to the increasing unsaturation on silver-loaded loaded packed columns (Demirbüker and Blomberg, 1990, 1991; Demirbüker et al., 1992, 1993). The difficulties in the combination of SFC and MS are related to the large gas

volumes produced as a result of the expansion of the carrier fluid at the SFC restrictor outlet. In addition, the rapid expansion of the mobile phase from high to ambient, or even lower, pressures considerably cools the outlet. This cooling effect has to be compensated to prevent the formation of cluster ions. Although SFC-MS has utilised several interfaces, such as moving-belt (e.g. Ramsay *et al.*, 1989; Perkins *et al.*, 1991), particle-beam (Randall and Wahrhaftig, 1981; Jedrzejewski and Taylor, 1994), and direct fluid introduction (DFI) (e.g. Smith *et al.*, 1982a,b; Smith and Udseth, 1983; Wright *et al.*, 1986; Kalinoski *et al.*, 1987; Huang *et al.*, 1988), only ionisation at atmospheric pressure has shown real potential for SFC-MS (e.g. Huang *et al.*, 1990; Anacleto *et al.*, 1991; Matsumoto *et al.*, 1992; Tyrefors *et al.*, 1993; Thomas *et al.*, 1994; Pinkston and Baker, 1995; Manninen and Laakso, 1997a). SFC-MS has been reviewed recently by several authors (Arpino, 1990; Olesik, 1991; Arpino *et al.*, 1993; Arpino and Haas, 1995; Pinkston and Chester, 1995).

Electron ionisation.   In general, the most common interfacing technique has been DFI, favouring the use of capillary columns because of their low column flow rate. The DFI interfaces are self-designed and self-constructed probes or modified GC-MS interfaces, which are connected to the mass spectrometer through direct insertion probe or GC-MS inlet. Heated air circulation along a DFI probe and/or installation of heating wires is required to ensure proper heating of the SFC restrictor. The disadvantages of DFI are unstable ionisation conditions and a loss in sensitivity as a result of the increasing mobile phase flow rate during SFC density programming, i.e. as the density of the mobile phase is increased (Cousin and Arpino, 1987; Pinkston *et al.*, 1992; Mertens *et al.*, 1996).

Pure carbon dioxide or carbon dioxide modified with a small amount of organic solvent, such as methanol, is most often used as a mobile phase in SFC. SFC-(EI)MS is affected by $CO_2^{+\cdot}$, resulting in charge-exchange spectra. Although these spectra resemble EI spectra, they are not always comparable to GC-(EI)MS spectra owing to the differences in the fragmentation and ion abundances (Jablonska *et al.*, 1993).

In the literature there is only one example of SFC-(EI)MS of triacylglycerols (Kallio *et al.*, 1989). Owing to the low sensitivity of the system, overloading of the capillary column and the use of single-ion monitoring (SIM) was required to detect M^+ ions of butterfat triacylglycerols. In the same study, the regiospecific position of the fatty acyl residue of reference compounds was examined by detecting $[M-RCOOCH_2]^+$ ions, which are not formed if the fatty acyl residue is located at the *sn*-2 position (Ryhage and Stenhagen, 1960; Lauer *et al.*, 1970).

Chemical ionisation. Proton transfer and $CO_2^{+\cdot}$ charge-exchange ionisation compete in SFC-(CI)MS when carbon dioxide is used as mobile phase, but charge-exchange ionisation becomes more dominant during the SFC programming (Cousin and Arpino, 1987; Houben *et al.*, 1991). The resulting mixed charge-exchange/CI spectra usually provide molecular ions, molecular adduct ions and fragment ions. The ionisation process can be controlled by applying a relatively high reactant gas pressure to the ion source. The sensitivity of CI is approximately one order of magnitude higher than that of EI (Lee *et al.*, 1988). SFC-MS in CI mode can also provide mass spectra similar to EI spectra by the use of a reactant gas that does not contain protons (Houben *et al.*, 1991).

SFC-(CI)MS of triacylglycerols has been demonstrated only with reference compounds. Methane as reactant gas yields a characteristic fragment ion $[MH-RCOOH]^+$ for tristearin (Wright *et al.*, 1986). The relative abundance of the protonated molecule $[M+H]^+$ was only $\sim 4\%$. Triacylglycerols have also been reported to yield low abundance ($\sim 2\%$ relative abundance) pseudo-molecular ions $[M-H]^+$ rather than $[M+H]^+$ ions under methane SFC-(CI)MS conditions (Hawthorne and Miller, 1987). Cousin and Arpino (1987) have demonstrated the spectra of a mixture of tricaproin and tricaprylin as a result of pure $CO_2^{+\cdot}$ charge-exchange ionisation in CI mode and mixed ammonia $CI/CO_2^{+\cdot}$ charge-exchange ionisation. In the first case, only fragment ions corresponding to the loss of one acyl group from undetected $M^{+\cdot}$ ions were formed. In the latter case, relatively abundant $[M+NH_4]^+$ adduct ions were also detected. The formation of $[M+NH_4]^+$ adduct ions of tristearin with ammonia SFC-(CI)MS has also been reported by Pinkston *et al.* (1988).

Atmospheric pressure ionisation. Atmospheric pressure chemical ionisation is preferred in SFC-MS owing to the use of carbon dioxide as mobile phase. Carbon dioxide readily evaporates at ambient pressure and therefore can be removed efficiently before MS. The sensitivity and baseline are independent of the column flow rate, which allows the use of both packed and capillary columns. The addition of a reactant ion solvent is recommended to achieve maximum sensitivity, stable ions and symmetrical peaks as well as efficient ionisation of analyte molecules (Tyrefors *et al.*, 1993). In capillary SFC-(APCI)MS, reactant ions are commonly produced via introduction of reactant solvent vapour in the ionisation region by leading the nebulising gas flow through a reactant ion solvent container. This procedure has greater flexibility in the selection of the reactant ions and their concentration at the ionisation region and, therefore, allows better controllability of the ionisation process compared with HPLC-(APCI)MS. In most cases, additional heating of the restrictor end is applied to compensate the cooling effect of the expanding carrier fluid. Adequate restrictor temperature can also be achieved by utilising the heating capacity of the APCI vaporiser and using pre-heated APCI gases. Two different examples of interfacing capillary SFC with (APCI)MS are described in detail by Tyrefors et al. (1993) and Manninen and Laakso (1997a). SFC-(APCI)MS can yield molecular ions, molecular adduct ions and fragment ions depending on the sample components and the reactant ions. In most cases, the ionisation and fragmentation of molecular species can be controlled by careful selection of the reactant ion solvent. Although SFC-(APCI)MS has been successfully applied to the analysis of various compounds, only a few applications concerning SFC-(APCI)MS of triacylglycerols have been published.

Tyrefors et al. (1993) demonstrated the ionisation of trilaurin by capillary SFC-(APCI)MS using water as reactant ion source, which yielded abundant $[M+H]^+$ and $[M+18]^+$ ions and less abundant $[M-RCOO]^+$ ions. The utilisation of APCI in packed capillary SFC-MS of triacylglycerols has been described by Schmeer et al. (1996). They reported the formation of $[M+H]^+$ and $[M-RCOO]^+$ ions using carbon dioxide as the mobile phase and heated air as a nebulising gas. They also detected $[M+NH_4]^+$ ions, the origin of which was not explained. More detailed studies concerning capillary SFC-(APCI)MS of triacylglycerols in natural mixtures, such as milk fat (Laakso and Manninen, 1997), berry oils (Manninen and Laakso, 1997a,b) and fish oil (Manninen and Laakso, 1997c) have been reported.

The flow rate of the APCI gases affects the overall sensitivity of the analysis, whereas the APCI vaporiser and heated capillary (see Figure 6.1b) temperatures affect principally the fragmentation of triacylglycerols (Manninen and Laakso, 1997a). Methanol, water and isopropanol as reactant ion solvents in SFC-(APCI)MS provided abundant $[M+H]^+$ and $[M-RCOO]^+$ ions for cloudberry seed oil triacylglycerols at identical capillary SFC-(APCI)MS conditions (Manninen and Laakso, 1997a). In general, methanol results in the formation of abundant protonated molecular ions of most triacylglycerols with the exception of saturated molecules and those having one to two double bonds in the acyl chains. Therefore, methanol as a reactant ion solvent is recommended for use in the analysis of relatively unsaturated oils such as seed and fish oils (Figure 6.12). The disadvantage of water as a reactant ion solvent is very weak fragment ion abundances for triacylglycerols having four or fewer double bonds (Manninen and Laakso, 1997a). In general, the ion abundances provided with the three above-mentioned reactant ion solvents follow the order methanol ≥ water > isopropanol.

The molecular environment has a strong effect on the formation of protonated molecular ion and fragmentation as reported with methanol as a reactant ion solvent (Manninen and Laakso, 1997a). The abundance

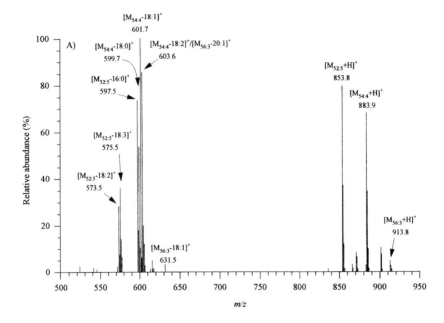

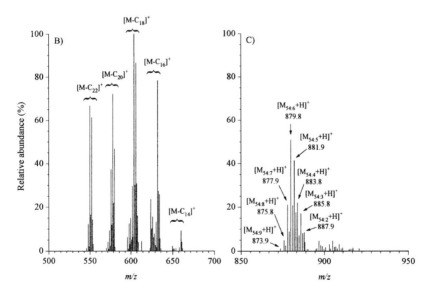

Figure 6.12 Capillary SFC-(APCI)MS spectra of (a) sea buckthorn seed oil triacylglycerols with 54 acyl carbons; (b) the region of $[M-RCOO]^+$ ions, and (c) the region of $[M+H]^+$ ions of fish oil triacylglycerols with 54 acyl carbons. Triacylglycerols of sea buckthorn seed oil were separated according to ACN + 2n on a 25% cyanopropylphenyl stationary phase (Manninen and Laakso (1997a). Copyright (1997) The American Oil Chemists' Society, reprinted with permission. Triacylglycerols of fish oil were separated according to ACN on an octyl stationary phase (Manninen and Laakso, 1997c).

of the $[M+H]^+$ ion increases with the increasing number of double bonds in the triacylglycerol acyl chains (Manninen and Laakso, 1997a). A fatty acid moiety in the sn-2 position forms less abundant $[M-RCOO]^+$ ion than in the sn-1/3 positions. The ion abundances are also affected by the acyl chain length and the number of double bonds of the fatty acid moieties. The abundance of $[M-RCOO]^+$ ion formed by the loss of α-linolenic acid moiety of a triacylglycerol was found to be clearly lower than that of γ-linolenic acid in the identical regiospecific position (Manninen and Laakso, 1997b). This indicates that the proximity of the double bonds to the glycerol skeleton makes the chain a better leaving group.

A solution of 0.5% ammonia in methanol as a reactant ion solvent provides abundant molecular adduct ions, $[M+18]^+$, and characteristic fragment ions $[M-RCOO]^+$, for less unsaturated and saturated triacylglycerols (Laakso and Manninen, 1997). Figure 6.13 shows an example of the difference in the molecular ion abundances of milk fat triacylglycerols with 54 acyl carbons produced either with methanol or with ammonia in methanol as a reactant ion solvent. In addition to $[M+18]^+$ ions, ammonia produces less abundant $[M+H]^+$ ions for triacylglycerols with five or more double bonds. For example, the ratio of $[M+18]^+$ and $[M+H]^+$ ions decreased approximately from 6 to 3 for triacylglycerols having 54 acyl carbons and from 5 to 9 double bonds. The abundance of the molecular ions also decreases rapidly with triacylglycerols having 8 or more double bonds and those with 56 or more acyl carbons. Therefore, ammonia in methanol is recommended to be used in the analysis of milk fat and other relatively saturated fats and oils (Figure 6.14).

6.4 Conclusions

In most cases (EI)MS is not a method to be recommended for the analysis of triacylglycerols: EI spectra are often difficult to interpret owing to extensive fragmentation of molecules. In addition, the molecular ion is present in a very low abundance or is entirely absent. (CI)MS, depending on the CI reactant gas, is a softer technique often providing ions containing the intact molecule as well as fragment ions. Most often, determination of triacylglycerols can be greatly improved by conducting a chromatographic separation step prior to MS detection. HT-GC can be combined on-line with (EI)MS and (CI)MS; however, the stationary phases often bleed to some extent at the high temperatures required to volatilise triacylglycerols. In addition, highly unsaturated triacylglycerols cannot be analysed by HT-GC owing to their thermal degradation.

Instead of HT-GC, triacylglycerols are more reliably separated by HPLC and SFC. The most practical approach for combining HPLC and

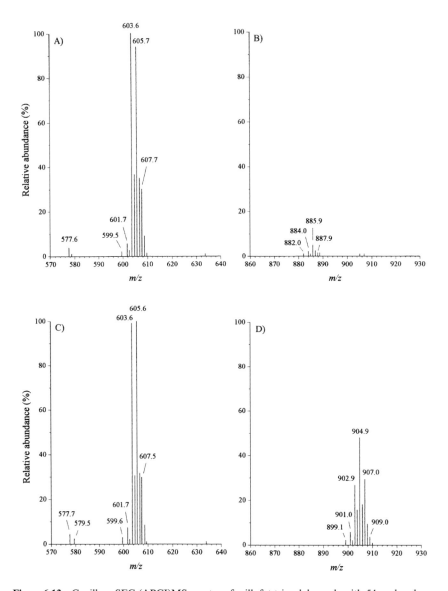

Figure 6.13 Capillary SFC-(APCI)MS spectra of milk fat triacylglycerols with 54 acyl carbons produced with methanol (a, b) or 0.5% ammonia in methanol (c, d) as reactant ion solvent: (a) the region of [M−RCOO]$^+$ ions, and (b) the region of [M + H]$^+$ ions produced by methanol; (c) the region of [M−RCOO]$^+$ ions, and (d) the region of [M + 18]$^+$ ions produced by 0.5% ammonia in methanol. Identifications of the [M + H]$^+$ ions in (b): m/z 882.0 corresponded to triacylglycerol 54:5, m/z 884.0 to 54:4, m/z 885.9 to 54:3, and m/z 887.9 to 54:2. Identifications of the [M + 18]$^+$ ions in D): m/z 899.1 corresponded to triacylglycerol 54:5, m/z 901.0 to 54:4, m/z 902.9 to 54:3, m/z 904.9 to 54:2, m/z 907.0 to 54:1, and m/z 909.0 to 54:0. Triacylglycerols were separated according to the ACN on an octyl stationary phase (Laakso and Manninen (1997). Copyright (1997) The American Oil Chemists' Society, reprinted with permission.)

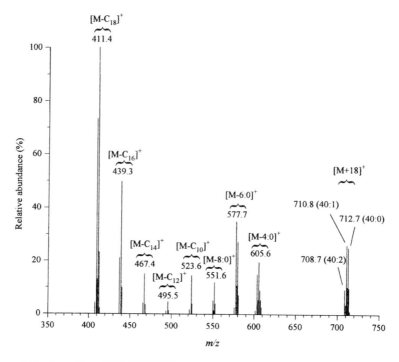

Figure 6.14 A capillary SFC-(APCI)MS spectrum of milk fat triacylglycerols with 40 acyl carbons produced by 0.5% ammonia in methanol as a reactant ion solvent: The $[M+18]^+$ ions defined the molecular mass species of triacylglycerols (the number of acyl carbons and double bonds in the acyl chains of a triacylglycerol) whereas the $[M-RCOO]^+$ fragment ions provided information on the acyl chains. Unambiguous identification of the loss of different C_{10} to C_{18} fatty acids from a single triacylglycerol was not possible: e.g. the ion m/z 411.4 may be formed by the loss of 18:0 from triacylglycerol 40:0, or by the loss of 18:1 from 40:1, or by the loss of 18:2 from 40:2. Triacylglycerols were separated according to the ACN on an octyl stationary phase (Laakso and Manninen (1997). Copyright (1997) The American Oil Chemists' Society, reprinted with permission.)

SFC with MS is to utilise interfacing techniques based on atmospheric pressure ionisation. Instruments equipped with API sources are robust in use and allow easy change between APCI and ESI modes. HPLC-(APCI)MS and SFC-(APCI)MS of most triacylglycerols yield abundant molecular ions and fragment ions. HPLC-(ESI)-CID-MS may also produce both molecular weight and fatty acid information, but ESI is not such a straightforward approach: typically splitting of the flow and addition of ionic species to the liquid flow in order to facilitate ionisation are required. In some cases, additional information concerning the fatty acid distribution between the *sn*-2 and the *sn*-1/3 positions may be achieved with (API)MS techniques. Quantitative determination is still a problem, because the mass spectrometric responses of triacylglycerols depend strongly on the molecular structure.

Depending on the information needed, direct probe sample introduction may be the method of choice: it allows fast analysis and, depending on the ionisation technique, provides data on molecular weight species and/or fragment ions. Ammonia negative-ion CI followed by CID of deprotonated molecular ions offers an efficient technique for determining molecular species compositions (fatty acid combinations as well as regiospecific distribution). However, isomeric triacylglycerols, containing for example *cis*- and *trans*-fatty acyl residues or α- and γ-linolenic acid moieties, cannot be distinguished with techniques based solely on the m/z separation with a quadrupole analyser. So far, ammonia negative-ion CI MS/MS has not been combined on-line with a chromatographic separation.

Table 6.2 presents a comparison of mass spectrometric techniques for the determination of triacylglycerols in fats and oils. In the following, some general remarks of practical value are made for different kinds of triacylglycerol samples.

6.4.1 Milk fat and other relatively saturated fats

Milk fat is characterised by a wide range of molecular weight species of triacylglycerols (ACN 26–54) having typically three or fewer double bonds in the acyl chains of a molecule. Milk fat contains several fatty acid constituents, the chain lengths of which vary from 4 to 18 carbon atoms. In addition, the presence of *trans*-fatty acids is a characteristic feature of all ruminant fats. The major drawback in most studies concerning MS analysis of milk fat triacylglycerols has been the nearly complete absence of molecular ions in the mass spectra. Molecular weight information on chromatographically separated saturated triacylglycerols as well as informative fragment ions has been produced exclusively with ionisation techniques based on the action of NH_4^+ ions, resulting in the formation of $[M + NH_4]^+$ ions (e.g. GC-(CI)MS, HPLC coupled with a particle-beam interface to (CI)MS and SFC-(APCI)MS). Chloride-attachment CI, ammonia CI and methane–nitrous oxide CI in negative-ion mode result in the formation of abundant ions containing the intact molecules without fragmentation. In addition, positive-ion CI with isobutane has been reported to provide both protonated molecules and fragment ions of relatively saturated triacylglycerols using direct probe sample introduction.

Milk fat is such a complex mixture of triacylglycerols that it is not possible to separate all molecular species with a single chromatographic technique. In most cases, interpretation of the results can be greatly simplified by fractionation of triacylglycerols according to unsaturation

Table 6.2 Comparison of mass spectrometric techniques in the analysis of triacylglycerols

Mass spectrometric technique	Milk fat and other relatively saturated fats			Vegetable and seed oils			Fish oils and other polyunsaturated oils		
	MW[a]	FA[b]	sn-Position[c]	MW	FA	sn-Position	MW	FA	sn-Position
GC-(EI)MS	−	+	−	not recommended			not recommended		
GC-(positive ion CI)MS	++	++	−	not recommended			not recommended		
GC-(negative ion CI)MS	+[d]	++	−	not recommended			not recommended		
HPLC-(APCI)MS, API	−	+++	+−	+++	+++	+−	+++	+++	+−
HPLC-(ESI)MS, API	++	−	−	++	−	−	++	−	−
HPLC-(ESI)-CID-MS, API	++	++	+−	++	++	+−	++	++	+−
HPLC-(positive ion CI)MS, DL	−	++	+−	++	++	+−	++	++	+−
HPLC-(negative ion CI)MS, DL	++	−	−	++	−	−	++	−	−
SFC-(APCI)MS, methanol	−	+++	+−	+++	+++	+−	+++	+++	+−
SFC-(APCI)MS, ammonia	+++	+++	+−	++−	+++	+−	−	+++	+−
DEP, positive ion CI	++	++	−	++	++	−	+[e]	+[e]	−
DEP, positive ion CI and CID	++	++	−	++	++	−	+[e]	+[e]	−
DEP, negative ion CI	+++	−	−	+++	−	−	+++	−	−
DEP, negative ion CI and CID	+++	+++	++	+++	+++	++	+++	+++	++

Abbreviations: APCI, atmospheric pressure chemical ionisation; API, atmospheric pressure ionisation; CI, chemical ionisation; CID, collision-induced dissociation; DEP, direct exposure probe; DLI, direct liquid inlet; EI, electron ionisation; ESI, electrospray ionisation; GC, gas chromatography; HPLC, high-performance liquid chromatography; MS, mass spectrometry; SFC, supercritical fluid chromatography. Information achieved on [a]molecular weight, [b]fatty acyl residues, and [c]distribution of fatty acyl residues between the sn-2 and the sn-1/3 positions:−, information not provided; +, information achieved to some extent; ++, information achieved with some restrictions or difficulties; +++, best available information achieved; +−, information available, but quality depends on the complexity of the sample; [d]Not reported with milk fat. [e]No published data available.

by silver-ion chromatography followed by separation according to molecular mass differences: e.g. silver-ion HPLC combined with reversed-phase HPLC-(APCI)MS or non-polar capillary SFC-(APCI)MS. Owing to the relatively saturated nature of milk fat triacylglycerols, they can also be analysed by GC-(CI)MS. Saturated triacylglycerols should yield both adduct ions and fragments with positive-ion ammonia CI combined with GC-MS, but the technique has not been utilised for milk fat analysis since the study of Murata (1977).

6.4.2   Vegetable and seed oils

The most abundant fatty acids in vegetable and seed oils contain 16, 18 or 20 acyl carbons and up to 3 double bonds. Thus, the triacylglycerols typically contain from 50 to 58 acyl carbons and up to 9 double bonds in the acyl chains. Most ionisation techniques combined with HPLC and SFC separation, except EI, yield informative mass spectra of vegetable and seed oils providing both molecular weight and fatty acid data. Adequate resolution of vegetable and seed oil triacylglycerols can often be achieved with a single chromatographic system depending on its separation mechanism and efficiency. For example, reversed-phase HPLC-(APCI)MS, silver-ion HPLC-(APCI)MS, and capillary SFC-(APCI)MS on a 25% cyanopropylphenyl stationary phase are efficient techniques in determination of molecular species compositions of vegetable and seed oils. GC-MS is not recommended for samples containing highly unsaturated triacylglycerols.

6.4.3   Fish oils and other highly unsaturated oils

Fish oils are the most complex samples of natural triacylglycerol mixtures. The typical feature of fish oils is a wide range of fatty acids having from 14 to 24 acyl carbons and up to 6 double bonds in one acyl chain. For example, eicosapentaenoic acid (20:5n−3) and docosahexaenoic acid (22:6n−3) are characteristic fatty acids for fish oils. The versatile fatty acid composition results in extremely complex triacylglycerol mixtures; the number of acyl carbons may vary from 44 to 68 and the number of double bonds per molecule may be even greater than 15. GC-MS is useless for the analysis of fish oil triacylglycerols owing to the high temperatures required. Fractionation before chromatography or two complementary chromatographic methods coupled with MS are required to achieve interpretable data; e.g. silver-ion HPLC followed by reversed-phase HPLC-(APCI)MS or non-polar capillary SFC-(APCI)MS. MS with most ionisation techniques, except EI, provides information on both the molecular masses and the fatty acyl residues of fish oil triacylglycerols.

References

Aasen, A.J., Lauer, W.M. and Holman, R.T. (1970) Mass spectrometry of triglycerides: II. Specifically deuterated triglycerides and elucidation of fragmentation mechanisms. *Lipids*, **5** 869-877.

Aitzetmüller, K. and Grönheim, M. (1992) Separation of highly unsaturated triacylglycerols by reversed phase HPLC with short wavelength UV detection. *J. High Resolut. Chromatogr.*, **15** 219-226.

Anderson, M.A., Collier, L., Dilliplane, R. and Ayorinde, F.O. (1993) Mass spectrometric characterization of *Vernolia galamensis* oil. *J. Am. Oil Chem. Soc.*, **70** 905-908.

Anacleto, J.F., Ramaley, L., Boyd, R.K., Pleasance, S., Sim, P.G. and Benoit, F.M. (1991) Analysis of polycyclic aromatic compounds by supercritical fluid chromatography/mass spectrometry using atmospheric pressure chemical ionization. *Rapid Commun. Mass Spectrom.*, **5** 149-155.

Arpino, P. (1990) Coupling techniques in LC/MC and SFC/MS. *Fresenius J. Anal. Chem.*, **337** 667-685.

Arpino, P.J. and Haas, P. (1995) Recent developments in supercritical fluid chromatography–mass spectrometry coupling. *J. Chromatogr. A*, **703** 479-488.

Arpino, P.J., Sadoun, F. and Virelizier, H. (1993) Reviews of recent trends in chromatography/mass spectrometry coupling. Part IV. Reasons why supercritical fluid chromatography is not so easily coupled with mass spectrometry as originally assessed. *Chromatographia*, **36** 283-288.

Baiocchi, C., Saini, G., Cocito, C., Giacosa, D., Roggero, M.A., Marengo, E. and Favale, M. (1993) Analysis of vegetable and fish oils by capillary supercritical fluid chromatography with flame ionization detection. *Chromatographia*, **37** 525-533.

Barber, M., Merren, T.O. and Kelly, W. (1964) The mass spectrometry of large molecules I. The triglycerides of straight chain fatty acids. *Tetrahedron Lett.*, 1063-1067.

Borch-Jensen, C., Staby, A. and Mollerup, J. (1993) Supercritical fluid chromatographic analysis of fish oil of the sand eel (*Ammodytes* sp.). *J. High Resolut. Chromatogr.*, **16** 621-623.

Bruins, A.P. (1991) Mass spectrometry with ion sources operating at atmospheric pressure. *Mass Spectrom. Rev.*, **10** 53-77.

Brühl, L., Schulte, E. and Thier, H.-P. (1993) Fraktionierung der Triglyceride von Muttermilch durch HPLC an einer Silberionensäule und an RP-18-Material mit dem Lichtstreudetektor. *Fat Sci. Technol.*, **95** 370-376.

Byrdwell, W.C. (1998) APCI-MS for lipid analysis. *Inform.*, **9** 986-997.

Byrdwell, W.C. and Emken, E.A. (1995) Analysis of triglycerides using atmospheric pressure chemical ionization mass spectrometry. *Lipids*, **30** 173-175.

Byrdwell, W.C., Emken, E.A., Neff, W.E. and Adlof, R.O. (1996) Quantitative analysis of triglycerides using atmospheric pressure chemical ionization–mass spectrometry. *Lipids*, **31** 919-935.

Christie, W.W. (1988) Separation of molecular species of triacylglycerols by high-performance liquid chromatography with a silver ion column. *J. Chromatogr.*, **454** 273-284.

Christie, W.W. (1991) Fractionation of the triacylglycerols of evening primrose oil by high-performance liquid chromatography in the silver ion mode. *Fat Sci. Technol.*, **93** 65-66.

Christie, W.W. (1995) Silver-ion high-performance liquid chromatography, in *New Trends in Lipid and Lipoprotein Analyses* (eds J.-L. Sebedio and E.G. Perkins), AOCS Press, Champaign, IL, pp 59-74.

Cousin, J. and Arpino, P.J. (1987) Construction of a supercritical fluid chromatograph–mass spectrometer instrument system using capillary columns, and a chemical ionization source accepting high flow-rates of mobile phase. *J. Chromatogr.*, **398** 125-141.

Currie, G. and Kallio, H. (1993) Triacylglycerols of human milk: rapid analysis by ammonia negative ion tandem mass spectrometry. *Lipids*, **28** 217-222.

Demirbüker, M. and Blomberg, L.G. (1990) Group separation of triacylglycerols on micropacked argentation columns using supercritical media as mobile phases. *J. Chromatogr. Sci.*, **28** 67-72.

Demirbüker, M. and Blomberg, L.G. (1991) Separation of triacylglycerols by supercritical-fluid argentation chromatography. *J. Chromatogr.*, **550** 765-774.

Demirbüker, M., Blomberg, L.G., Olsson, N.U., Berqvist, M., Herslöf, B.G. and Jacobs, F.A. (1992) Characterization of triacylglycerols in the seeds of *Aquilegia vulgaris* by chromatographic and mass spectrometric methods. *Lipids*, **27** 436-441.

Demirbüker, M., Anderson, P.E. and Blomberg, L.G. (1993) Miniaturized light scattering detector for packed capillary supercritical fluid chromatography. *J. Microcol. Sep.*, **5** 141-147.

Dobson, G., Christie, W.W. and Nikolova-Damyanova, B. (1995) Silver ion chromatography of lipids and fatty acids. *J. Chromatogr. B*, **671** 197-222.

Duffin, K.L., Henion, J.D. and Shieh, J.J. (1991) Electrospray and tandem mass spectrometric characterization of acylglycerol mixtures that are dissolved in nonpolar solvents. *Anal. Chem.*, **63** 1781-1788.

Evans, C., Traldi, P., Bambagiotti-Alberti, M., Giannellini, V., Coran, S.A. and Vincieri, F.F. (1991) Positive and negative fast atom bombardment mass spectrometry and collision spectroscopy in the structural characterization of mono-, di- and triglycerides. *Biol. Mass Spectrom.*, **20** 351-356.

Evershed, R.P. (1994) Application of modern mass spectrometric techniques to the analysis of lipids. *Spec. Publ. R. Soc. Chem.*, **160** 123-160.

Evershed, R.P. (1996) High-resolution triacylglycerol mixture analysis using high-temperature gas chromatography/mass spectrometry with a polarizable stationary phase, negative ion chemical ionization, and mass-resolved chromatography. *J. Am. Soc. Mass Spectrom.*, **7** 350-361.

Evershed, R.P., Prescott, M.C. and Goad, L.J. (1990) High-temperature gas chromatography/mass spectrometry of triacylglycerols with ammonia negative-ion chemical ionization. *Rapid Commun. Mass Spectrom.*, **4** 345-347.

Fales, H.M., Milne, G.W.A., Winkler, H.U., Beckey, H.D., Damico, J.N. and Barron, R. (1975) Comparison of mass spectra of some biologically important compounds as obtained by various ionization techniques. *Anal. Chem.*, **47** 207-219.

Finnigan MAT (1993) *TSQ/SSQ 7000 Series—Atmospheric Pressure Ionization, Operator's and Service Manual*, Revision A, September 1993, 70001-97056, Finnigan MAT, San Jose, CA.

Fukatsu, M. and Tamura, T. (1988) Analysis of triglyceride by GC-MS, in *Proceedings of Session Lectures and Scientific Presentations on JSF-JOCS World Congress 1988*, vol. II, Japan Oil Chemists' Society, pp 956-962.

Games, D.E. (1978) Soft ionization mass spectral methods for lipid analysis. *Chem. Phys. Lipids*, **21** 389-402.

Geeraert, E. and Sandra, P. (1985) Capillary GC of triglycerides in fats and oils using a high temperature phenylmethylsilicone stationary phase, Part I., *J. High Resolut. Chromatogr. Chromatogr. Commun.*, **8** 415-422.

Geeraert, E. and Sandra, P. (1987) Capillary GC of triglycerides in fats and oils using a high temperature phenylmethylsilicone stationary phase, Part II. The analysis of chocolate fats. *J. Am. Oil Chem. Soc.*, **64** 100-105.

Geeraert, E., Sandra, P. and De Shepper, D. (1983) On-column injection in the capillary gas chromatographic analysis of fats and oils. *J. Chromatogr.*, **279** 287-295.

Grob, K., Jr (1979) Evaluation of injection techniques for triglycerides in capillary gas chromatography. *J. Chromatogr.*, **178** 387-392.

Grob, K., Jr (1981) Degradation of triglycerides in gas chromatographic capillaries: studies by reversing the column. *J. Chromatogr.*, **205** 289-296.

Hawthorne, S.B. and Miller, D.J. (1987) Analysis of commercial waxes using capillary supercritical fluid chromatography–mass spectrometry. *J. Chromatogr.*, **388** 397-409.

Hites, R.A. (1970) Quantitative analysis of triglyceride mixtures by mass spcetrometry. *Anal. Chem.*, **42** 1736-1740.

Hites, R.A. (1975) Mass spectrometry of triglycerides. *Methods Enzymol.*, **35** 348-359.

Hogge, L.R., Taylor, D.C., Reed, D.W. and Underhill, E.W. (1991) Characterization of castor bean neutral lipids by mass spectrometry/mass spectrometry. *J. Am. Oil Chem. Soc.*, **68** 863-868.

Hori, M., Sahashi, Y., Koike, S., Yamaoka, R. and Sato, M. (1994) Molecular species analysis of polyunsaturated fish triacylglycerol by high-performance liquid chromatography/fast atom bombardment mass spectrometry. *Anal. Sci.*, **10** 719-724.

Houben, R.J., Leclercq, P.A. and Cramers, C.A. (1991) Ionization mechanisms in capillary supercritical fluid chromatography–chemical ionization mass spectrometry. *J. Chromatogr.*, **554** 351-358.

Huang, A.S., Robinson, L.R., Gursky, L.G., Profita, R. and Sabidong, C.G. (1994) Identification and quantification of SALATRIM 23CA in foods by the combination of supercritical fluid extraction, particle beam LC-mass spectrometry, and HPLC with light-scattering detector. *J. Agric. Food Chem.*, **42** 468-473.

Huang, A.S., Robinson, L.R., Pelluso, T.A., Gursky, L.G., Pidel, A., Manz, A., Softly, B.J., Templeman, G.J., Finley, J.W. and Leville, G.A. (1997) Quantification of generic salatrim in foods containing salatrim and other fats having medium- and long-chain fatty acids. *J. Agric. Food Chem.*, **45** 1770-1778.

Huang, E., Covey, T.R. and Henion, J. (1990) Packed-column supercritical fluid chromato-graphy–mass spectrometry and supercritical fluid chromatography–tandem mass spectro-metry with ionization at atmospheric pressure. *J. Chromatogr.*, **511** 257-270.

Huang, E.C., Jackson, B.J., Markides, K.E. and Lee, M.L. (1988) Direct heated interface probe for capillary supercritical fluid chromatography/double focusing mass spectrometry. *Anal. Chem.*, **60** 2715-2719.

Ikonomou, M.G., Blades, A.T. and Kebarle, P. (1991) Electrospray–ion spray: a comparison of mechanisms and performance. *Anal. Chem.*, **63** 1989-1998.

Jablonska, A., Hansen, M., Ekeberg, D. and Lundanes, E. (1993) Determination of chlorinated pesticides by capillary supercritical fluid chromatography–mass spectrometry with positive- and negative-ion detection. *J. Chromatogr.*, **647** 341-350.

Jedrzejewski, P.T. and Taylor, T.L. (1994) Evaluation of the particle beam interface for packed-column supercritical fluid chromatography–mass spectrometry with pure and modified CO_2. *J. Chromatogr.* A, **677** 365-376.

Jensen, N.J. and Gross, M.L. (1988) A comparison of mass spectrometry methods for structural determination and analysis of phospholipids. *Mass Spectrom. Rev.*, **7** 41-69.

Johansson, A.K., Kuusisto, P.H., Laakso, P.H., Derome, K.K., Sepponen, P.J., Katajisto, J.K. and Kallio, H.P. (1997a) Geographical variations in seed oils from *Rubus chamaemorus* and *Empetrum nigrum*. *Phytochemistry*, **44** 1421-1427.

Johansson, A., Laakso, P. and Kallio, H. (1997b) Molecular weight distribution of the triacylglycerols of berry seed oils analyzed by negative-ion chemical ionization mass spectrometry. *Z. Lebensm. Unters. Forsch.* A, **204** 308-315.

Johnstone, R.A.W. and Rose, M.E. (1996) *Mass Spectrometry for Chemists and Biochemists*, 2nd edn, Cambridge University Press, Cambridge.

Kalinoski, H.T., Udseth, H.R., Wright, B.W. and Smith, R.D. (1987) Analytical applications of capillary supercritical fluid chromatography–mass spectrometry. *J. Chromatogr.*, **400** 307-316.

Kallio, H. (1992) Tandem mass spectrometric analysis of triacylglycerols of low unsaturation level of Baltic herring (*Clupea harengus membras*) flesh oil, in *Contemporary Lipid Analysis, Proceedings of the 2nd meeting* (eds N.U. Olsson and B.G. Herslöf), Järna Tryckeri AB, Järna, Sweden, pp 48-62.

Kallio, H. and Currie, G. (1991) Triacylglycerols of turnip rapeseed oil. A new method of stereospecific analysis of triacylglycerols, in *The Short Course on HPLC of Lipids*, 9–11 May, AOCS, Bloomingdale.

Kallio, H. and Currie, G. (1993a) Analysis of low erucic acid turnip rapeseed oil (*Brassica campestris*) by negative ion chemical ionization tandem mass spectrometry. A method giving information on the fatty acid composition in positions sn-2 and sn-1/3 of triacylglycerols. *Lipids*, **28** 207-215.

Kallio, H. and Currie, G. (1993b) Analysis of natural fats and oils by ammonia negative ion tandem mass spectrometry—triacylglycerols and positional distribution of their acyl groups, in *CRC Handbook of Chromatography—Analysis of Lipids* (eds K.D. Mukherjee, N. Weber and J. Sherma), CRC Press, Boca Raton, FL, pp 435-458.

Kallio, H. and Rua, P. (1994) Distribution of the major fatty acids of human milk between sn-2 and sn-1,3 positions of triacylglycerols. *J. Am. Oil Chem. Soc.*, **71** 985-992.

Kallio, H., Currie, G., Gibson, R. and Kallio, S. (1997) Mass spectrometry of food lipids: negative ion chemical ionization/collision induced dissociation analysis of oils containing γ-linolenic acid as an example. *Ann. Chim. (Rome)*, **87** 187-198.

Kallio, H., Laakso, P., Huopalahti, R., Linko, R.R. and Oksman, P. (1989) Analysis of butter fat triacylglycerols by supercritical fluid chromatography/electron impact mass spectrometry. *Anal. Chem.*, **61** 698-700.

Kalo, P. and Kemppinen, A. (1993) Mass spectrometric identification of triacylglycerols of enzymatically modified butterfat separated on a polarizable phenylmethyl-silicone column. *J. Am. Oil Chem. Soc.*, **70** 1209-1217.

Kebarle, P. and Tang, L. (1993) From ions in solution to ions in gas phase: The mechanism of electrospray mass spectrometry. *Anal. Chem.*, **65** 972A-986A.

Kim, H.Y. and Salem, N. (1987) Application of thermospray high-performance liquid chromatography/mass spectrometry for the determination of phospholipids and related compounds. *Anal. Chem.*, **59** 722-726.

Kim, H.Y. and Salem, N. (1993) Liquid chromatography–mass spectrometry of lipids. *Prog. Lipid Res.*, **32** 221-245.

Kostiainen, R. and Bruins, A.P. (1994) Effect of multiple sprayers on dynamic range and flow rate limitations in electrospray and ionspray mass spectrometry. *Rapid Commun. Mass Spectrom.*, **8** 549-558.

Kuksis, A. (1994) GLC and HPLC of neutral glycerolipids, in *Lipid Chromatographic Analysis* (ed. T. Shibamoto), Marcel Dekker., New York, pp 177-222.

Kuksis, A. (1997) Mass spectrometry of complex lipids, in *Lipid Analysis in Oils and Fats* (ed. R.J. Hamilton), Blackie Academic and Professional, London, pp 181-249.

Kuksis, A. and Myher, J.J. (1995) Application of tandem mass spectrometry for the analysis of long-chain carboxylic acids. *J. Chromatogr. B*, **671** 35-70.

Kuksis, A., Marai, L. and Myher, J.J. (1983) Strategy of glycerolipid separation and quantitation by complementary analytical techniques. *J. Chromatogr.*, **273** 43-66.

Kuksis, A., Marai, L., Myher, J.J., Cerbulis, J. and Farrell, H.M., Jr (1986) Comparative study of the molecular species of chloropropanediol diesters and triacylglycerols in milk fat. *Lipids*, **21** 183-190.

Kuksis, A., Marai, L. and Myher, J.J. (1991a) Reversed-phase liquid chromatography–mass spectrometry of complex mixtures of natural triacylglycerols with chloride-attachment negative chemical ionization. *J. Chromatogr.*, **588** 73-87.

Kuksis, A., Marai, L. and Myher, J.J. (1991b) Plasma lipid profiling by liquid chromatography with chloride-attachment mass spectrometry. *Lipids*, **26** 240-246.

Kusaka, T., Ishihara, S., Sakaida, M., Mifune, A., Nakano, Y., Tsuda, K., Ikeda, M. and Nakano, H. (1996) Composition analysis of plant triacylglycerols and hydroperoxidized rac-1-stearoyl-2-oleoyl-3-linoleoyl-sn-glycerols by liquid chromatography–atmospheric pressure chemical ionization mass spectrometry. *J. Chromatogr. A*, **730** 1-7.

Laakso, P. (1995) Molecular species analysis of triacylglycerols by mass spectrometry using ammonia negative ion chemical ionization. Presented at *the 21st World Congress of the International Society for Fat Research (ISF)*, 1–6 October, The Hague.

Laakso, P. (1996) Analysis of triacylglycerols—approaching the molecular composition of natural mixtures. *Food Rev. Int.*, **12** 199-250.

Laakso, P. (1997) Characterization of α- and γ-linolenic acid oils by reversed-phase high-performance liquid chromatography–atmospheric pressure chemical ionization mass spectrometry. *J. Am. Oil Chem. Soc.*, **74** 1291-1300.

Laakso, P. and Kallio, H. (1993) Triacylglycerols of winter butterfat containing configurational isomers of monoenoic fatty acyl residues. I. Disaturated monoenoic triacylglycerols. *J. Am. Oil Chem. Soc.*, **70** 1161-1171.

Laakso, P. and Kallio, H. (1996) Optimization of the mass spectrometric analysis of triacylglycerols using negative-ion chemical ionization with ammonia. *Lipids*, **31** 33-42.

Laakso, P. and Manninen, P. (1997) Identification of milk fat triacylglycerols by capillary supercritical fluid chromatography–atmospheric pressure chemical ionization mass spectrometry. *Lipids*, **32** 1285-1295.

Laakso, P. and Voutilainen, P. (1996) Analysis of triacylglycerols by silver-ion high-performance liquid chromatography–atmospheric pressure chemical ionization mass spectrometry. *Lipids*, **31** 1311-1322.

Laakso, P., Manninen, P., Mäkinen, J. and Kallio, H. (1996) Postparturition changes in the triacylglycerols of cow colostrum. *Lipids*, **31** 937-943.

Lamberto, M. and Saitta, M. (1995) Principal component analysis in fast atom bombardment–mass spectrometry of triacylglycerols in edible oils. *J. Am. Oil Chem. Soc.*, **72** 867-871.

Lauer, W.M., Aasen, A.J., Graff, G. and Holman, R.T. (1970) Mass spectrometry of triglycerides: I. Structural effects. *Lipids*, **5** 861-868.

Lee, E.D., Hsu, S.-H. and Henion, J.D. (1988) Electron-ionization-like mass spectra by capillary supercritical fluid chromatography/charge exchange mass spectrometry. *Anal. Chem.*, **60** 1990-1994.

Maniongui, C., Gresti, J., Bugaut, M., Gauthier, S. and Bezard, J. (1991) Determination of bovine butterfat triacylglycerols by reversed-phase liquid chromatography and gas chromatography. *J. Chromatogr.*, **543** 81-103.

Manninen, P. and Laakso, P. (1997a) Capillary supercritical fluid chromatography–atmospheric pressure chemical ionization mass spectrometry of triacylglycerols in berry oils. *J. Am. Oil Chem. Soc.*, **74** 1089-1098.

Manninen, P. and Laakso, P. (1997b) Capillary supercritical fluid chromatography–atmospheric pressure chemical ionization mass spectrometry of γ- and α-linolenic acid containing triacylglycerols in berry oils. *Lipids*, **32** 825-831.

Manninen, P. and Laakso, P. (1997c) Capillary supercritical fluid chromatography–(APCI)MS in the analysis of triacylglycerols, presented in *The 14th International Mass Spectrometry Conference*, 25–29 August, Tampere, Finland.

Manninen, P., Laakso, P. and Kallio, H. (1995a) Method for characterization of triacylglycerols and fat-soluble vitamins in edible oils and fats by supercritical fluid chromatography, *J. Am. Oil Chem. Soc.*, **72** 1001-1008.

Manninen, P., Laakso, P. and Kallio, H. (1995b) Separation of triacylglycerols γ- and α-linolenic acid containing triacylglycerols in berry oils fats by supercritical fluid chromatography, *Lipids*, **30** 665-671.

Marai, L., Kuksis, A. and Myher, J.J. (1994) Reversed-phase liquid chromatography–mass spectrometry of the unknown triacylglycerol structures generated by randomization of butteroil. *J. Chromatogr. A*, **672** 87-99.

Marai, L., Myher, J.J. and Kuksis, A. (1983) Analysis of triacylglycerols by reversed-phase high pressure liquid chromatography with direct liquid inlet mass spectrometry. *Can. J. Biochem. Cell. Biol.*, **61** 840-849.

Mareš, P. (1988) High temperature capillary gas liquid chromatography of triacylglycerols and other intact lipids, *Prog. Lipid Res.*, **27** 107-133.

Mareš, P., Rezanka, T. and Novak, M. (1991) Analysis of human blood plasma triacylglycerols using capillary gas chromatography, silver ion thin-layer chromatographic fractionation and desorption chemical ionization mass spectrometry. *J. Chromatogr.*, **568** 1-10.

Matsubara, T. and Hayashi, A. (1991) FAB/Mass spectrometry of lipids. *Prog. Lipid Res.*, **30** 301-322.

Matsumoto, K., Nagata, S., Hattori, H. and Tsuge, S. (1992) Development of directly coupled supercritical fluid chromatography with packed capillary column-mass spectrometry with atmospheric pressure chemical ionization. *J. Chromatogr.*, **605** 87-94.

McLafferty, F.W. and Turecek, F. (1993) *Interpretation of Mass Spectra*, 4th edn, University Science Books, Mill Valley, CA.

Merritt, C. Jr., Vajdi, M., Kayser, S.G., Halliday, J.W. and Bazinet, M.L. (1982) Validation of computational methods for triglyceride composition of fats and oils by liquid chromatography and mass spectrometry, *J. Am. Oils Chem. Soc.*, **59** 422-432.

Mertens, M.A.A., Janssen, H.-G.M., Gramers, C.A., Genuit, W.J.L., van Velzen, G.J., Dirkzwager, H. and van Binsbergen, H. (1996) Development and evaluation of an interface for coupled capillary supercritical fluid chromatography/magnetic sector mass spectrometry. *J. High Resolut. Chromatogr.*, **19** 17-22.

Mottram, H.R. and Evershed, R.P. (1996) Structure analysis of triacylglycerol positional isomers using atmospheric pressure chemical ionisation mass spectrometry. *Tetrahedron Lett.*, **37** 8593-8596.

Murata, T. (1977) Analysis of triglycerides by gas chromatography/chemical ionization mass spectrometry. *Anal. Chem.*, **49** 2209-2213.

Murata, T. and Takahashi, S. (1973) Analysis of triglyceride mixtures by gas chromatography–mass spectrometry. *Anal. Chem.*, **45** 1816-1823.

Murata, T. and Takahashi, S. (1977) Qualitative and quantitative chemical ionization mass spectrometry of triglycerides. *Anal. Chem.*, **49** 728-731.

Murphy, R.C. (1993) *Mass Spectrometry of Lipids*, Plenum Press, New York.

Myher, J.J. and Kuksis, A. (1995a) Electrospray-MS for lipid identification. *INFORM*, **6** 1068-1072.

Myher, J.J. and Kuksis, A. (1995b) General strategies in chromatographic analysis of lipids. *J. Chromatogr. B*, **671** 3-33.

Myher, J.J., Kuksis, A., Marai, L. and Manganaro, F. (1984) Quantitation of natural triacylglycerols by reversed-phase liquid chromatography with direct liquid inlet mass spectrometry. *J. Chromatogr.*, **283** 289-301.

Myher, J.J., Kuksis, A., Marai, L. and Sandra, P. (1988) Identification of the more complex triacylglycerols in bovine milk fat by gas chromatography–mass spectrometry using polar capillary columns. *J. Chromatogr.*, **452** 93-118.

Myher, J.J., Kuksis, A. and Marai, L. (1993) Identification of the less common isologous short-chain triacylglycerols in the most volatile 2.5% molecular distillate of butter oil. *J. Am. Oil Chem. Soc.*, **70** 1183-1191.

Myher, J.J., Kuksis, A. and Park, P.W. (1997) Stereospecific analysis of docosahexaenoic acid-rich triacylglycerols by chiral-phase HPLC with online electrospray mass spectrometry, in *New Techniques and Applications in Lipid Analysis* (eds R.E. McDonald and M.M. Mossoba), Champaign, IL, pp 100-120.

Neff, W.E. and Byrdwell, W.C. (1995) Soybean oil triacylglycerol analysis by reversed-phase high-performance liquid chromatography coupled with atmospheric pressure chemical ionization mass spectrometry. *J. Am. Oil Chem. Soc.*, **72** 1185-1191.

Niessen, W.M.A. and van der Greef, J. (1992) *Liquid Chromatography–Mass Spectrometry, Principles and Applications*, Marcel Dekker, New York.

Nikolova-Damyanova, B. (1992) Silver ion chromatography and lipids, in *Advances in Lipid Methodology—One* (ed. W.W. Christie), The Oily Press, Ayr, pp 181-237.

Nikolova-Damyanova, B. (1997) Reversed-phase high-performance liquid chromatography: general principles and application to the analysis of fatty acids and triacylglycerols, in *Advances in Lipid Methodology—Four* (ed. W.W. Christie), The Oily Press, Dundee, pp 193-251.

Nurmela, K.V.V. and Satama, L.T. (1988) Quantitative analysis of triglycerides by high-performance liquid chromatography using non-linear gradient elution and flame ionization detection. *J. Chromatogr.*, **435** 139-148.

Ohshima, T., Yoon, H.-S. and Koizumi, C. (1989) Application of selective ion monitoring to the analysis of molecular species of vegetable oil triacylglycerols separated by open-tubular column GLC on a methylphenylsilicone phase at high temperature. *Lipids*, **24** 537-544.

Olesik, S.V. (1991) Recent advances in supercritical fluid chromatography/mass spectrometry. *J. High Resolut. Chromatogr.*, **14** 5-9.

Perkins, J.R., Games, D.E., Startin, J.R. and Gilbert, J. (1991) Analysis of sulphonamides using supercritical fluid chromatography and supercritical fluid chromatography–mass spectrometry. *J. Chromatogr.*, **540** 239-256.

Perrin, J.-L., Prevot, A., Traitler, H. and Bracco, U. (1987) Analysis of triglyceride species of blackcurrant seed oil by HPLC via a laser light scattering detector. *Rev. Franc. Corps Gras*, **34** 221-223.

Phillips, F.C., Erdahl, W.L., Nadenicek, J.D., Nutter, L.J., Schmit, J.A. and Privett, O.S. (1984) Analysis of triglyceride species by high-performance liquid chromatography via a flame ionization detector. *Lipids*, **19** 142-150.

Pinkston, J.D. and Baker, T.R. (1995) Modified ionspray interface for supercritical fluid chromatography/mass spectrometry: Interface design and initial results. *Rapid Commun. Mass Spectrom.*, **9** 1087-1094.

Pinkston, J.D. and Chester, T.L. (1995) Putting opposites—guidelines for successful SFC/MS. *Anal. Chem.*, **67** 650A-656A.

Pinkston, J.D., Delaney, T., Morand, K.L. and Cooks, R.G. (1992) Supercritical fluid chromatography/mass spectrometry using a quadrupole mass filter/quadrupole ion trap hydrid mass spectrometer with external ion source. *Anal. Chem.*, **64** 1571-1577.

Pinkston, J.D., Owens, G.D., Burkes, L.J., Delaney, T.E., Millington, D.S. and Malthy, D.A. (1988) Capillary supercritical fluid chromatography-mass spectrometry using a "high mass" quadrupole and splitless injection. *Anal. Chem.*, **60** 962-966.

Proot, M., Sandra, P. and Geeraert, E. (1986) Resolution of triglycerides in capillary SFC as a function of column temperature. *J. High Resolut. Chromatogr. Chromatogr. Commun.*, **9** 189-192.

Ramsay, E.D., Perkins, J.R., Games, D.E. and Startin, J.R. (1989) Analysis of drug residues in tissue by combined supercritical-fluid extraction–supercritical–fluid chromatography–mass spectrometry–mass spectrometry. *J. Chromatogr.*, **464** 353-364.

Randall, L.G. and Wahrhaftig, A.L. (1981) Direct coupling of a dense (supercritical) gas chromatograph to a mass spectrometer using a supersonic molecular beam interface. *Rev. Sci. Instrum.*, **52** 1283-1295.

Rezanka, T. and Mareš, P. (1991) Determination of plant triacylglycerols using capillary gas chromatography, high-performance liquid chromatography and mass spectrometry. *J. Chromatogr.*, **542** 145-159.

Rezanka, T., Mareš, P., Hušek, P. and Podojil, M. (1986) Gas chromatography–mass spectrometry and desorption ionization mass spectrometry of triacylglycerols from the green alga *Clorella kessleri*. *J. Chromatogr.*, **355** 265-271.

Richter, B.E., Andersen, M.R., Knowles, D.E., Cambell, E.R., Porter, N.L., Nixon, L. and Later, D.W. (1988) Capillary supercritical fluid chromatography—use for the analysis of food components and contaminants, in *Supercritical fluid Extraction and Chromatography* (eds B.A. Charpentier and M.R. Sevenants), American Chemical Society, Washington, DC, pp 179-190.

Ruiz-Gutiérrez, V. and Barron, L.J.R. (1995) Methods for the analysis of triacylglycerols. *J. Chromatogr. B*, **671** 133-168.

Ryhage, R. and Stenhagen, E. (1960) Mass spectrometry in lipid research. *J. Lipid Res.*, **1** 361-390.

Sadoun, F., Virelizier, H. and Arpino, P.J. (1993) Packed-column supercritical fluid chromatography coupled with electrospray ionization mass spectrometry. *J. Chromatogr.*, **647** 351-359.

Schmeer, K., Nicholson, G., Zhang, S., Bayer, E. and Bohning-Gaese, K. (1996) Identification of the lipids and the ant attractant 1,2-dioleoylglycerol in the arils of *Commiphora guillaumini* Perr. (*Burseraceae*) by supercritical fluid chromatography–atmospheric pressure chemical ionisation mass spectrometry. *J. Chromatogr. A*, **727** 139-146.

Schulte, E., Höhn, M. and Rapp, U. (1981) Mass spectrometric determination of triglyceride patterns of fats by direct chemical ionization technique (DCI). *Fresenius Z. Anal. Chem.*, **307** 115-119.

Schulten, H.-R., Murray, K.E. and Simmleit, N. (1987) Natural waxes investigated by soft ionization mass spectrometry. *Z. Naturforsch.*, **42c** 178-190.

Schuyl, P.J.W., de Joode, T. Duchateau, G.S.M.J.E. and Vasconcellos, M.A. (1995) HPLC-MS analysis of triacylglycerides, in *Proceedings of the 43rd ASMS Conference on Mass Spectrometry and Allied Topics*, 21–26 May, Atlanta, GA, p 1143.

Smith, K.W., Perkins, J.M., Jeffrey, B.S.J. and Phillips, D.L. (1994) Separation of molecular species of *cis*- and *trans*-triacylglycerols in *trans*-hardened confectionery fats by silver-ion high-performance liquid chromatography. *J. Am. Oil Chem. Soc.*, **71** 1219-1222.

Smith, R.D. and Udseth, H.R. (1983) Mass spectrometry with direct supercritical fluid injection. *Anal. Chem.*, **55** 2266-2272.

Smith, R.D., Felix, W.D., Fjeldsted, J.C. and Lee, M.L. (1982a) Capillary column supercritical fluid chromatography/mass spectrometry. *Anal. Chem.*, **54** 1883-1885.

Smith, R.D., Fjeldsted, J.C. and Lee, M.L. (1982b) Direct fluid injection interface for capillary supercritical fluid chromatography–mass spectrometry. *J. Chromatogr.*, **247** 231-243.

Spanos, G.A., Schwartz, S.J., Breemen, R.B. and Huang, C.-H. (1995) High-performance liquid chromatography with light-scattering detection and desorption chemical-ionization tandem mass spectrometry of milk fat triacylglycerols. *Lipids*, **30** 85-90.

Spitzer, V., Tomberg, W., Hartmann, R. and Aichholz, R. (1997) Analysis of the seed oil of *Heisteria silvanii* (Olacaceae)—a rich source of a novel C_{18} acetylenic fatty acid. *Lipids*, **32** 1189-1200.

Sprecher, H.W., Maier, R., Barber, M. and Holman, R.T. (1965) Structure of an optically active allene-containing tetraester triglyceride isolated from the seed oil of *Sapium sebiferum*. *Biochemistry*, **4** 1856-1863.

Staby, A., Borch-Jensen, C., Balchen, S. and Mollerup, J. (1994) Supercritical fluid chromatographic analysis of fish oils. *J. Am. Oil Chem. Soc.*, **71** 355-359.

Stroobant, V., Rozenberg, R., Bouabsa, M., Deffence, E. and Hoffmann, E. (1995) Fragmentation of conjugate bases of esters derived from multifunctional alcohols including triacylglycerols. *J. Am. Soc. Mass Spectrom.*, **6** 498-506.

Taylor, D.C., Giblin, E.M., Reed, D.W., Hogge, L.R., Olson, D.J. and MacKenzie, S.L. (1995) Stereospecific analysis and mass spectrometry of triacylglycerols from *Arabidopsis thaliana* (L.) Heynh. Columbia seed. *J. Am. Oil Chem. Soc.*, **72** 305-308.

Taylor, D.C., Weber, N., Barton, D.L., Underhill, E.W., Hogge, L.R., Weselake, R.J. and Pomeroy, M.K. (1991) Triacylglycerol bioassembly in microspore-derived embryos of *Brassica napus* L. cv Reston. *Plant Physiol.*, **97** 65-79.

Thomas, D., Sim, P.G. and Benoit, F.M. (1994) Capillary column supercritical fluid chromatography/mass spectrometry of polycyclic aromatic compounds using atmospheric pressure chemical ionization. *Rapid Commun. Mass Spectrom.*, **8** 105-110.

Tillman-Sutela, E., Johansson, A., Laakso, P., Mattila, T. and Kallio, H. (1995) Triacylglycerols in the seeds of northern Scots pine, *Pinus sylvestris* L., and Norway spruce, *Picea abies* (L.) Karst. *Trees*, **10** 40-45.

Tyrefors, L.N., Moulder, R.X. and Markides, K.E. (1993) Interface for open tubular column supercritical fluid chromatography/atmospheric pressure chemical ionization mass spectrometry. *Anal. Chem.*, **65** 2835-2840.

Valeur, A., Michelsen, P. and Odham, G. (1993) On-line straight-phase liquid chromatography/plasmaspray tandem mass spectrometry of glycerolipids. *Lipids*, **28** 255-259.

Van Oosten, H.J., Klooster, J.R., Vandeginste, B.G.M. and De Galan, L. (1991) Capillary supercritical fluid chromatography for the analysis of oils and fats. *Fat Sci. Technol.*, **93** 481-487.

Wakeham, S.G. and Frew, N.M. (1982) Glass capillary gas chromatography–mass spectrometry of wax esters, steryl esters and triacylglycerols. *Lipids*, **17** 831-841.

Wright, B.W., Kalinoski, H.T., Udseth, H.R. and Smith, R.D. (1986) Capillary supercritical fluid chromatography–mass spectrometry. *J. High Resolut. Chromatogr. Chromatogr. Commun.*, **9** 145-153.

7   Gas chromatography-mass spectrometry of lipids

John A.G. Roach, M.P. Yurawecz, M.M. Mossoba
and K. Eulitz

7.1   Introduction: the process of GC-MS analysis

Mass spectrometry examines the gaseous ion chemistry of the lipid
substance rather than the physical properties of the lipid that are
nondestructively studied by infrared, nuclear magnetic resonance,
ultraviolet and X-ray spectroscopy (Murphy, 1993). The information in
a mass spectrum is in the ionized pieces of the molecule that was
destroyed to provide clues to its molecular structure and identity. The
interpretation of a mass spectrum is based on knowledge that
the pathways of ion decomposition mirror the chemistry involved in
the formation of the molecule. The salient feature of mass spectrometry is
that the sensitivity of the technique often provides considerable
information from very little material.

Lipids constitute several groups of compounds (Mangold, 1984)
including free fatty acids (FFA), waxes (n-alkyl esters), mono-, di-, and
triacylglycerides (TG), phospholipid esters (choline, serine, inositol),
sterols, sterol esters, tocopherols, ether lipids, fatty acid polymers, and
smaller molecules such as aldehydes, ketones, di-acids, and alkanes. Gas
chromatography–mass spectrometry is often used to identify derivatives
of various lipid components including the fatty acid portions of lipid
material after suitable clean-up. Confirmation and quantitation by gas
chromatography–mass spectrometry are possible for known components.

A short chapter on the topic of the gas chromatography–mass
spectrometry of lipids is not the place adequately to reference the
majority of the techniques or researchers involved in lipid research.
Practical approaches to lipid analysis (Hamilton and Hamilton, 1992)
and advances in lipid methodology (Christie, 1992) are recurrent themes
in texts on lipids that reflect the perspectives and experiences of the
authors. The material in this chapter is drawn from the current work of
the authors on the identification of lipids and lipid oxidation products in
foods. A general approach is presented for the acquisition and
interpretation of GC and MS data. Examples are provided for the use
of 4,4-dimethyloxazoline derivatives in the structural elucidation of fatty
acid species by mass spectrometry.

Gas chromatography–mass spectrometry (GC-MS) as it is used today
is the merger of the analytical techniques of gas chromatography, mass

spectrometry and high-speed electronic data processing. Each technique was flourishing on its own before their combination into a single automated instrument. The instrument is a 'hyphenated' analytical tool that makes it difficult to discuss separately GC, MS, data processing, or the resulting information without an awareness of the close interdependence of the component parts of the system. The interdependence within the system is more apparent in a stepwise examination of the analytical process depicted in Figure 7.1.

7.1.1   Analyte preparation

Analyte preparation will generally provide an ester or another functional group at the carboxylic portion(s) of the molecule. In order to effect this derivatization, the use of acid or base is required. In some instances, this reaction also alters other functional groups in the molecule. Batna and Spiteller report intramolecular aldol condensation under alkaline conditions and formation of isomeric species by recyclization with introduction of a conjugated double bond under acidic conditions in their study of the oxidation products of tetra-alkyl-substituted furan fatty acids (Batna and Spiteller, 1994).

Acid-catalyzed methylation procedures adversely affect analysis for conjugated *cis* and *trans* isomers of unsaturated fatty acids in several ways. First, conjugated *cis,trans* fatty acids are isomerized to *trans,trans* isomers during acid-catalyzed methylation (Kramer *et al.*, 1997). The conversion of functional groups in other molecules to form artifacts of the intended analyte is a second problem. A comparison of acid- and base-catalyzed methylation procedures in analyses for conjugated linoleic acid (CLA) isomers led to a determination that acid-catalyzed methylation converted allylic hydroxy oleates to CLA (Yurawecz *et al.*, 1994). A third problem is loss of intended analyte through conversion to artifacts during analyte preparation. Allylic methoxy artifacts are formed from conjugated dienes by acid-catalyzed methylation of CLA (Kramer *et al.*, 1997).

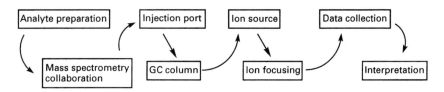

Figure 7.1   Flow chart of the process of GC-MS analysis of a lipid sample.

7.1.2   Mass spectrometry collaboration

The task of getting useful data from a test portion is a collaborative effort in which good communication is essential. A mass spectrometer is an information-rich detector that must be appropriately set up in order for an analysis to provide useful data. If you want GC-MS information, share your GC conditions with the mass spectrometrist so that they can be adapted to the GC-MS analysis. Identify the origins of your test portions and explain the purpose of your analytical request. After examining the initial GC-MS results, you may find that other features of the GC-MS data may have greater significance than the original GC peak of interest, so be prepared to change your request.

The GC-MS analysis is an iterative process adjusted by all of the chemists involved in the work to obtain optimal results. For example, in a study of the oxidation products of conjugated linoleic acid and furan fatty acids (Yurawecz et al., 1997), test portions mixed with methyl stearate internal standard are placed in capped test tubes in the presence of air and heated in an oil bath. Individual tubes are removed from the bath and examined by GC with flame ionization detection (FID) to monitor the progress of the oxidation reaction. The portions submitted for GC-MS analysis are drawn from the starting material and tubes thought to contain components of interest based on the FID data. The first GC-MS analyses locate the known compounds in the test portions, tentatively identify some components, and determine the range of molecular masses present in the reaction mixtures. At this point, the team examines the data and discusses what other information would be useful. The mass spectrometrist then uses the locations of the identified compounds to mark retention time windows for other components of interest. The molecular mass information is used to select mass scan parameters to acquire data in the mass range of interest and the GC-MS analyses are repeated as necessary until the mass spectrometrist is satisfied that the MS data sufficiently characterize all of the components of interest.

7.1.3   The injection port

A GC-MS analysis begins with an injection. For some compounds, the injection port is the last step in their journey to the mass spectrometer. For example, allylic hydroxyoleates may convert to CLA when injected into a gas chromatograph (Yurawecz et al., 1997). Hot metal surfaces are frequently the culprit in such reactions (Morris et al., 1960), so all-glass systems are suggested for GC-MS analyses. Figure 7.2 lists some compounds that are reported (Ishii et al., 1988; Morris et al., 1960;

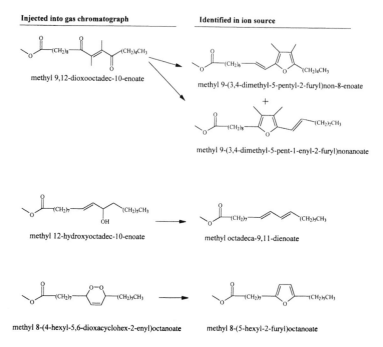

Figure 7.2 Some compounds that decompose during gas chromatography and the resulting products detected by mass spectrometry. Data are from the reports of Bascetta *et al.* (1984), Ishii *et al.* (1988) and Morris *et al.* (1960).

Bascetta *et al.*, 1984) to decompose thermally in the injection port. Alternative modes of chromatographic introduction (liquid chromatography (LC), supercritical fluid chromatography (SFC)), use of a cold, on-column injector, or analyte introduction by direct insertion probe with MS/MS separation may be used to study thermally labile species.

Methyl 9,12-dioxooctadec-10-enoate was tentatively identified in an MS examination of the singlet oxidation products of 9,12-epoxy-9,11-octadecadienoic acid by direct insertion probe introduction. Probe introduction was used to avoid possible thermal alteration of this suspected product in the gas chromatograph (Ishii *et al.*, 1988). Additional products were identified by GC-MS (Sehat *et al.*, 1998).

Our GC-MS system is used for a variety of analyses; for those lipid-related compounds that do gas chromatograph intact, the first concern is to return the Hewlett Packard 5890 series II gas chromatograph to an initial setup for lipid analysis. The high vacuum of the mass spectrometer aspirates material through the GC column. This effect increases the apparent linear velocity of the column compared to a column with its exit at atmospheric pressure. The pull of the vacuum system requires that GC conditions observed with FID must be adapted for GC-MS.

Modern gas chromatographs equipped with electronic pressure control compensate for increasing resistance to flow with increasing column temperature and for the pull of a vacuum system on the end of the column. Direct comparisons of electronic and analog pressure control for GC-MS analyses in this laboratory have found that the most visible improvements for GC-MS are reduced run times and narrower peaks for late-eluting components. The suggestions presented here are for analog control of the GC pressure.

We use the split/splitless injector in the splitless mode for most injections because the injector tolerates a wider range of injection volume than the on-column injector and other members of the team do not have access to an on-column injector. The glass injection port liner is a 4 mm i.d. double gooseneck (Hewlett Packard 5181-3315, Palo Alto, CA, USA). Minor peak tailing or analyte loss will occur if the end of the capillary column is not properly positioned near the exit of the glass injection liner. Use a graphite/vespel ferrule and experimentally determine the best position for the column before the ferrule fuses to the column. Fusion of the ferrule to the column in the optimal position simplifies subsequent placement of the column in the injector. The helium carrier gas head pressure is set at 14 psi. The septum purge flow is set at 6–$10\,ml\,min^{-1}$ and the split purge is set at $100\,ml\,min^{-1}$. For a $50\,m \times 0.25\,mm$ i.d. capillary column, this will provide a measured linear velocity of around $35\,cm\,s^{-1}$ with helium carrier gas for an injection of argon gas with the column oven set at an initial temperature of 75°C.

The 14 psi head pressure approximates the separation obtained by GC with FID at a head pressure of 30 psi. If the purpose of the analysis requires optimal chromatographic separation, vary the head pressure and examine the separation and peak shape. Optimal flow provides the best chromatographic peak shape as well as the best separation.

7.1.4   The GC column

Constantly changing the capillary column to analyze various types of analytes by GC-MS provides multiple opportunities for water and air to damage the column. Gas fittings should be carefully assembled and checked for air leaks. Use the best grade of helium that you can obtain, clean it with a heated gettering metal gas purifier (Supelco 2-3800-U, Bellefonte, PA, USA) to remove traces of water and oxygen and then patiently outgas a newly installed column before heating it.

The end of the capillary column should pass through the heated interface of the mass spectrometer to the edge of the ion source. Maintain the interface temperature at not much more than the final temperature of the oven temperature program during use, and reduce the interface temperature when the system is not in use to prolong the life of the column.

Gas chromatograph–mass spectrometer instruments are purchased as systems rather than as parts. The gas chromatograph of our Finnigan TSQ-46 GC-MS system was designed for packed-column chromatography. Its injection ports and MS interface included provisions for packed and capillary columns that compromised its performance. The Micromass Autospec Q high-resolution mass spectrometer was coupled to a Hewlett Packard Series II capillary GC by a simple heated tube that provided a capillary column with a straight line of entry from the GC oven to the ion source. We chose to replace the gas chromatograph of the TSQ-46 GC-MS with a Hewlett Packard Series II gas chromatograph and to convert the GC inlet of the Finnigan movable transfer line assembly (Finnigan 40006-60320, San Jose, CA, USA) into a heated tube. The changes resulted in an efficiently heated line-of-sight interface that permitted placement of the end of a GC column at the edge of the ion source (Roach, 1998). The success of the modification was exemplified by the GC-MS separation of *trans* monounsaturated C_{18} fatty acid positional isomers described as in section 7.3.2. The separation was developed with the TSQ-46 using multiple ion detection. The column was then transferred to the Autospec Q and the isomers were identified by their full mass spectra (Mossoba *et al.*, 1997).

Chromatographic reproducibility is of paramount importance in the characterization of unknown mixtures. The work starts by locating known compounds and relies on reproducible chromatography to find other compounds. The initial low-resolution MS data are supplemented with accurate mass measurements to determine the elemental compositions of ions of interest. Switching from low-resolution to high-resolution MS reduces the intensity of the analyte signal and limits the portion of the mass spectrum that can be recorded during a single GC-MS analysis. In order to obtain high-resolution data over a mass range of 300 daltons, the mass range is broken down into three or four smaller windows that are examined in separate GC-MS analyses. This would be a taxing task in the absence of reproducible chromatography.

7.1.5   The ion source

Ions are generated by electron ionization (EI) in an ion source by bombarding gaseous molecules with 70 eV electrons. Loss of an electron to form a radical cation through the close approach of a 70 eV electron deposits some additional kinetic energy in the radical cation. The energy range associated with chemical bonds is on the order of 10 eV. Kinetic energy in excess of what is required to dissociate a typical chemical bond may be imparted to the newly formed radical cation during ionization. If this singly-charged positive ion can accommodate the additional energy obtained during ionization, it will be detected as the molecular $[M]^{+\cdot}$ ion and will appear as the highest mass ion in the positive ion EI mass spectrum of that substance (Hamilton and Hamilton, 1992). If the molecular ion cannot reduce its newly acquired excess internal energy through collisions with other molecules or accommodate it within permissible vibrational levels, it flies apart to provide additional degrees of freedom in which to disperse the energy. One of the pieces has the positive charge resulting from the loss of the electron. This positive ion can be directed by the electrostatic and magnetic fields of the mass spectrometer to the detector. Neutrals, radical species and negative ions produced by fragmentations in the ion source can only be inferred from the EI data provided by the detected positive ions because they are not positively charged species.

A positive charge equilibrates to favorable sites such as the non-bonding orbitals of oxygen and π electron systems. A site of charge localization is one way to account for the rearrangements and cleavages that are observed in a mass spectrum because breakage of an adjacent bond is considered to arise by electron movement towards the positive charge. The lowest ionization potential of any electron in a saturated fatty acid methyl ester is a non-bonding electron on the carboxyl oxygen atom, making this a good site for localization of the positive charge in the molecular ion. The mass spectra of saturated fatty acid methyl esters suggest an orderly process of successive losses from the alkyl chain and rearrangement ions indicative of the carbomethoxy functionality. However, transfer of a hydrogen atom from a remote site in the molecular ion to the carbonyl oxygen moves the radical site from the oxygen to the chain, facilitating isomerization as well as fragmentation. Detailed labeling studies revealed fragmentation of the molecular ion via several competing pathways including expulsions of intermediate portions of the chain and radical site-driven alkyl migration to the carbonyl oxygen (McCloskey, 1970).

The usual ions observed for a saturated fatty acid methyl ester include the molecular ion, loss of a methoxyl radical ($M - 31$), a γ hydrogen rearrangement to the carbonyl oxygen with cleavage β to the carbonyl to form the most abundant ion (base peak) with a mass-to-charge ratio (m/z) of 74, and a series of ions at m/z ($59 + 14n$) typically starting with m/z 87. Deuterium labeling on carbon atoms 2, 3, 4 and 6 (Dinh-Nguyen, 1961) has revealed extensive hydrogen interchange principally between position 2 and positions 5, 6, 7. Labeling data also show that ions of m/z 87, 101, 115 and 129 arise from systematic cleavage of the chain, while

143 and higher members of the series are formed by expulsion of intermediate portions of the chain plus one hydrogen.

Esters with functional groups in the chain afford additional sites for localization of the charge that compete with charge-driven and radical-driven fragmentation pathways derived from initial placement of the charge on the carbomethoxy oxygen. The presence of an oxygen atom in the fatty acid chain has a dramatic influence on the EI mass spectrum (Murphy, 1993). A keto function will reveal its presence with adjacent α cleavages of the chain directed by localization of the charge on the keto oxygen.

A lack of the usual ions observed for saturated fatty acid methyl esters and the appearance of ions characteristic of a functional group may suggest the presence of that functionality, but a determination by MS of the location of the group in the chain is not assured. For example, the location of a double bond in the chain cannot be determined from the EI mass spectrum of an unsaturated fatty acid methyl ester because of the tendency of the molecular ion to isomerize.

Location of the charge on the oxygen permits the loss of stable neutral molecules such as water to move the charge site onto the alkyl chain, where isomerization is more facile than fragmentation (Gross, 1992). The reaction of the fatty acid group with nitrogen-containing compounds to form picolinyl, piperidyl, and morpholinyl esters or pyrrolidides, triazolopyridines, 2-alkenylbenzoazoles and 2-alkenyl-4,4-dimethyloxa-zolines shifts the preferred site of ionization to a nitrogen atom. Unlike oxygen, the nitrogen atoms of the derivatives do not tend to leave the ion through formation of neutral species. This affects the resultant EI spectra in two ways: (1) preferential ionization at the nitrogen reduces the likelihood of ionization of unsaturated systems in the hydrocarbon chain, resulting in less bond migration along the hydrocarbon chain; (2) fragmentation at a radical site far removed from the location of the charge becomes a preferred fragmentation pathway.

Retention of the charge on the nitrogen atom allows hydrogen atoms to migrate to the charge site. The probability of migration to the nitrogen atom for any hydrogen in the chain is thus equal to the probability of a fragmentation arising from the radical site formed at the former location of the hydrogen in the chain (Spitzer, 1997).

The radical site-directed fragmentation simplifies interpretation of the observed spectrum because it tends to be more structurally specific, but it does not entirely preclude the possibility of isomerization. For example, shown in Figure 7.20 is the EI mass spectrum of a GC peak recorded in a study of cyclic fatty acid monomer structures described later in this chapter. The EI mass spectrum contains features that are consistent with the isomeric structures 2-[8-(2-prop-*trans*-1-enyl-cyclohex-*cis*-1-enyl)oc-tyl]-4,4-dimethyloxazoline (Figure 7.20, top) and a corresponding structure containing a cyclohexadienyl ring with an *n*-propyl substituent (Figure 7.20, bottom). It cannot be stated with certainty whether the data represent the coelution of these compounds or isomerization of a molecular ion in the ion source, but GC–infrared analysis of the same material showed only a *cis* double bond in a cyclohexenyl ring and a *trans* double bond on a hydrocarbon chain at the retention time of this GC peak (Mossoba *et al.*, 1995).

7.1.6   *Ion focusing*

Some ions formed in the ion source may decompose beyond the source to produce metastable ions that are useful clues that may support a proposed fragmentation. Not all instruments can detect a metastable ion, but there are MS/MS instruments that can collisionally dissociate ions to yield a product ion spectrum that can be used to determine the possible origins or fragmentations of selected ions. A third aid to the interpretation of mass spectra is the determination of the elemental compositions of ions of interest with a high-resolution mass spectrometer. The interpretations of spectra in this text are largely based on knowledge of the elemental compositions of the ions in the spectra obtained through accurate mass measurements with a high-resolution instrument.

The mass analyzer portion of a mass spectrometer separates ions according to their mass-to-charge ratios to produce a mass spectrum. The separation between adjacent ions in a mass spectrum is described as the instrument resolution used to obtain that mass spectrum. An instrument operating at unit resolution is able to detect separately ions differing in mass by one mass unit. Thus an instrument need only operate at a resolution of 278 to separately detect ions with mass-to-charge ratios of 277 and 278. A high-resolution instrument operating at more than unit resolution can distinguish between ions of the same nominal molecular mass that differ in elemental composition. For example, dibutyl phthalate ($C_{16}H_{22}O_4$) and sulfamethazine ($C_{12}H_{14}N_4O_2S$) have the same nominal molecular mass of 278 but differ in their exact molecular masses (278.1518 and 278.0839) because of the differences in their elemental compositions. The resolution required to detect these two molecular ions separately is $(278)/(278.1518 - 278.0839) = 4094$. Low-resolution data are typically acquired at a resolution of 1000 for compounds with molecular masses of less than 500. Accurate mass measurements in this laboratory are recorded at a resolution in excess of 10 000.

The resolution of the high-resolution instrument is increased by limiting the energy spread in the ions and by restricting divergence of the ions from an optimal path through the instrument. The trade-off in using this limited set of ions for accurate mass measurement is a corresponding loss in signal intensity. Changing from a resolution of 1000 to a resolution of 10 000 decreases the intensity of the measured ion beam to 10% of its value at 1000 in an Autospec Q mass spectrometer.

A technique known as voltage scanning is used to scan rapidly a small portion of the mass range at high resolution in order to acquire rapidly accurate mass data for the eluate during a GC-MS analysis. Small windows of accurate mass data are recorded in successive GC-MS analyses in order to obtain data for all of the ions in the mass spectra of the compounds in the test portion.

7.1.7   Data collection

An ion current that is greater than the maximum signal that can be measured by the detector is a saturated signal. Acquisition of saturated signals is not productive. Background subtraction of low-resolution saturated signals will not provide representative spectra. Saturated high-resolution signals cannot be used to determine accurately the centroid of an ion.

Chromatograms for GC-MS analyses are derived from the ion intensity data recorded for each scan of the GC eluate. Hence, the scan rate influences the appearance of the chromatograms in the same way as the sampling rate of any other type of chromatographic detector. Low- and high-resolution data acquired at a scan rate of 0.8 seconds per scan provide chromatograms that compare favorably with flame ionization detector (FID) data.

Chromatograms in the form of ion profiles are often the first clues to the classes of compounds present in a test portion. A reconstructed ion profile for m/z 74 will contain a peak at the retention time of every saturated fatty acid methyl ester detected during the analysis. Use known or proposed ions of mass spectra of target compounds to search for the retention time(s) and then subtract signal adjacent to a chromatographic response to obtain its mass spectrum.

Automated routines for subtracting signals contributed by adjacent chromatographic components do exist, but they are not infallible, so a close examination of ion profiles for some ions in the spectrum is in order. The ions to check are of two types: those you may use to interpret the spectrum and those that seem out of place in the spectrum. Once you have winnowed the list of suspect ions down to those that co-elute at the retention time of the compound of interest, interpret its spectrum.

7.1.8   Interpretation

Fatty acids and their derivatives were among the first compounds studied extensively by mass spectrometry. A wealth of information exists regarding the behavior of these compounds in a mass spectrometer. The EI mass spectra of many compounds associated with the methyl esters of fatty acids are available in mass spectral databases. A library search, or automated comparison of an unknown spectrum against a database, can frequently provide spectra of similar compounds. A range of chain lengths and functionalities is usually present in a mixture of reaction products. Any library spectra of short-chain congeners can provide a starting point for the tentative identification of longer-chain members of the same class of compounds.

The purchase of a library search capability with an instrument usually includes one or more databases of spectra and the ability to create a custom library of spectra for known compounds that are frequently encountered by the analyst. Creation of a custom library simplifies subsequent lipid analyses, but a library search is a starting point rather than an ending point in the identification of compounds that are not in any available database.

Prior reviews, chapters and texts on the mass spectrometry of lipids are the foundation for current reports in the primary literature. McCloskey's summary of information in his 1970 chapter (McCloskey, 1970) is still relevant to an understanding of the mass spectrometry of fatty acid derivatives. An earlier work by Budzikiewicz et al. on the mass spectrometry of organic compounds (Budzikiewicz et al., 1967) is still a useful resource in any mass spectrometry laboratory. For example, Figure 7.3 shows the EI mass spectra of methyl 8-oxo-9,12-epoxy-9,11-octadecadienoate (Figure 7.3a) and methyl 13-oxo-9,12,-epoxy-9,11-octadecadienoate (Figure 7.3b) tentatively identified as oxidation products of 9,12-epoxy-9,11-octadecadienoic acid (Sehat, 1998).

The structures of these isomeric compounds are such that competing processes may give rise to isobaric products. Distinguishing between the two isomers requires an understanding of the origins of the fragment ions that are unique to each structure. Many of the abundant ions in the mass spectra appear to arise from localization of the charge on the keto and furan functions. Budzikiewicz et al., suggest that the distinctive processes that may occur at these functionalities are: (a) cleavage α to the carbonyl, (b) cleavage β to the ring, (c) γ hydrogen rearrangements with cleavage β to the carbonyl, and (d) cleavage γ to the carbonyl where the driving force may be allylic bond rupture in the enol form or formation of a cyclic oxonium ion (Budzikiewicz et al., 1967).

Cleavage of the carbon chain of the 8-oxo isomer α to the carbonyl and cleavage β to the ring of the 13-oxo isomer gives rise to isobaric ions. The possible ion structures for m/z 179 are (i) for the 8-oxo isomer and (ii) for the 13-oxo isomer (Figure 7.4). Cleavage of the carbon chain α to the carbonyl would give rise to m/z 251 for the 13-oxo isomer but cleavage β

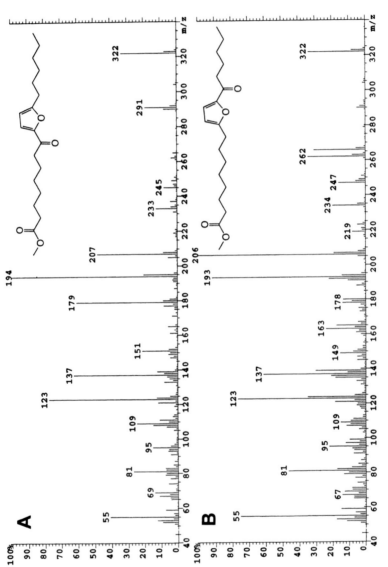

Figure 7.3 EI mass spectra of (a) methyl 8-oxo-9,12-epoxy-9,11-octadecadienoate and (b) methyl 13-oxo-9,12-epoxy-9,11-octadecadienoate.

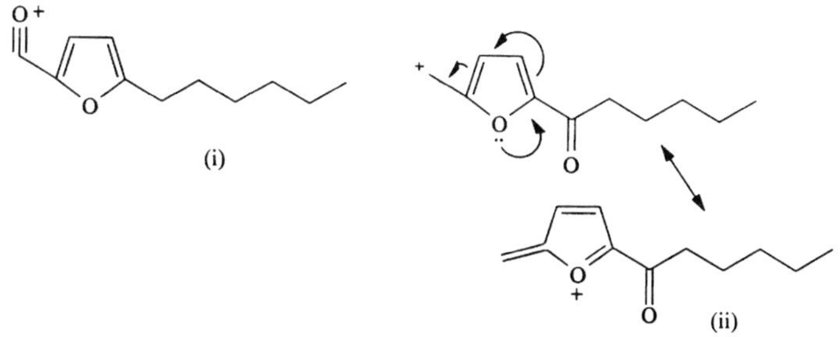

Figure 7.4 Possible ion structures for m/z 179 arising from (i) cleavage α to the carbonyl of the 8-oxo isomer, and (ii) cleavage β to the ring of the 13-oxo isomer.

to the ring of the 8-oxo isomer would also give rise to m/z 251. It is interesting to note that in these spectra, and the spectra of similar furan fatty acid methyl esters containing a furan or a keto function in the chain, fragment ions arising from cleavages α to a carbonyl or β to the ring that retain the ester function are of low relative abundance. This may be due to subsequent decompositions of these ions to other products or to the preferred loss of the ester portion as the larger alkyl group to form an ion such as m/z 179 or 251. Thus, it is possible that cleavage of the carbon chain α to the carbonyl in the 8-oxo isomer may result in the most abundant ion of this set, but it is still necessary to consider other prominent ions in the spectra to distinguish the two isomers.

An elemental composition of $C_{12}H_{17}O_2$ for m/z 193 in Figure 7.3b is consistent with cleavage of the carbon chain of the 13-oxo isomer γ to the furan ring. Hydrogen rearrangement to the furan oxygen with radical-directed dissociation of the bond between carbons 6 and 7 of the chain may yield m/z 193 (Figure 7.5a). Gamma hydrogen rearrangement to a carbonyl oxygen with cleavage β to the carbonyl can occur at the ester or keto function. This process at the ester carbonyl is the origin of m/z 74 in the EI mass spectra of the methyl esters of saturated fatty acids. Gamma hydrogen rearrangement with cleavage β to the carbonyl at the keto function will give rise to m/z 194 for the 8-oxo isomer (Figure 7.5b) and m/z 266 for the 13-oxo isomer (Figure 7.5c).

Gamma hydrogen rearrangement in the molecular ion and neutral losses of the elements of methanol and carbon monoxide may account for the presence of m/z 290 and 262 in the spectra. Corresponding losses from the ion at m/z 266 for the 13-oxo isomer would result m/z 234 and 206 in Figure 7.3b.

For the 8-oxo compound, the driving force for cleavage γ to the ketone may correspond to allylic bond rupture in an enol form to give rise to ion structure (**iii**) (Figure 7.6), or to the formation of a cyclic oxonium ion

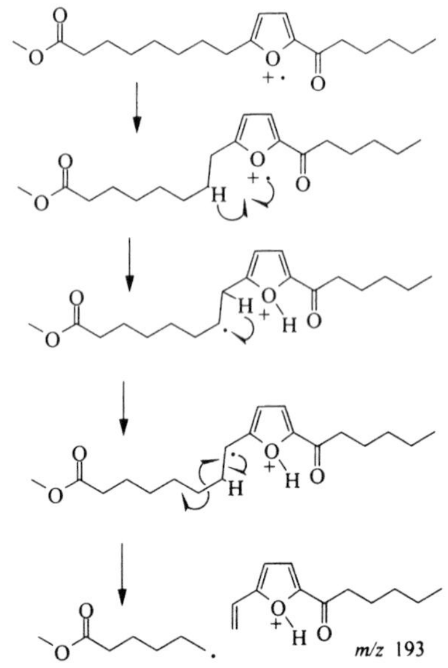

Figure 7.5 Proposed fragmentations consistent with (a) m/z 193 in Figure 7.3b, (b) m/z 194 in Figure 7.3a, and (c) m/z 266 in Figure 7.3b.

such as ion structure (**iv**). Formation of a cyclic oxonium ion at the furan oxygen would correspond to ion structure (**v**). All of these possible ion structures have an identical elemental composition and m/z of 207. The corresponding ions for the 13-oxo isomer would have m/z of 279.

The influence of a furan with an adjacent carbonyl function in the alkyl chain was examined further by a comparison with 5-hexylfuran-2-carboxyaldehyde. High-resolution data were acquired for all ions in the EI mass spectrum of this compound. Hypothetical ion structures were proposed from the elemental composition data and used to devise corresponding ion structures for the 8-oxo and 13-oxo isomers (Figure 7.7). In some cases, the structures devised were in agreement with previously proposed structures. In the case of m/z 207, an additional ion structure was suggested for the 13-oxo isomer by the comparison.

On the basis of these data, the mass spectrum in Figure 7.3a is tentatively identified as that of the 8-oxo isomer and the mass spectrum in Figure 7.3b is tentatively identified as that of the 13-oxo isomer. Like all identifications based on an interpretation of a mass spectrum, final proof of structure must rely on an analytical comparison with authentic reference material.

(b)

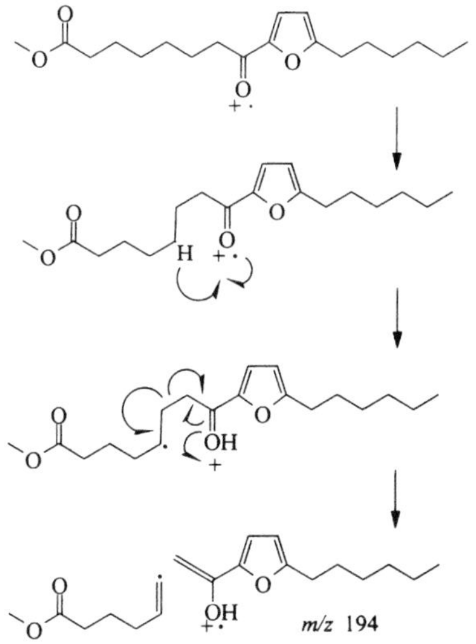

(c)

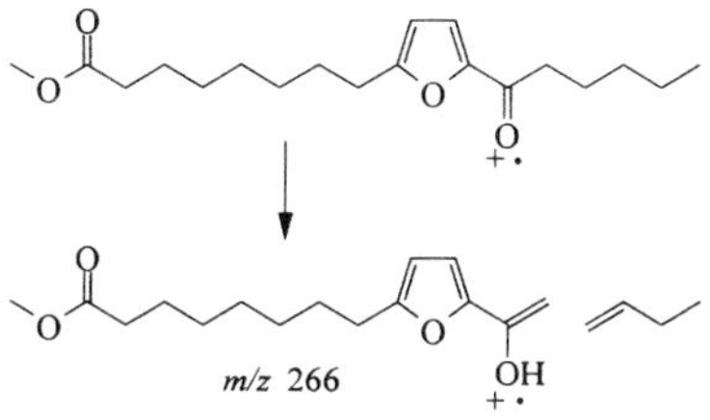

Figure 7.5   (Continued).

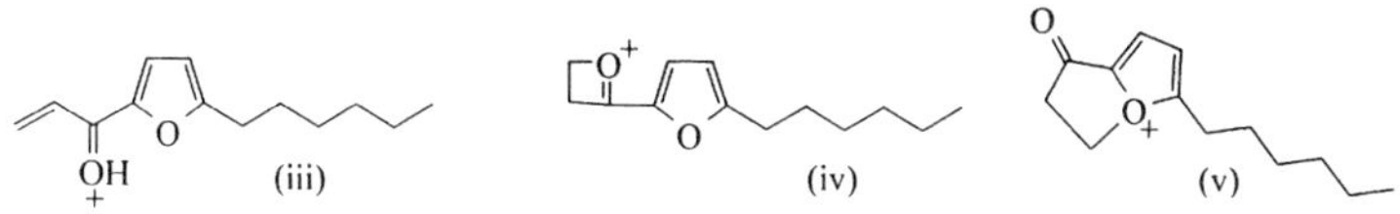

Figure 7.6   Possible ion structures for m/z 207 derived for methyl 8-oxo-9,12-epoxy-9,11-octade-cadienoate arising by (iii) allylic bond cleavage of enol form, (iv) cyclization to keto oxygen, (v) cyclization to furan oxygen.

7.2   Mass spectrometry of 4,4-dimethyloxazoline derivatives of fatty acids

The GC-MS of lipids has been reviewed in detail by Spitzer (1997). The methyl esters are of limited utility in the determination of the structure of unsaturated fatty acids by MS owing to the migration of multiple bonds in the ion source of the mass spectrometer (Bieman, 1962). The 2-alkenyl-4,4-dimethyloxazoline (DMOX) derivative localizes the positive charge

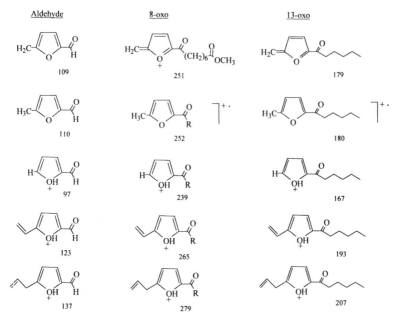

Figure 7.7 Hypothetical product ion structures for the 8-oxo and 13-oxo isomers modeled after corresponding structures based on elemental composition data for ions in mass spectrum of 5-hexylfuran-2-carboxyaldehyde.

on the nitrogen, thereby reducing bond migration. Included here is a general approach to the identification and examination of DMOX derivatives of the fatty acid moieties found in lipids.

7.2.1   Preparation of 4,4-dimethyloxazoline derivatives

The 4,4-dimethyloxazolines (DMOX) were first introduced by Zhang *et al.* in 1988 (Zhang *et al.*, 1988a). These derivatives are readily prepared by condensation of the free fatty acids with an excess of 2-amino-2-methyl-propanol (AMP). They gas chromatograph as well as or better than the fatty acid methyl esters (FAMEs). Elution temperatures are approximately 10°C higher than those for FAMEs and the gas chromatographic resolution is comparable or better. The DMOX EI mass spectra are more structurally specific than the EI mass spectra of FAMEs. Therefore, DMOX derivatives are widely used to elucidate the positions of multiple bonds, alkyl branching, hydroxy and oxo functions and the presence and position of cyclopropane and cyclopropene groups (Spitzer *et al.*, 1997).

There are several procedures for preparing DMOX derivatives from the free fatty acids (Figure 7.8). Zhang *et al.* (1988a) suggest heating one part free fatty acid with five parts of 2-amino-2-methyl-propanol at 170–180°C for 0.5–1 hour to obtain a greater than 95% yield. A variation of this procedure using a 2-hour reaction time at 210°C has been proposed (Liebich *et al.*, 1994) that yields even less remaining free acids. The free fatty acids can be mixed with equal parts of dicyclohexylcarbodiimide and AMP in dichloromethane at room temperature for 1–4 hours, followed by a treatment with thionylchloride for 0.5–1 hour to avoid heating (Yu *et al.*, 1991).

DMOX derivatives can also be obtained from FAMEs by adding an excess of AMP to the FAME mixture and heating overnight at 180°C (Fay and Richli, 1991). It is our experience (Mossoba *et al.*, 1994) that it is possible successfully to reduce the heating time to 6 hours. Garrido and Medina (1994) describe the direct conversion of fatty acids from the total lipids into their DMOX derivatives by heating the lipids with excess AMP for 18 hours at 180°C.

7.2.2   Interpretation of fatty acid DMOX mass spectra

In most mass spectra of DMOX derivatives, ions at m/z 113 and m/z 126 dominate and one of these is the base peak. Both ions are directly related to the DMOX group with 113 formed by a McLafferty rearrangement and 126 by a cyclization-displacement reaction as shown in Figure 7.9.

In addition to these two ions, there is an ion series starting at m/z 126 of m/z $126 + 14n$ for saturated straight-chain fatty acids containing n additional methylene groups. The peaks at m/z 140 and m/z 154 are typically weak and are followed by a more intense ion at m/z 168. The higher mass ions in the series decrease steadily in intensity with increasing m/z, the $M - 15$ ion often shows a local maximum (Figure 7.10).

Departures from these general rules for saturated straight-chain fatty acids are indicators for multiple bonds or alkyl, hydroxyl or oxo substituents along the methylene chain. A listing of some of the more specific differences reviewed by Spitzer (1997) follows.

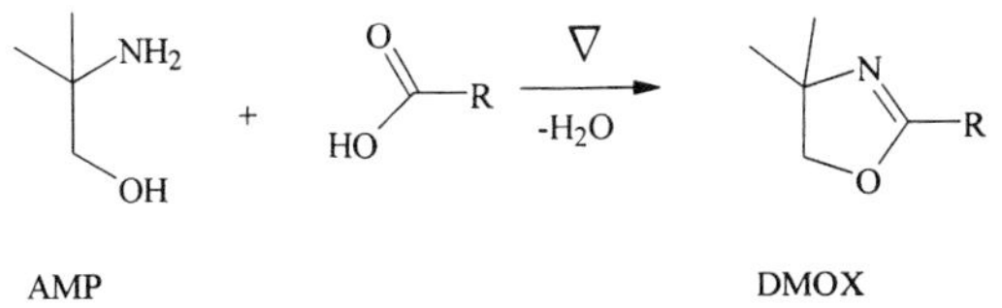

Figure 7.8   Preparation of 4,4-dimethyloxazoline derivatives.

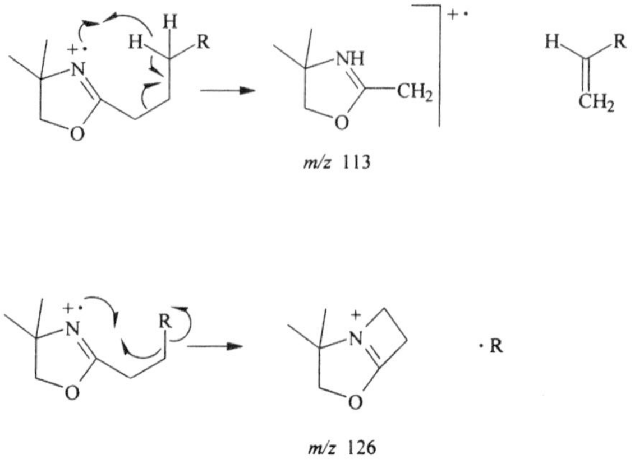

m/z 113

m/z 126

Figure 7.9 Formation of *m/z* 113 by McLafferty rearrangement and *m/z* 126 by cyclization–displacement.

7.2.2.1 Spectra with an unusual intensity distribution

An unusual intensity distribution in the m/z $126 + 14n$ series might indicate branching. The peak representing the cleavage at the branched point (m/z X) is weak and surrounded by more intense peaks m/z X − 14 and m/z X + 14, respectively. Using these guidelines even dimethyl and trimethyl branched fatty acids have been identified (Yu *et al.*, 1988).

7.2.2.2 Spectra with 12-dalton gaps

A 12-dalton (Da) gap between two peaks in the usual m/z $126 + 14n$ series indicates a double bond equivalent. Unfortunately, this can be a double bond or a cyclopropane ring. Consideration of the intensity of the surrounding peaks can provide valuable information to distinguish between them. When the 12 Da difference occurs between the peaks belonging to ions with $n − 1$ and n carbon atoms, the intensity of the peaks representing fragments with $n − 2$, $n + 2$ and $n + 3$ carbon atoms can be used.

   These three comparably intense ions (relative to the pair separated by 12 Da) suggest the presence of a double bond (Zhang *et al.*, 1988a). The relatively intense ions representing $n − 2$ and $n + 2$ ions are due to allylic cleavages. On the other hand, when only the $n + 3$ peak is intense and those for the $n − 2$ and $n + 2$ fragments are comparable in intensity to those with $n − 1$ and n carbons, the presence of a cyclopropane ring is suggested (Zhang *et al.*, 1987). In both cases, the double bond or the cyclopropane ring is located between the n and $n + 1$ positions on the fatty acid chain.

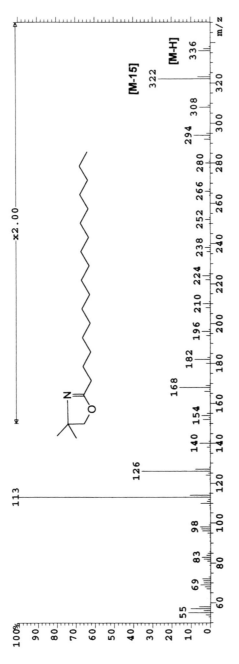

Figure 7.10 EI mass spectrum of the DMOX derivative of 18:0 fatty acid. The molecular mass is 337. Formation of [M−H] in greater abundance than the molecular ion is common in the EI spectra of saturated fatty acid DMOX derivatives examined in this laboratory.

Sometimes the peaks showing the 12 Da gap are too weak to be observed. This is especially true for fatty acids with a double bond between C-3 and C-6. In these cases, the gap of 40 Da between the neighboring, more intense peaks caused by the $n-2$ and $n+2$ carbon fragments can be used to identify the double bond position.

7.2.2.3   Spectra with 10-dalton gaps
A gap of 10 Da in the usual m/z $126 + 14n$ series indicates a triple bond equivalent, which can be a triple bond or a cyclopropene ring. Similar to the 12 Da gap discussed above, the intensity of the surrounding peaks should be considered to distinguish between these functionalities. If the 10 Da gap is between ions corresponding to $n-1$ and n carbon atoms, and the $n-2$, $n+2$ and $n+3$ ions are of comparable intensity, then there is a triple bond between the carbon atoms n and $n+1$ (Zhang et al., 1988a). On the other hand, if only the $n+3$ fragment ion is intense, then there is a cyclopropene ring between the positions n and $n+1$ (Spitzer et al., 1991b).

7.2.2.4   Spectra with 12- and 10-dalton gaps
The rules for identification of the position of multiple bonds can also be used for molecules with several conjugated or unconjugated multiple bonds. Patterns of intense ions separated by gaps of 40 Da or 38 Da, and relatively less intense ions with a gap between them of 12 Da or 10 Da, respectively, suggest that a methylene-interrupted multiple bond system is present. This is typical of most natural polyunsaturated fatty acids. It is noteworthy that in these compounds, the $n+2$ and $n+3$ ions from the first multiple bond and those for the $m-1$ and m ions of the following multiple bond are similar, and of lower abundance than those of a monounsaturated system.

A single intense ion, with several following ions showing gaps of 12 Da or 10 Da that are not interrupted by a single intense ion, indicates the presence of a conjugated multiple bond system. The spectra of conjugated polyunsaturated fatty acids also show relatively intense ions at the $n+2$ and $n+3$ positions with respect to the last multiple bond in the chain (Spitzer et al., 1991a). The positions of the double bonds in conjugated linoleic acid and its positional isomers can also be determined using these rules (Yurawecz et al., 1994).

7.2.2.5   Spectra with a gap of 28 daltons
A gap of 28 Da in the usual m/z $126 + 14n$ series suggests an oxo-group (Classen et al., 1994). In addition to the intense ions at $X - 28$ (or $n-1$) and X (or n), that are formed by α-cleavage on both sides of the oxo-group, an intense peak at $X + 15$ due to a McLafferty rearrangement is present.

7.2.2.6 Spectra with a gap of 29 daltons

A gap of 29 Da in the usual m/z $126 + 14n$ series indicates a hydroxyl group (Zhang et al., 1988b). The two ions forming the 29 Da gap are due to α cleavage on both sides of the hydroxyl group after migration of the hydroxyl hydrogen atom to the heterocyclic portion of the molecule. If one of the peaks forming the 29 Da gap is weak, conjugation of the hydroxyl group with a multiple bond system is indicated (Spitzer, 1997).

7.2.2.7 Spectra with large gaps

Large gaps between neighboring peaks in the usual m/z $126 + 14n$ series indicate the presence of rings within the fatty acid chain (Mossoba et al., 1994). Typical gaps are 66 Da (cyclopentenyl), 68 Da (cyclopentyl), 78 Da (di-unsaturated 6-membered ring), 80 Da (cyclohexenyl) or 82 Da (cyclohexyl), etc. The gap directly indicates the size and saturation of the ring system.

7.2.2.8 Spectra with an intense peak at m/z 127

The 2 and 6-methyl branched fatty acids usually show an intense ion at m/z 127. If the ion at m/z 127 is the base peak rather than the ion at m/z 113, the spectrum is that of a 2-methyl branched fatty acid. On the other hand, if m/z 113 is the base peak, and m/z 127 is accompanied by a comparably intense ion at m/z 126, then the presence of a 6-methyl branched fatty acid (Yu, 1988) is suggested.

7.2.2.9 Spectra with intense ions at m/z 139, 153 or 167

Intense ions at m/z 139, 153 or 167 indicate double bonds at positions C-4, C-5 or C-6, respectively. These ions are diagnostic for double bonds because they do not normally appear in the spectra of DMOX derivatives (Zhang et al., 1988a) and accompany the expected ions at m/z 138, 152 or 166. These ions may be explained by ionization of the nitrogen atom followed by cyclization (Fay and Richli, 1991) using the remaining non-bonding electron of the nitrogen atom and an electron from one bond of the double bond in the alkyl chain. The nitrogen atom is now positively charged and the remaining electron of the former double bond is a radical site on the chain. An electron shift driven by the location of the radical site on the chain cleaves the chain, resulting in a neutral loss of the portion of the alkyl chain that is not part of the newly formed ring. The unpaired electron left behind on the ion by the departing alkyl chain is now a radical site. The odd-electron product ion may then shift the unpaired electron back to the more favorable location of the non-bonding orbital of the nitrogen atom and reform a double bond in the alkyl chain by pairing one electron of the recently formed nitrogen–carbon bond with the unpaired electron on the alkyl chain. A hydrogen atom shifts from the carbon atom at one end of this new double bond to the terminal carbon of the double bond at the end of the recreated alkyl chain to complete the sequence.

7.3 Mass spectrometry of complex fatty acid mixtures

The usefulness of DMOX derivatives for the structural analysis of fatty acid species by mass spectrometry can be truly appreciated when the GC-MS analysis of *complex mixtures* of fatty acids and their isomers is required. Many other derivatives that are potentially useful for MS analysis degrade GC resolution. Eliminating this deterioration is the single most important prerequisite for the successful detection of GC effluents of complex mixtures by spectral methods.

7.3.1   Cyclic fatty acid monomers

An excellent example of the application of GC-MS to the complex analysis of fatty acid DMOX derivatives has been the elucidation of the structures of cyclic fatty acid monomers (CFAMs). It required discriminating between cyclic and bicyclic ring sizes and double bond positions and configurations for closely related fatty acids with subtle structural differences. Gas chromatography–electron ionization mass spectrometry (GC-EIMS) of DMOX derivatives was used to confirm the identities of a complex mixture of C_{18} diunsaturated CFAMs isolated from heated flaxseed (linseed) oil (Mossoba *et al.*, 1994). The molecular masses and *positions* of double bonds and 1,2-disubstituted unsaturated 5- and 6-membered rings along the fatty acid hydrocarbon chains were established by this method (Mossoba *et al.*, 1995). The double bond *configuration* for the methyl ester and DMOX derivatives of these CFAMs was confirmed by using GC–matrix isolation–Fourier transform infrared (MI-FTIR) spectroscopy (Mossoba, 1996b). An additional reason for using the DMOX derivative for CFAM analysis was that it was previously used for the structural determination of fatty acids containing terminal (*mono*-substituted) 5-membered rings (Zhang *et al.*, 1989).

Intramolecular cyclization of polyunsaturated fatty acids occurs when fat and oil frying operations take place under high temperature, for a long period or under other abusive conditions. CFAM products, which are easily absorbed by the digestive system, are of concern from a dietary-toxicity point of view. Hence, efforts to synthesize CFAMs or isolate them from heated oils and characterize their structures by chemical, chromatographic and spectrometric methods have been pursued for several decades. When this research was initiated in our laboratory, the classical problem of locating double bonds for a complex mixture of diunsaturated CFAM methyl esters isolated from heated oils had been only partially solved by mass spectrometry.

The GC separation of CFAM DMOX derivatives did not deteriorate the chromatographic resolution; for some peaks resolution was even enhanced relative to that found for methyl ester derivatives. As expected, the FID profile for the methyl ester derivatives of the unsaturated CFAM mixture isolated from heated flaxseed oil was qualitatively similar to that of the same mixture of oxazoline derivatives (Figure 7.11).

Low-resolution GC-EIMS analyses were obtained with a Hewlett-Packard 5890 series II gas chromatograph coupled to a Micromass (Wythenshawe, UK) Autospec Q mass spectrometer and OPUS 2000 data system. The GC-MS system used version 1.6C software. The capillary GC column was CP-Sil-88 (Chrompack, Inc., Bridgewater, NJ, USA), $50\,m \times 0.22\,mm$ (i.d.), with a $0.19\,mm$ stationary phase film. Adjusting the capillary GC column head pressure to $10\,psi$ gave chromatographic profiles comparable to those obtained with a flame ionization detector. The GC-MS conditions were as follows: splitless injection with helium sweep restored 1 min after injection; injector and transfer lines held at $230^{\circ}C$; oven temperature program, $75^{\circ}C$ for 2 min after injection, $20^{\circ}C\,min^{-1}$ to $185^{\circ}C$, hold for 15 min, $4^{\circ}C\,min^{-1}$ to $225^{\circ}C$, hold for 5 min. The mass spectrometer was tuned to a resolution

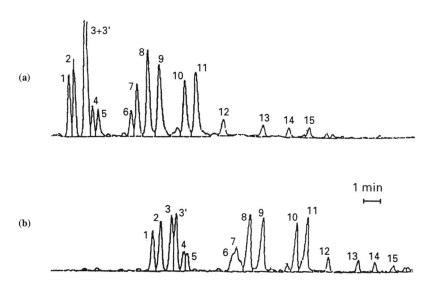

Figure 7.11   Flame ionization detection gas chromatograms for unsaturated mixtures of cyclic fatty acid monomer (CFAM) methyl esters (a) and CFAM oxazolines (b).

of 1000 (5% valley) by observing the ion at m/z 305 in the EI mass spectrum of perfluorokerosene (PFK). The mass scale was calibrated with PFK for magnet scans from 440 to 44 daltons at 1 s per decade. The trap current was 200 μA at 70 eV. The ion source temperature was 250°C.

As stated above, the homologous ion series with a pattern of peaks 14 Da apart is due to the cleavage of methylene groups along the hydrocarbon chain. For oxazoline derivatives of CFAMs, this fragmentation at each skeletal carbon–carbon bond was interrupted when a double bond and/or a ring was present along the fatty acid chain. This structural tool was applied to unsaturated and saturated (hydrogenated or deuterated) CFAM mixtures whose major components had one ring along the hydrocarbon chain.

7.3.1.1 Cyclic fatty acid monomers containing cyclopentenyl rings

The elucidation of CFAM structures in complex mixtures by MS requires the complementary use of vibrational spectroscopy as demonstrated with selected examples.

The EI mass spectrum for GC peak 1 in Figure 7.11b, exhibited successive mass intervals of 14 Da with interruptions of 66 and 12 Da (Figure 7.12). This spectrum is due to a C_{18} CFAM structure with a cyclopentenyl ring having a $2'$-n-butene substituent. As shown in the structure in Figure 7.12, the double bond is on C-1 of the butene group according to the empirical rule by Zhang *et al.* (1988a) discussed above. The configuration of the double bond of this butene substituent was found to be *trans* by IR (Mossoba *et al.*, 1995). Double bonds in 5- (or 6-) membered rings are usually stable in the *cis* configuration. In general, more than a single position is possible for a double bond with two *cis* hydrogen atoms in a 1,2-disubstituted 5- (or 6-) membered ring structure, which suggests that positional CFAM isomers may exist. However, the CFAM that is more likely to occur is the one in which the *cis* double bond corresponds to the C-12 position in the parent (linolenic) fatty acid.

While an identical IR spectrum was observed for the CFAM that gave rise to GC peak 2 in Figure 7.11b, the positions of the double bond and the 5-membered ring in this case were distinguished by MS. The mass spectrum obtained for this compound (Figure 7.13), indicated the presence of a double bond between C-8 and C-9 and a cyclopentenyl ring having a $2'$-n-propyl substituent in this molecule. Carbon atoms C-8 and C-9 along the 2-alkenyl chain of DMOX correspond to C-9 and C-10, respectively, in linolenic acid. It is noted that the observed EI mass spectra for unsaturated (or saturated) disubstituted 5- (or 6-) membered ring oxazoline species did not indicate that the rings in these CFAMs are *ortho*-disubstituted. However, the ring 1,2-disubstitution pattern was well established for several documented CFAM structures having 5-mem-

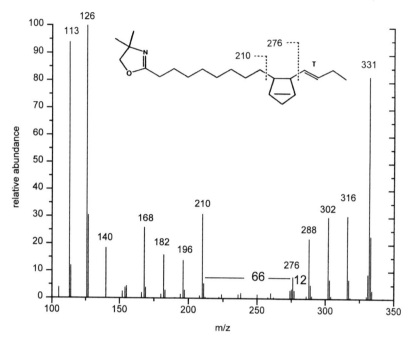

Figure 7.12   EI mass spectrum of GC peak 1 (Figure 7.11b), 2-[8-(2-but-*trans*-1-enyl-cyclopent-enyl) octyl]-4,4-dimethyloxazoline.

bered (Awl *et al.*, 1988; Rojo and Perkins, 1989a) or 6-membered (Le Quere *et al.*, 1991; Gast *et al.*, 1963) rings isolated from heated linseed oil, or synthesized (Hutchison and Alexander, 1963; MacDonald, 1956; McInnes *et al.*, 1961; Saito and Kameda, 1976; Potteau *et al.*, 1978).

The homologous series of ions, m/z $126 + 14n$, in the mass spectrum for GC peak 2 can be traced from m/z 126 to the 12 Da interruption at m/z 196. The relative abundance of m/z 154 ($n = 2$) was low in this spectrum and in those of several other CFAMs in this mixture including linoleic acid. The low relative abundance of $n = 2$ in this series (Zhang *et al.*, 1988a) did not interfere with the structural determination of unsaturated C_{18} CFAMs because the more abundant adjacent ions in the series reliably indicated the absence of a double bond in this portion of the molecule.

To verify MS assignments, different portions of the CFAM methyl ester mixture were hydrogenated and deuterated (Mossoba *et al.*, 1996b), and subsequently analyzed as DMOX derivatives by GC-EIMS. A portion of the isolated mixture of CFAM methyl esters was catalytically hydrogenated over platinum oxide with a microhydrogenator (Rojo and

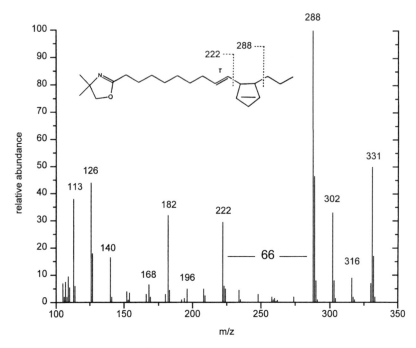

Figure 7.13 EI mass spectrum of GC peak 2 (Figure 7.11b), 2-[9-(2-propyl-cyclopentenyl)non-*trans*-8-enyl]-4,4-dimethyloxazoline. A double bond between carbons *n* and *n*+1 in the original fatty acid chain corresponds to an unsaturation site between carbons *n*−1 and *n*, respectively, along the 2-alkenyl chain of the oxazoline derivative.

Perkins, 1987). Unsaturated FAMEs were deuterated by the method of Rakoff and Emken (Rakoff and Emken, 1978) using a Wilkinson's catalyst [(Ph$_3$P)$_3$RhCl(I)]. In general, the chromatograms obtained for the hydrogenated and deuterated species were qualitatively similar (Figure 7.14), and the components of the latter mixture eluted about 0.1 min sooner than those of the former under our experimental conditions. In Figure 7.14, the major components, labeled 1–6, are saturation products of major and/or minor diunsaturated 5- (or 6-) membered ring CFAM mixture components. Similar pairs of mass spectra were found for the pairs of CFAMs that exhibited each of GC peaks 1 and 3, 2 and 5, as well as 4 and 6, and were probably cyclic stereoisomers in which only the configurations of the two *ortho* substituents on the ring are different: *trans* and *cis* for the early- and late-eluting CFAM, respectively. This elution sequence was previously determined (Awl and Frankel, 1982, Rojo and Perkins, 1989b) on the

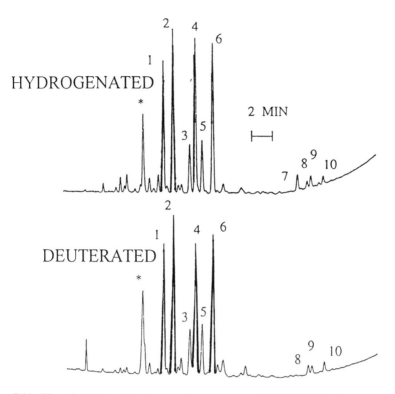

Figure 7.14 Flame ionization detection gas chromatograms of DMOX derivatives portions of the CFAM methyl ester mixture that were hydrogenated (top) and deuterated (bottom). The asterisks denote saturated non-cyclic C_{18} species.

basis of the characterization of synthesized 1,2-disubstituted CFAM isomers.

The determination of the sites of deuterium atoms along the chains of saturated CFAMs deuterated GC peaks 1 and 2 in Figure 7.14 further confirmed the previously assigned locations of double bonds along the carbon backbone of the corresponding diunsaturated CFAMs (peaks 1 and 2 in Figure 7.11). Typical mass spectral evidence is presented in Figures 7.15 and 7.16 for these deuterated CFAMs. Consecutive fragments separated by two adjacent intervals of 15 Da each (or a total of 30 Da) in the mass spectra of deuterated CFAMs indicated the presence of a —CHD—CHD— moiety along the hydrocarbon chain. An interruption of the 14 Da pattern by a mass interval of 70 Da confirmed the identity and location of the cyclopentyl ring.

Although the mass spectra for the pair of GC peaks 1 and 6 (Figure 7.11) were identical, IR spectroscopy discriminated between them

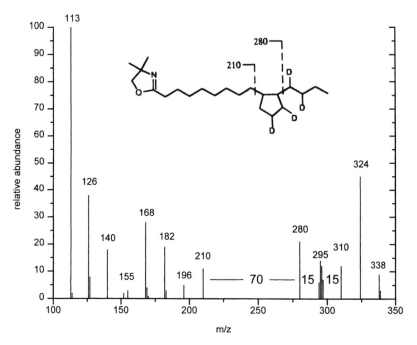

Figure 7.15   EI mass spectrum and structure for deuterated analyte of GC peak 1 (Figure 7.14, bottom).

and indicated that the configuration of the double bond in the hydrocarbon chain was different: *trans* for peak 1, and *cis* for peak 6. Other examples are discussed next for mass spectra obtained for diunsaturated CFAM oxazolines having 6-membered rings. Their spectra were found to be more complex than those of the cyclopentenyl CFAMs.

7.3.1.2   Cyclic fatty acid monomers containing cyclohexenyl rings

Unlike mass spectra of C_{18} CFAMs with a saturated 5- or 6-membered ring or a cyclopentenyl ring, which exhibit mostly large mass intervals, the mass spectrum of a compound containing a cyclohexenyl ring appears to have characteristic intense even and odd mass fragments at m/z 248 and 277, respectively (Figure 7.17). Fragments at m/z 208 and 288 with a mass interval of 80 Da (Figure 7.17), consistent with a cyclohexenyl structure, were observed for the compounds corresponding to GC peaks 8 and 9 in Figure 7.11b. The fragment at m/z 277 may arise via a retro Diels–Alder reaction resulting in the loss of C_4H_6 from the cyclohexenyl ring. The odd-electron product ion could give rise to m/z 248 with an electron shift resulting in resonance stabilization of the positive charge and loss of a C_2H_5 radical (Figure 7.18).

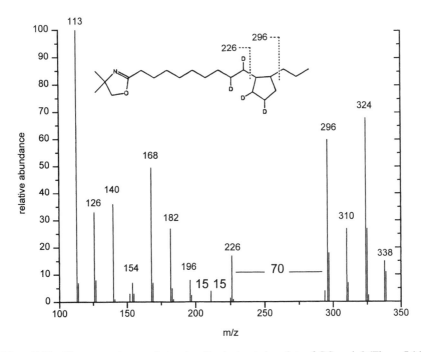

Figure 7.16 EI mass spectrum and structure for deuterated analyte of GC peak 2 (Figure 7.14, bottom).

These spectra also suggest that these two CFAMs have a 2'-*n*-propyl substituent on the ring, as well as a double bond on the aliphatic chain between C-7 and C-8. A *trans* double bond configuration was determined by IR for these two GC peaks. The complexity of the observed cyclohexenyl CFAM mass spectra found for each GC peak does not rule out the presence of other co-eluting species. Hence, the m/z 248 peak may alternatively be due to the presence of a component with a different ring structure. The similarity of the mass spectra of the compounds for GC peaks 8 and 9 suggests that the corresponding CFAMs are probably ring stereoisomers. The corresponding deuterated structure exhibited characteristic ions at m/z 212 and 296 for a saturated ring with two deuterium atoms. Other ions (discussed below) were due to co-eluting deuterated CFAMs having the same carbon skeleton. Isolation of a CFAM that exhibited only the ions that are consistent with the chemical structure shown in Figure 7.19 was recently achieved by silver ion HPLC (Dobson *et al.*, 1995).

The mass spectra (Figure 7.20) obtained for the two compounds corresponding to GC peaks 10 and 11 (Figure 7.11b) were less useful than

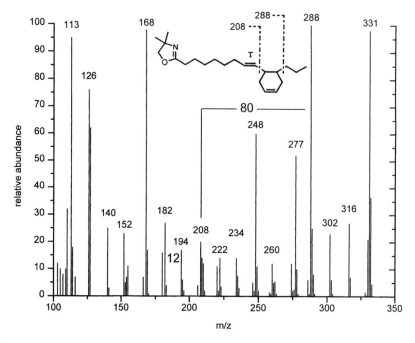

Figure 7.17 EI mass spectrum of GC peak 8 (Figure 7.11b). The tentative structure is 2-[8-(2-propyl-cyclohex-*cis*-4-enyl)oct-*trans*-7-enyl]-4,4-dimethyloxazoline.

those shown above. They indicated a common fragmentation pattern consisting of a series of 14 Da mass intervals and fragments at m/z 210 and 288 with a difference of 78 Da, consistent with a CFAM having a cyclohexadienyl ring with a 2′-*n*-propyl substituent. This cyclohexadienyl structure would preclude formation of m/z 277 by the pathway proposed for a cyclohexenyl ring. However, the MS data also provide support for a cyclohexenyl ring in these compounds in the form of an abundant ion at m/z 277 and a gap of 80 Da (m/z 290 and 210) at reduced relative abundance (Figure 7.20).

These mass spectra point to the presence of more than a single CFAM for each GC peak, and do not necessarily exclude the presence of other CFAM structures. IR data were more reliable in this case and confirmed that each of these two compounds has a *cis* double bond on a cyclohexenyl ring and a *trans* double bond on a hydrocarbon chain (Mossoba *et al.*, 1995). Support for this assignment was found in the mass spectrum for the corresponding deuterated structure (Figure 7.21) with characteristic ions at m/z 210 and 294.

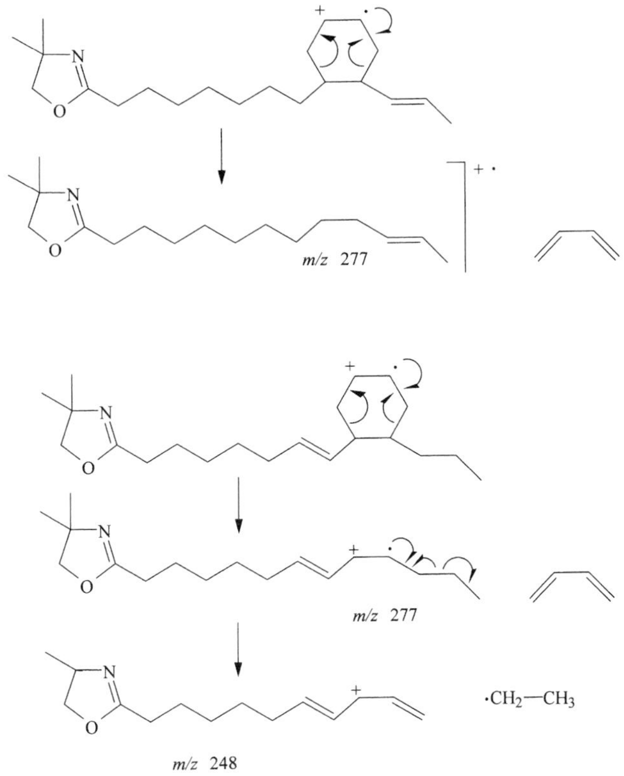

Figure 7.18 Proposed retro Diels–Alder fragmentation to m/z 277 and loss of a C_2H_5 radical from m/z 277 to form m/z 248. Note that the charge is in the cyclohexenyl ring rather than the nitrogen atom.

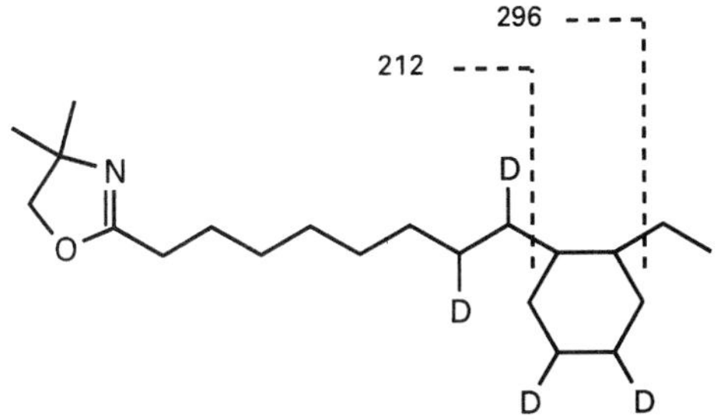

Figure 7.19 Deuterated structure consistent with a $2'$-n-propyl substituent on the ring.

7.3.1.3   Minor cyclic fatty acid monomer components

Similar inconclusive mass spectra (see Figure 7.20) were also found for each of the *minor* components that gave rise to GC peaks 12 and 13 in

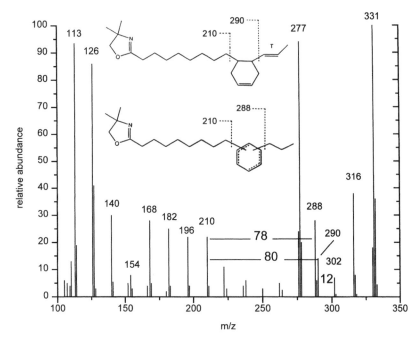

Figure 7.20 EI mass spectrum and possible structures for GC peak 10 (Figure 7.11b).

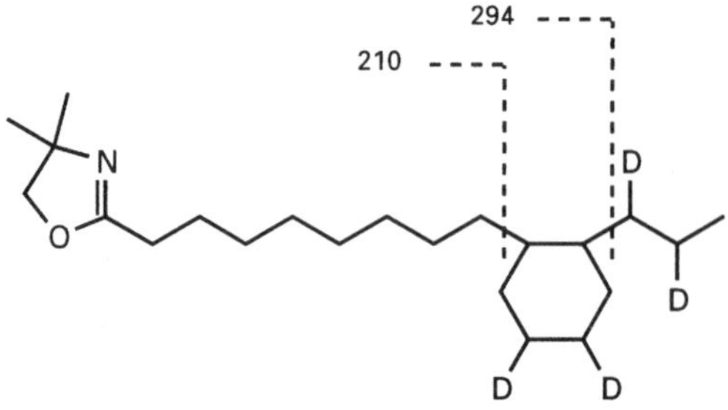

Figure 7.21 Deuterated structure corresponding to GC peak 10 (Figure 7.11b).

Figure 7.11. Once more, useful information was obtained by IR. These two CFAMs exhibited the IR band (664 cm^{-1}) characteristic of a double bond in a 6-membered ring, and another new one near 723 cm^{-1} that was attributed to a second ethylenic bond in the ring (Mossoba *et al.*, 1995). The relative location of the double bonds in the ring is not known; and of the several possible unsaturation sites, one could perhaps be at a substituted ring carbon. A UV band with a maximum at 233 nm (E$_{1\%}$ (absorptivity of a 1% solution) = 32.8), characteristic of conjugated dienes, was observed for the CFAM methyl ester mixture. Since the UV spectrum was not obtained for each of the compounds of interest, the observed data were not sufficient to confirm the presence of a conjugated diene system in these CFAMs (GC peaks 12 and 13). However, further evidence for a cyclohexadienyl ring structure was obtained from the spectrum of a minor deuterated species that exhibited a gap of 86 Da (between m/z 210 and 296, Figure 7.22) that is consistent with a 6-membered ring having 4 deuterium atoms.

We do not have an explanation for the presence of features indicative of both cyclohexenyl and cyclohexadienyl structures in the MS data for GC peaks 10–13 in Figure 7.11. The previously proposed explanation of double bond migration during ionization (Andersson and Holman, 1974) could not be tested with the available MS data in the present case.

The mass spectra observed for two other *minor* compounds (GC peaks 14 and 15 in Figure 7.11) showed a large gap of 120 Da due to the loss of a sizable fragment in each case. The identity of a methyl or an ethyl ring substituent, respectively, for these compounds is indicated by the mass intervals of 15 Da (GC peak 14), and 14 and 15 Da (GC peak 15) at the high-mass end of the corresponding spectra. These mass spectra are

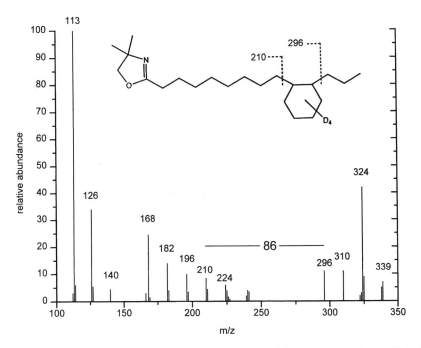

Figure 7.22   EI mass spectrum for minor deuterated species with 86 Da gap consistent with a 6 membered ring containing 4 deuterium atoms.

consistent with structures of bicyclic monoenoic fatty acid derivatives. In published mechanisms (Potteau *et al.*, 1978; Mikusch and Sagredos, 1971) for the formation of cyclic and bicyclic fatty acids from heated linolenic acid, monounsaturated bicyclic structures were proposed consisting of a 6-membered ring fused to a 5-membered ring. Similar fatty acid structures originating from octadecatrienoic acids were reportedly formed during tall oil distillation (Hase *et al.*, 1992). These tentative structures were also consistent with the observed IR data (Mossoba *et al.*, 1995). Supporting evidence was found in the mass spectra observed for the saturated components that correspond to GC peak 8 (Figure 7.14), which indicate the presence of large gaps of 122 (Figure 7.23a) and 124 Da (Figure 7.23b), respectively, due to the loss of a bicyclic fragment.

The gaps found for the hydrogenated and deuterated components are 2 and 4 Da higher, respectively, than the 120 Da gap obtained for the corresponding unsaturated CFAM and are consistent with the saturation of a single double bond. Mass spectra with similar gaps, between m/z 196 and 318 (hydrogenated) and m/z 196 and 320 (deuterated), were also

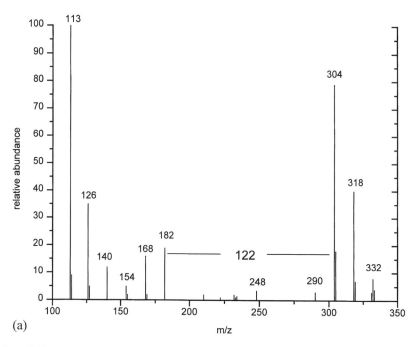

Figure 7.23 EI mass spectra of (a) hydrogenated and (b) deuterated compounds corresponding to GC peak 8 (Figure 7.14).

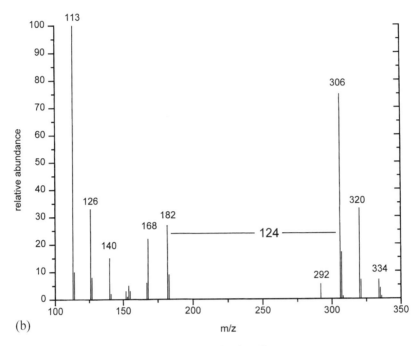

Figure 7.23 (Continued).

obtained for the saturated components that gave rise to GC peaks 9 and 10 (Figure 7.14). A methyl substituent on the bicyclic ring structure was indicated by a mass interval of 15 Da at the high-mass end, between m/z 318 and 333 and between 320 and 335, for the hydrogenated and deuterated compounds, respectively.

Thus, it is possible to elucidate the structures of diunsaturated monocyclic and monounsaturated bicyclic CFAMs in a complex mixture by using the complementary techniques GC-MI-FTIR and GC-EIMS.

7.3.2   Trans *monounsaturated C_{18} fatty acid positional isomers*

Trans fatty acids occur naturally in animal fats as a result of biohydrogenation and as *complex isomeric mixtures* in many food products containing partially hydrogenated vegetable oils (Firestone and Sheppard, 1992). The nutritional significance of *trans* fatty acids has led to increased interest in accurate and rapid methods for quantifying the total *trans* fatty acid content of foods (Mossoba *et al.*, 1996a). There is also interest in procedures (Christie and Breckenridge, 1989) that can separate and confirm the double bond configuration and position for *individual trans* fatty acid positional isomers.

Capillary GC has been the standard tool for the separation of fatty acids, usually as fatty acid methyl esters (FAMEs) (Christie, 1990). However, the peaks of several *trans* and *cis* octadecenoic acid (18:1) positional isomers overlap, even with GC columns that have polar stationary phases (Duchateau *et al.*, 1996). Therefore, in order to confirm unequivocally the double bond *position* for *individual trans* 18:1 positional isomers by mass spectrometry, the complete separation of 18:1 *trans* from *cis* geometric isomers was carried out first by using silver-ion high-performance liquid chromatography (HPLC) (Christie and Breckenridge, 1989; Adlof *et al.*, 1995). The double bond configuration for the FAMEs in the *trans* 18:1 HPLC fraction was confirmed by using GC–direct deposition–Fourier transform infrared (DD-FTIR) spectroscopy (Mossoba *et al.*, 1997). A portion of this HPLC fraction was converted to the DMOX derivative, and used for GC-EIMS analysis.

HPLC separations were performed utilizing a Waters 600E solvent delivery system (Waters Associates), a Rheodyne 7125 injector (Rheodyne, Inc., Cotati, CA, USA) with a 20 µl injection loop, and a Waters Model 996 Photodiode Array Detector (Waters Associates). A Chrom-Spher Lipids column (4.6 mm i.d. × 250 mm stainless steel; 5 µm particle size; silver impregnated) was acquired from Chrompack, Inc., Bridgewater, NJ, USA. Solvent flow was set at 1.0 µl min^{-1}. The mobile phase was 0.3% acetonitrile in hexane (isocratic). The test portion size was 100 µg FAMEs injected with each 20 µl loop injection. Semi-preparative HPLC fractionations were made using similar conditions with a 10 mm i.d. × 250 mm ChromSpher Lipids column and a flow rate of 5 ml min^{-1}.

An unusually low temperature was used to achieve the separation of positional 18:1 DMOX isomers. Low-resolution GC-EIMS analyses were performed with a Hewlett-Packard (Avondale, PA, USA) 5890 series II gas chromatograph coupled to a Micromass (Wythenshawe, UK) Autospec Q mass spectrometer and OPUS 4000 data system. The GC-MS system used version 2.1BX software. The capillary GC columns used were the CP-Sil 88 (Chrompack, Inc., Bridgewater, NJ, USA), 100 m × 0.22 mm i.d., with a 0.19 mm stationary phase film, and the SP2560 (Supelco, Bellefonte, PA, USA), 100 m × 0.25 mm i.d. with a 0.20 mm film thickness. The GC-MS conditions were as follows: hydrogen carrier gas; split injection; head pressure 20 psi; oven temperature 140°C (isothermal); split vent flow rate 30 ml min^{-1}; septum purge flow rate 40 ml min^{-1}; linear velocity 32 cm s^{-1}; split ratio 23:1; and injector and transfer lines held at 220 and 200°C, respectively. The same chromatographic resolution was obtained with both columns. The mass spectrometer was tuned to a resolution of 1400 (5% valley) by observing m/z 281 in the EI mass spectrum of perfluorokerosene (PFK). The mass scale was calibrated with PFK for magnet scans from 360 to 60 Da at 0.5 s per decade. The trap current was 200 μA at 70 eV. The ion source temperature was 250°C.

Under our reaction conditions, the conversion from FAME to DMOX derivatives was not quantitative. The presence of both derivatives in the same mixture allowed the monitoring and comparison of their GC chromatographic characteristics (in particular, resolution) under identical GC conditions. The *trans* 18:1 FAME eluted about 1 h before the DMOX derivatives at 140°C (isothermal). These data (Figure 7.24) are for two separate GC runs that differed only in the injection volume, 0.4 and 1.0 μl, shown in the top and bottom traces, respectively. For each of these two GC runs, the *m/z* 264 and 113 ion profiles, recorded for the *trans* 18:1 FAME and DMOX positional isomers, appeared in this same sequence, respectively. Inspection of the bottom GC profile (1.0 μl injection) shows evidence of overload in the early part of the *m/z* 113 DMOX trace, while

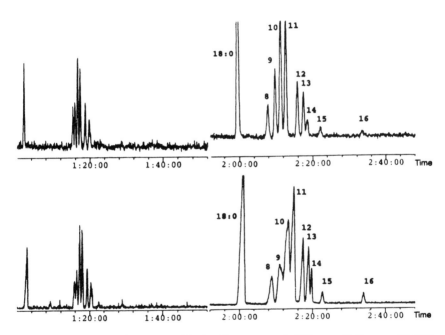

Figure 7.24 GC-EIMS chromatographic data for two separate GC runs with the same SP-2560, 100 m capillary column. The top and bottom traces were for two GC runs in which 0.4 and 1.0 μl portions of the test sample were injected, respectively. The *m/z* 264 (for FAME) and 113 (for DMOX) ion profiles were recorded for *trans* 18:1 positional isomer FAME and 4,4-dimethyloxazoline (DMOX) derivatives present in the same test portions. The DMOX eluted about 1 h after the FAME derivatives under 140°C isothermal GC conditions. The numerical labels 8 through 16 next to the GC peaks denote the Δ8 through Δ16 *trans* 18:1 positional isomers.

the response of the Δ 13 and Δ 14 pair of positional isomers (in the second half of this trace) appears to be enhanced. This was the first capillary GC report of the separation of these two *trans* 18:1 DMOX positional isomers on polar columns.

As stated above, fatty acid DMOX mass spectra consist of a series of *even*-mass ions separated by a mass difference of 14 due to the loss of a methylene group (Zhang *et al.*, 1988a). A gap of 12 Da between ions containing $n - 1$ and n carbon atoms indicates the presence of a double bond between carbons n and $n + 1$ of the parent fatty acid. More abundant peaks due to allylic cleavage, corresponding to ions containing $n - 2$, $n + 2$, and $n + 3$ carbons of the parent fatty acid, flank the pair of diagnostic ions that are separated by a gap of 12 Da (Zhang *et al.*, 1988a). Deviations from these rules are found in DMOX spectra (Figure 7.25) when a double bond is closer to the extremities of the fatty acid molecule (e.g. Δ 6 and Δ 7; see below). The mass spectral data confirmed the double bond position along the hydrocarbon chain for nine individual *trans* 18:1 positional isomers, each having a double bond at a carbon located

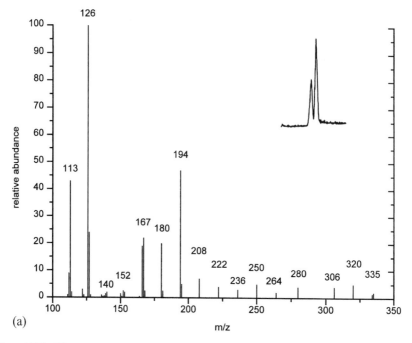

Figure 7.25   EI mass spectra of (a) Δ6 and (b) Δ7 *trans* 18:1 DMOX positional isomer standards. The inset shows m/z 113 chromatographic trace obtained for a co-injected qualitative test mixture of these two standards.

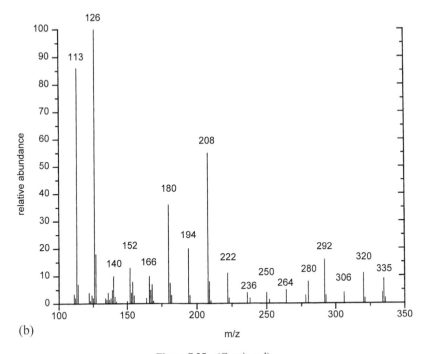

(b)

Figure 7.25   (Continued).

between C-8 and C-16. The identities of the $\Delta 9$, $\Delta 11$, $\Delta 12$, $\Delta 13$, and $\Delta 15$ *trans* 18:1 positional isomers were further confirmed by comparison with standards.

For partially hydrogenated fats, 18:1 positional isomers are usually produced as a distribution in which the abundance of an isomer generally decreases as its double bond gets closer to either end of the fatty acid chain (Marchand, 1982). Accordingly, the levels of the $\Delta 4$ through $\Delta 7$ *trans* 18:1 isomers are expected to be low and difficult to identify. The $\Delta 6$ and $\Delta 7$ *trans* 18:1 FAME positional isomer standards were commercially acquired and analyzed by GC under identical experimental conditions. As FAMEs, the $\Delta 6$ and $\Delta 7$ *trans* 18:1 positional isomers co-eluted, but they were well resolved as DMOX derivatives (Figure 7.25). The mass spectra of D6 and D7 *trans* 18:1 DMOX positional isomers were unique and permitted identification of these isomers, even though their DMOX mass spectra did *not* exhibit the expected diagnostic ions. When co-injected with the partially hydrogenated soybean oil test portion analyzed, the $\Delta 6$ and $\Delta 7$ DMOX isomers co-eluted with the $\Delta 8$ and $\Delta 9$ DMOX isomers, respectively. Therefore, it was not possible to confirm by GC-EIMS the identity of the minor $\Delta 6$ (or $\Delta 7$) isomer expected to be found at a much lower level than that of the $\Delta 8$ (or $\Delta 9$, respectively) isomer with which it would be co-eluting. However, their presence was recently confirmed by nuclear magnetic resonance (Mazzola *et al.*, 1997).

Once more, DMOX proved to be an important derivative that can confirm the double bond *position* for individual 18:1 fatty acid isomers by GC-EIMS. Better GC resolution was obtained for the DMOX than for the FAME derivatives for the two pairs of $\Delta 6$ and $\Delta 7$, and $\Delta 13$ and $\Delta 14$ *trans* 18:1 fatty acid positional isomers.

7.4 Conclusions

Localization of charge on the nitrogen atom of a DMOX derivative of a fatty acid begins with the preferential ionization of the non-bonding electrons of the nitrogen atom versus ionization of the non-bonding electrons of an oxygen atom or the π electrons of a double or triple bond elsewhere in the molecule. Unlike oxygen, the nitrogen atom in the DMOX derivative does not tend to leave the ion through formation of a neutral species. Retention of charge on the nitrogen atom allows hydrogen migration to the charge site with fragmentation directed by the radical site at the former location of the hydrogen atom in the chain. Radical site-directed fragmentation simplifies interpretation of the observed spectrum because it tends to be more structurally specific.

The DMOX derivative is an invaluable tool in the structural elucidation of many fatty acids by EIMS. However, it is not a universal tool that will independently answer every question. A mass spectrum is the result of competing processes in which the ease of ionization and the relative stabilities of the products limit, but cannot eliminate, all possible routes to the observed set of ions that is a mass spectrum.

Functional groups in the chain afford additional sites for localization of the charge that compete with localization of the charge on the nitrogen atom. A rationale for the origin of m/z 277 and 248 in Figure 7.18 that places the charge on the cyclohexenyl ring rather than the nitrogen atom suggests that the ring is an effective competitive site. A corresponding retro Diels–Alder fragmentation is reported for cyclohexenyl CFAM methyl esters (Awl and Frankel, 1982), but there is a need for more DMOX data because there are significant differences in the fragmentation pathways of DMOX and methyl ester derivatives of fatty acids. However, a comparison of the EI mass spectra of the DMOX derivatives of α-licanic acid and α-eleostearic acid provides a convincing example of a competitive site because the functional difference in the two molecules is a carbonyl group (Spitzer, 1991a).

The statement that the presence of an oxygen atom in the fatty acid chain has a dramatic influence on the EI mass spectrum (Murphy, 1993)

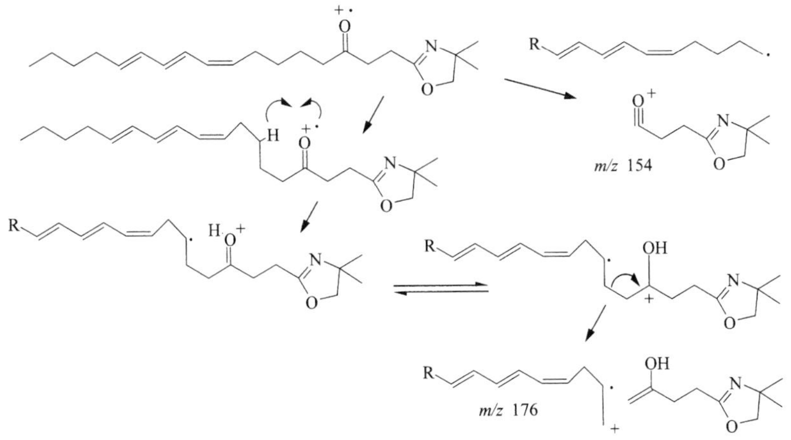

Figure 7.26   Proposed formation of m/z 154 via α cleavage adjacent to the carbonyl and m/z 176 by McLafferty rearrangement to the keto group with charge migration to the alkyl chain.

is as applicable to the DMOX derivatives as it is to the methyl ester derivatives of fatty acids. The principal ions observed in the EI mass spectrum of the DMOX derivative of α-licanic are m/z 154 and 176. Both ions may be interpreted as resulting from localization of charge on the oxygen of the C-4 keto function (Spitzer *et al.*, 1991a). Alpha cleavage with retention of charge on the oxygen results in the base peak at m/z 154. A McLafferty hydrogen rearrangement to the keto group with charge migration to the unsaturated carbon chain (McLafferty, 1967) gives rise to m/z 176 (Figure 7.26).

The related compound α-eleostearic acid lacks the keto function of α-licanic acid, but is otherwise identical to α-licanic acid. The locations of the double bonds of α-eleostearic acid are clearly evident in the EI mass spectrum of its DMOX derivative. In contrast, the keto group of α-licanic acid controls fragmentation to the extent that information about double bond location cannot be obtained from the EI mass spectrum of its DMOX derivative (Spitzer *et al.*, 1991a).

Thus, a DMOX derivative may be an invaluable tool, but final proof of structure should continue to rely on corroborative evidence provided by independent means of structural determination.

Acknowledgements

The authors thank Martha L. Gay and James A. Sphon for their review and comments.

References

Adlof, R.O., Copes, L.C. and Emken, E.A. (1995) Analysis of the monoenic fatty acid distribution in hydrogenated vegetable oils by silver ion HPLC. *J. Am. Oil Chem. Soc.*, **72** 571-574.

Andersson, B.A. and Holman, R.T. (1974) Pyrrolidides for mass spectrometric determination of the position of the double bond in monounsaturated fatty acids. *Lipids*, **9** 185-190.

Awl, R.A. and Frankel, E.N. (1982) Cyclic fatty esters: synthesis and characterization of methyl ω-(6-alkyl-3-cyclohexenyl) alkenoates. *Lipids*, **17** 414-426.

Bascetta, E., Gunstone, F.D. and Scrimgeour, C.M. (1984) Synthesis, characterization, and transformations of lipid cyclic peroxide. *J. Chem. Soc., Perkin Trans.*, **1** 2199-2205.

Batna, A. and Spiteller, G. (1994) Effects of soybean lipoxygenase-1 on phosphatidylcholines containing furan fatty acids. *Lipids*, **29** 397-403.

Bieman, K. (1962) *Mass Spectrometry Organic Chemical Application*, McGraw-Hill, New York.

Budzikiewicz, H., Djerassi, C. and Williams, D.H. (1967) *Mass Spectrometry of Organic Compounds*, Holden-Day, San Francisco.

Christie, W.W. (1990) *Gas Chromatography and Lipids: A Practical Guide*, The Oily Press, Ayr.

Christie, W.W. (ed.) (1992) *Advances in Lipid Methodology—One*, The Oily Press, Ayr.

Christie, W.W. and Breckenridge, G.McG. (1989) Separation of *cis* and *trans* isomers of unsaturated fatty acids by high performance liquid chromatography in the silver ion mode. *J. Chromatogr.*, **469** 261-269.

Classen, E., Marx, F. and Fabricius, H. (1994) Mass spectra of 4,4-dimethyloxazoline derivatives of oxooctadecanoic acids. *Fette Wissens. Tech.*, **96** 331-332.

Dinh-Nguyen, N., Ryhage, R., Stallberg-Stenhagen, S. and Stenhagen, E. (1961) Mass spectrometric studies. VIII. A study of the fragmentation of normal-chain methyl esters and hydrocarbons under electron impact with the aid of D-substituted compounds. *Arkiv foer Kemi*, **18** 393-399.

Dobson, G., Christie, W.W., Brechany, E.Y., Sebedio, J.L. and Le Quere, J.L. (1995) Silver ion chromatography and gas chromatography–mass spectrometry in the structural analysis of cyclic dienoic acids formed in frying oils. *Chem. Phy. Lipids*, **75** 171-182.

Duchateau, G.S.M.J.E., van Oosten, H.J. and Vascocellos, M.A. (1996) Analysis of *cis*- and *trans*-fatty acid isomers in hydrogenated and refined vegetable oils by capillary gas-liquid chromatography. *J. Am. Oil Chem. Soc.*, **73** 275-282.

Evershed, R.P. (1992) Mass spectrometry of lipids, in *Lipid Analysis A Practical Approach* (eds R.J. Hamilton and S. Hamilton), IRL Press, New York, pp 263-310.

Fay, L. and Richli, U. (1991) Location of double bonds in polyunsaturated fatty acids by gas chromatography–mass spectrometry after 4,4-dimethyloxazoline derivatization. *J. Chromatogr.*, **541** 89-98.

Firestone, D. and Sheppard, A. (1992) Determination of *trans* fatty acids, in *Advances in Lipid Methodology—One* (ed. W.W. Christie) The Oily Press, Ayr, pp 273-322.

Garrido, J.L. and Medina, I. (1994) One-step conversion of fatty acids into their 2-alkenyl-4,4-dimethyloxazoline directly from total lipids. *J. Chromatogr.*, **273** 101-105.

Gast, L.E., Schneider, W.J., Forest, C.A. and Cowan, J.C. (1963) Composition of methyl esters from heat-bodied linseed oils. *J. Am. Oil Chem. Soc.*, **40** 287-289.

Gross, M.L. (1992) Charge-remote fragmentations: method, mechanism and applications. *Int. J. Mass Spectrom. Ion Processes*, **118/119** 137-165.

Hamilton, R.J. and Hamilton, S. (eds) (1992) *Lipid Analysis, A Practical Approach*, IRL Press, Oxford.

Hase, A., Kaltia, S., Marikainen, J., Ala-Peijari, M. and Hase, T. (1992) Cyclopinolenic acids: synthesis, derivatives and thermal properties. *J. Am. Oil Chem. Soc.*, **69** 1027-1031.

Hutchison, R.B. and Alexander, J.C. (1963) The structure of a cyclic C_{18} acid from heated linseed oil. *J. Org. Chem.*, **28** 2522-2526.

Ishii, K., Okajima, H., Koyamatsu, T., Okada, Y. and Watanabe, H. (1988) The composition of furan fatty acids in the crayfish. *Lipids*, **23** 694-700.

Kramer, J.K.G., Fellner, V., Dugan, M.E.R., Sauer, F.D., Mossoba, M.M. and Yurawecz, M.P. (1997) Evaluating acid and base catalysts in the methylation of milk and rumen fatty acids with special emphasis on conjugated dienes and total *trans* fatty acids. *J. Am. Oil Chem. Soc.*, **32** 1219-1228.

Le Quere, J.L., Sebedio, J.L., Henry, R., Couderc, F., Demont, N. and Prome, J.C. (1991) Gas chromatography–mass spectrometry and gas chromatography–tandem mass spectrometry of cyclic fatty acid monomers isolated from heated fats. *J. Chromatogr.*, **562** 659-672.

Liebich, H.M., Schmieder, N., Wahl, H.G. and Wahl, J. (1994) Separation and identification of unsaturated fatty acid isomers in blood serum and therapeutic oil preparations in the form of their oxazolin derivatives by GC-MS. *J. High Resolut. Chromatogr.*, **17** 519-521.

MacDonald, J.A. (1956) Evidence of cyclic monomers in heated linseed oil. *J. Am. Oil Chem. Soc.*, **33** 394-396.

Mazzola, E.P., McMahon, J.B., McDonald, R.E., Yurawecz, M.P., Sehat, N. and Mossoba, M.M. (1997) [13]C Nuclear magnetic resonance spectral confirmation of Δ6- and Δ7-*trans*-18:1 fatty acid methyl ester positional isomers. *J. Am. Oil Chem. Soc.*, **74** 1335-1337.

Mangold, H.K. (1984) *CRC Handbook of Chromatography Lipids*, Vols I and II, CRC Press, Boca Raton, FL.

Marchand, C.M. (1982) Positional isomers of *trans*-octadecenoic acids in margarines. *J. Can. Inst. Food Sci. Technol.*, **15** 196-199.

McCloskey, J.A. (1970) Mass spectrometry of fatty acid derivatives, in *Topics in Lipid Chemistry 1* (ed. F.D. Gunstone), Logos Press, London, pp 369-440.

McInnes, A.G., Cooper, F.P. and MacDonald, J.A. (1961) Further evidence for cyclic monomers in heated linseed. *Can. J. Chem.*, **39** 1906-1914.

McLafferty, F.W. (1967) *Interpretation of Mass Spectra*, W.A. Benjamin, New York, p 127.

Mikusch, J.D. and Sagredos, A.N. (1971) Zur chemie der cyclischen fettsauren. *Fette Seifen Anstrichm.*, **73** 384-393.

Morris, L.J., Holman, R.T. and Fontell, K. (1960) Alteration of some long-chain esters during gas–liquid chromatography. *J. Lipid Rev.*, **1** 412-420.

Mossoba, M.M., Yurawecz, M.P., Roach, J.A.G., Lin, H.S., McDonald, R.E. and Flickinger, B.D. (1994) Rapid determination of double bond configuration and position along the hydrocarbon chain in cyclic fatty acid monomers. *Lipids*, **29** 893-896.

Mossoba, M.M., M.P. Yurawecz, M.P., Roach, J.A.G., Lin, H.S., McDonald, R.E., Flickinger, B.D. and Perkins, E.G. (1995) Elucidation of cyclic fatty acid monomer structures. Cyclic and bicyclic ring sizes and double bond position and configuration. *J. Am. Oil Chem. Soc.*, **72** 721-727.

Mossoba, M.M., Yurawecz, M.P. and McDonald, R.E. (1996a) Rapid determination of the total *trans* content of neat hydrogenated oils by attenuated total reflection spectroscopy. *J. Am. Oil Chem. Soc.*, **73** 1003-1009.

Mossoba, M.M., Yurawecz, M.P., Roach, J.A.G., McDonald, R.E. and Perkins, E.G. (1996b) Confirmatory mass-spectral data for cyclic fatty acid monomers. *J. Am. Oil Chem. Soc.*, **73** 1317-1321.

Mossoba, M.M., McDonald, R.E., Roach, J.A.G., Fingerhut, D.D., Yurawecz, M.P. and Sehat, N. (1997) Spectral confirmation of *trans* monounsaturated C_{18} fatty acid positional isomers. *J. Am. Oil Chem. Soc.*, **74** 125-130.

Murphy, R.C. (1993) *Handbook of Lipid Research 7, Mass Spectrometry of Lipids*, Plenum Press, New York, pp 71-130.

Potteau, B., Dubois, P. and Rigaud, J. (1978) Identification and determination of saturated monomeric cyclic acids obtained from thermopolymerized and thermally oxidized linseed oil. *Ann. de Technol. Agri.*, **27** 655-679.

Rakoff, H. and Emken, E.G. (1978) Syntheses of geometric isomers of di-, tetra- and hexadeuterated 12-octadecenoates and their triglycerides. *J. Labelled Comp. Radiopharm.*, **15** 233-252.

Roach, J.A.G. (1998) Application of GC/MS and SFC/MS to food analysis, in *Spectral Methods in Food Analysis: Instrumentation and Applications* (ed. M.M. Mossoba), Marcel Dekker, New York, pp 159-250.

Rojo, J.A. and Perkins, E.G. (1987) Cyclic fatty acid monomer formation in frying fats. I. Determination and structural study. *J. Am. Oil Chem. Soc.*, **64** 414-421.

Rojo, J.A. and Perkins, E.G. (1989a) Cyclic fatty acid monomer: isolation and purification with solid phase extraction. *J. Am. Oil Chem. Soc.*, **66** 1593-1595.

Rojo, J.A. and Perkins, E.G. (1989b) Chemical synthesis and spectroscopic characteristics of C_{18} 1,2-disubstituted cyclopentyl fatty acid methyl esters. *Lipids*, **24** 467-476.

Saito, M. and Kaneda, T. (1976) Studies on the relation between the nutritive value and the structure of polymerized oils. *Yukagaku (Oil Chem.)*, **25** 79-86.

Sehat, N., Yurawecz, M.P., Roach, J.A.G., Mossoba, M.M., Eulitz, K. and Ku, Y. (1998) Autoxidation of the furan fatty acid ester, methyl 9,12-epoxyoctadeca-9,11-dienoate. *J. Am. Oil Chem. Soc.*, **75** 1313-1319.

Spitzer, V. (1997) Structure analysis of fatty acids by gas chromatography–low resolution electron impact mass spectrometry of their 4,4-diemthyloxazoline derivatives—a review, *Progr. Lipids Res.*, **35** 387-408.

Spitzer, V., Marx, F., Maia, J.G.S. and Pfeilsticker, K. (1991a) Occurrence of conjugated fatty acids in the seed oil of *Couepia longipendula* (Chrysobalanaceae). *J. Ame. Oil Chem. Soc.*, **68** 440-442.

Spitzer, V., Marx, F., Maia, J.G.S. and Pfeilsticker, K. (1991b) GC-MS characterization (chemical ionization and electron impact modes) of the methyl esters and oxazoline derivatives of cyclopropenoid fatty acids. *J. Am. Oil Chem. Soc.*, **68** 963-969.

Yu, Q.T., Liu, B.N., Zhang, J.Y. and Huang, Z.H. (1988) Location of methyl branching in fatty acids: fatty acids in urpgial secretion of Shanghai duck by GC-MS of 4,4-dimethyloxazoline derivatives. *Lipids*, **23** 804-810.

Yu, Q.T., Liu, B.N., Zhang, J.Y. and Huang, Z.H. (1989) Location of double bonds in fatty acids of fish oil and rat testis lipids. Gas chromatography–mass spectrometry of the oxazoline derivatives. *Lipids*, **24** 79-83.

Yurawecz, M.P., Hood, J.K., Roach, J.A.G., Mossoba, M.M., Daniels, D., Ku, Y., Pariza, M.W. and Chin, S.F. (1994) Conversion of allylic hydroxy oleate to conjugated linoleic acid and methoxy oleate by acid-catalyzed methylation procedures. *J. Am. Oil Chem. Soc.*, **71** 1149-1155.

Yurawecz, M.P., Sehat, N., Mossoba, M.M. and Roach, J.A.G. (1997) Oxidation products of conjugated linoleic acid and furan fatty acids, in *New Techniques and Applications in Lipid Analysis* (eds. M.M. Mossoba and R.E. McDonald), AOCS Press, Champaign, IL, pp 183-215.

Zhang, J.Y., Yu, Q.T. and Huang, Z.H. (1987) 2-Substituted 4,4-dimethyloxazolines as useful derivatives for the localization of cyclopropane rings in long-chain fatty acids. *Shitzurjo Bunseki*, **35** 23-30.

Zhang, J.Y., Yu, Q.T., Liu, B.N. and Huang, Z.H. (1988a) Chemical modification in mass spectrometry IV: 2-alkenyl-4,4-dimethyloxazolines as derivatives for the double bond location of long-chain olefinic acids. *Biomed. Environ. Mass Spectrom.*, **15** 33-44.

Zhang, J.Y., Yu, Q.T., Yang, Y.M. and Huang, Z.H. (1988b) Chemical modification in mass spectrometry II. A study on the mass spectra of 4,4-dimethyloxazoline derivatives of hydroxy fatty acids. *Chem. Scr.*, **28** 357-363.

Zhang, J.Y., Wang, N.Y., Yu, Q.T., Yu, X.J., Liu, B.N. and Huang, Z.H. (1989) The structures of cyclopentenyl fatty acids in the seed oils of *Flacourtiaceae* species by GC-MS of their 4,4-dimethyloxazoline derivatives. *J. Am. Oil Chem. Soc.*, **66** 242-246.

8 Infrared spectroscopy of lipids: principles and applications

Ashraf A. Ismail, Antonio Nicodemo, Jacqueline Sedman, Frederick R. van de Voort and Inès Elizabeth Holzbaur

8.1 Introduction

Mid-infrared (IR) spectroscopy is a well-established technique for the identification and characterisation of chemical structures. The chemical bonds of a molecule absorb energy in the mid-IR region of the electromagnetic spectrum ($4000-400 \, cm^{-1}$) owing to excitation of the molecule's vibrational modes (involving bond stretching and deformation). As the wavelengths of the radiation absorbed depend not only on the type of bond (e.g., C—C, C=O, N—H), but on its position and the nature of the bonds surrounding it, the pattern of absorbed wavelengths for a molecule—its IR spectrum—can be regarded as its 'fingerprint'. Consequently, IR spectroscopy is a powerful qualitative analytical tool that can be employed for the identification of chemical compounds and the elucidation of molecular structure. It can also be applied in quantitative analysis because the amount of infrared energy absorbed by a compound is proportional to its concentration in the sample.

IR spectroscopy has played a major role in fundamental research on lipid systems (Chapman, 1965a,b). In terms of practical applications, it has been widely utilised by the fats and oils industry for one specific purpose, namely, the determination of *trans* isomers. In recent years, following the advent of Fourier transform infrared (FTIR) spectroscopy in the late 1960s, there has been increased interest in extending the application of IR spectroscopy in the routine analysis of fats and oils. This interest has arisen not only as a result of the superior performance of FTIR instruments in comparison with the dispersive IR spectrometers that they have largely replaced, but also because of the concomitant development of powerful chemometric tools that can be used to extract information from IR spectra that is not discernible by visual inspection. Major advances in sample handling have also been made. Accordingly, FTIR spectroscopy has the potential of becoming an important tool for the routine analysis of fats and oils.

Near-infrared (NIR) spectroscopy has played a much less important role than mid-IR spectroscopy in the study and the analysis of lipid systems, but it is likely to be increasingly utilised in fats and oils analysis within the next few years. Molecular absorption of energy in the near-infrared region of the electromagnetic spectrum (12 500–4000 cm^{-1}) results from the excitation of overtones and combinations of the fundamental vibrational modes that give rise to mid-IR absorption. This type of absorption is weak and is generally only observed for bonds involving hydrogen (e.g., C—H, O—H, N—H). In addition, the bands in NIR spectra are much broader and less well defined than bands in mid-IR spectra. Thus, NIR spectroscopy does not provide the same kind of detailed structural information as mid-IR spectroscopy and is of limited utility in qualitative analysis. However, during the past 20 years, it has emerged as a powerful technique for quantitative analysis, and its applications in routine analysis far exceed those of mid-IR spectroscopy. This is primarily because NIR spectroscopy is a more versatile technique in terms of the types of samples that can be analysed; in particular, NIR reflectance spectroscopy allows for the analysis of solid, opaque materials without any sample preparation beyond grinding to a powder. Thus, NIR spectroscopy is used extensively for the rapid quantitative determination of moisture, lipid, protein, carbohydrate and fibre in food products and agricultural commodities, such as grains and oilseeds (Williams and Norris, 1987; Osborne *et al.*, 1993). Another advantage of NIR spectroscopy, in relation to industrial utilisation, is the availability of low-cost fibre optics that are transparent to NIR radiation, making NIR spectroscopy useful for on-line process monitoring applications.

This chapter will review certain aspects of the IR spectroscopic analysis of fats and oils. A historical overview will be given first. We will then describe the instrumentation involved in FTIR spectroscopy and the characteristics of the two sample handling techniques that are most commonly employed in the analysis of fats and oils. Some of the mathematical tools used in data handling will also be discussed. Finally, following a brief discussion of the wet chemistry analytical techniques that can potentially be replaced by IR methods, the recent developments in the application of mid-IR and NIR spectroscopy in the analysis of fats and oils will be reviewed.

8.2 History of lipid analysis by infrared spectroscopy

The investigation of the infrared spectral characteristics of lipids dates back to the beginning of the 20th century, when the pioneering studies demonstrating the potential general utility of IR spectroscopy as a tool for qualitative analysis were performed. In fact, the first compilation of IR spectra, published by Coblentz in 1905, included the spectra of several fatty acids and vegetable oils, and a paper entitled 'The infra-red spectra of vegetable oils' also appeared in the early literature on IR spectroscopy (Gibson, 1920). However, major progress in the interpretation of the IR spectra of lipids was not made until the late 1940s and early 1950s, a period that represents the beginning of modern IR spectroscopy, as advances made during the Second World War led to the commercial availability of IR spectrometers and their widespread use in organic research laboratories. In this postwar period, a number of researchers conducted detailed studies of the IR spectra of pure reference materials in order to build the necessary knowledge base for the application of IR spectroscopy in fats and oils analysis. Following this, it became obvious that the potential applications of IR spectroscopy in fats and oils analysis were numerous. In the past fifty years, IR spectroscopy has been employed as a tool for qualitative and quantitative analysis of fats and oils, as well as for the elucidation of the chemical structure of fatty acids.

8.2.1 Determination of the geometric configuration of double bonds

An important result of early work on the IR spectra of lipids was the recognition of the potential use of IR spectroscopy to distinguish between *cis* and *trans* double bonds. In 1947, Rasmussen and co-workers had reported the IR spectra of seven octenes. Two bands characterised *trans* double bonds, a strong band at $966\,cm^{-1}$ ($10.3\,\mu m$) assigned to the $C{=}C{-}H$ deformation and a weak band at $1667\,cm^{-1}$ ($6.0\,\mu m$) assigned to the $C{=}C$ stretching vibration (Rasmussen *et al.*, 1947). The *trans* band at $966\,cm^{-1}$ was subsequently observed in the spectra of *trans*-monounsaturated fatty acids, esters, and triglycerides (Shreve *et al.*, 1950a) and has since been employed for both the qualitative and quantitative analysis of a variety of fats and oils.

In 1949, Lemon and Cross demonstrated that the band at $966\,cm^{-1}$ was present in the spectra of samples taken during hydrogenation of unsaturated oils, as a result of a *cis–trans* isomerisation occurring during the reaction. Subsequently, Feuge *et al.* (1953) employed IR spectroscopy to measure *trans* isomer formation during hydrogenation of methyl linolenate under various experimental conditions. Following this, IR spectroscopy was employed extensively in the study of *trans* isomer formation in catalytic hydrogenation processes as a function of various parameters, such as temperature, nature of the catalyst and the solvent, composition of the oil, and degree of dispersion of hydrogen gas in the oil (O'Connor, 1961).

Ahlers *et al.* (1953) noted small hypsochromic shifts of the *trans* absorption band in the spectra of fatty acids containing conjugated double bonds. These shifts varied with the number and the configuration of the conjugated bonds (Table 8.1), making it possible, for example, to distinguish between *cis,trans*- and *trans,trans*-conjugated bonds.

In the 1950s, many groups focused their attentions on the development of an IR method for the quantitation of non-conjugated *trans* double bonds based on measurement of the band at 966 cm^{-1} (Ahlers *et al.*, 1953; Jackson and Callen, 1951; Knight *et al.*, 1951; Shreve *et al.*, 1950b; Swern *et al.*, 1950) and in 1959 a tentative method was adopted by the American Oil Chemists' Society (AOCS) as Official Method Cd 14-61. In this method, which is schematically represented in Figure 8.1, a sample of oil or fat is dissolved in carbon disulfide, and its spectrum is recorded in a fixed path length (0.2–2.0 mm) transmission cell. The peak height at

Table 8.1 Characteristic C—H deformation frequencies of *trans* bonds

Type of *trans* bond	Frequency (cm^{-1})[a]
Isolated *trans*	967
Conjugated *cis–trans*	983
Conjugated *trans–trans*	988
Conjugated *cis–cis–trans*	989
Conjugated *cis–trans–trans*	991
Conjugated *trans–trans–trans*	994

[a]From Chapman (1965a,b).

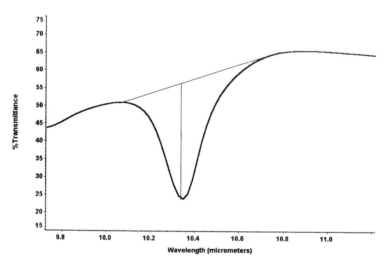

Figure 8.1 Measurement of the characteristic IR absorption band of *trans* carbon–carbon double bonds in accordance with Official Method Cd 14-61 of the American Oil Chemists' Society.

10.3 μm (966 cm^{-1}) is then measured relative to a baseline drawn between 10.05 μm (995 cm^{-1}) and 10.67 μm (937 cm^{-1}). Calibration involves recording the spectrum of trielaidin (a C18:1t triglyceride) under the same conditions. The *trans* content of the sample is then expressed as %*trans* relative to a value of 100% for trielaidin, using the following equation:

$$\%trans = (a_{\text{sample}}/a_{\text{trielaidin}}) \times 100 \qquad (8.1)$$

where $a = A/bc$, and A is absorbance (peak height), b is the cell path length in centimetres, and c is the concentration of the CS_2 solution used in the measurement, expressed in grams per litre. A modified procedure is required for samples containing low levels of *trans* isomers as their *trans* values are overestimated by the method because the contribution that a weakly overlapping triglyceride absorption makes to the measured peak height at 966 cm^{-1} becomes significant. The AOCS method requires that samples containing less than 15% isolated *trans* isomers be saponified and the fatty acids converted to methyl esters prior to analysis. For such cases, calibration is performed with methyl elaidate in place of trielaidin and slightly different baseline points are employed in calculating the *trans* peak height.

The method described above has been widely utilised in the fats and oils industry, particularly in the analysis of partially hydrogenated oils. Because of the importance of this method, it has been the subject of numerous investigations (e.g. Chapman, 1965b; Firestone and De La Luz Villadelmar, 1961; Firestone and LaBouliere, 1965; Huang and Firestone, 1971a,b; Madison *et al.*, 1982; Sreenivasan and Holla *et al.*, 1967). Much of this work has been reviewed by Firestone and Sheppard (1992), and the various suggested improvements of the method have been enumerated by Mosssoba *et al.* (1996). Recent modifications of the method will be discussed in Section 8.6.2.1.

8.2.2 *Studies of lipid autoxidation*

IR spectroscopy has also been used extensively as an analytical tool in the study of the autoxidation of lipids. Honn and co-workers (1949) monitored the autoxidation of linseed oil over time by IR spectroscopy and observed a 3–4 hour induction period following which a sharp band at 3450 cm^{-1} and a broad band at 980 cm^{-1} rapidly increased in intensity. The sharp band was attributed to the total hydroxyl content of the oxidised oil. Infrared spectroscopy alone did not allow for the separation and identification of the various hydroxy oxidation products. Other analytical tools were employed. Thus, the concentration of alcohols was determined by subtraction of the titrimetrically determined concentration of hydroperoxides and free fatty acids from the total hydroxy compound content measured by IR spectroscopy.

Dugan and co-workers (1949) conducted an extensive study of the influence of autoxidation on the IR spectrum of methyl linoleate. They recorded the IR spectra of samples having peroxide values (PV) ranging from $PV = 1$ to $PV = 940$ meq kg^{-1} and compared them to the spectra of aliquots of the same samples in which the PV was reduced to zero by treatment with KI. They detected major changes in two regions: 3550–3400 cm^{-1}, the hydroxy absorption region, and 1775–1650 cm^{-1}, the carbonyl absorption region. In particular, a band at 3470 cm^{-1} was predominant in highly oxidised samples and absent in samples of $PV < 1$. A double maximum, at 1724 cm^{-1} and 1733 cm^{-1}, present in all samples, and an additional small maximum at 1755 cm^{-1}, present in some samples, were observed. The authors explained the appearance of a split maximum by the presence in the autoxidised samples of at least two separate carbonyl compounds, formed in differing amounts. The band at lower frequency was attributed to keto groups conjugated to a double bond, and the higher frequency band to non-conjugated keto groups. The weaker 1755 cm^{-1} band was attributed to aldehydes, resulting from the scission of primary oxidation products. This study led to utilisation of these regions as the criteria for the purity of fats and oils with respect to oxidation.

Knight and co-workers (1951) further ascertained that the oxidation of fatty acids resulted not only in the increasing concentration of hydroperoxides, carbonyl and hydroxy compounds, but also in the increase of *trans* content. This suggested that the mechanism of peroxidation involves *cis–trans* isomerisation.

Concurrently, Lemon and co-workers (1951) studied the effects of incubation temperature and added iron on the oxidation of the esterified fatty acids of peanut oil. They observed that increased absorption due to oxidation occurred principally in three regions: 3730–3005 cm^{-1}, 1840–1630 cm^{-1}, and 1130–740 cm^{-1}. Along with observations that confirmed previous band assignments for hydroperoxides, carbonyl compounds and *trans* double-bonds, they assigned a band at 1710 cm^{-1} and another between 980 and 880 cm^{-1} to free fatty acids, which can also be formed during oxidation.

In 1951, Henick published the first IR study of a fat or oil since the work of Gibson in 1920, titled 'Detection of deterioration products of autoxidising milk fat by infrared spectrophotometry'. The course of the autoxidation of milk fat was monitored through the changes in three parameters: the IR spectra of the volatile components, such as carbonyl compounds, the acceptability of the flavour of the fat, and the peroxide value. The author's aim was to correlate the loss of flavour and appearance of 'off flavours' with the fluctuations in the intensity of specific bands. Henick (1951) focused on the changes, over time, in the hydroperoxide region ($\sim$3450 cm^{-1}) and in the carbonyl region ($\sim$1755–1725 cm^{-1}) and compared the IR spectral data with the opinions of a taste panel. The appearance of a sharp band at 1754 cm^{-1} correlated with the detected loss of flavour, and that of one at 1740 cm^{-1}, with the level of freshness of the fat. The findings of this study showed that the region between 1820 and 1610 cm^{-1} could be used as an indicator of flavour and fat acceptability.

8.2.3 Mid-IR band assignments

By the 1950s, many IR band assignments were listed in the literature. Shreve and co-workers (1950a) described the various bands present in the spectra of four groups of long-chain fatty acids and glycerides. They outlined common spectral features as well as the spectral differences between each group. This study stressed the potential utility of IR spectroscopy in characterisation and identification of fats and oils. O'Connor and co-workers (1951) identified the bands that follow Beer's law from the spectra of seven fatty acids and their methyl and ethyl esters. These data had a large impact on further studies of the quantitative assessment of fatty acids by IR spectroscopy. Other groups confirmed the assignment of bands using the spectra of different saturated and unsaturated fatty acids (Ahlers et al., 1953; Sinclair et al., 1952a,b). Mono-, di- and triglycerides were also characterised, in chloroform and as pure melts (Chapman, 1960).

As a result of the fundamental studies of the IR spectra of fatty acids and esters and triglycerides conducted in the 1950s, the assignments of the major bands were established. Figure 8.2 shows the FTIR spectrum of a partially hydrogenated vegetable oil in a 0.025 mm transmission cell; as the bands in the 3000–2850 cm^{-1} region are off-scale in this spectrum, the inset shows this spectral region recorded with a path length of 0.010 mm. The corresponding band assignments are tabulated in Table 8.2. Small variations in the band positions listed in Table 8.2 occur depending on the triglyceride composition of the sample. For example, for pure triglycerides, the position of the band due to the C—H stretching vibration of cis double bonds depends on the number of double bonds in the fatty acid chain, shifting from 3005 cm^{-1} for triolein (18:1c) to 3009 cm^{-1} for trilinolein (18:2c) and 3011 cm^{-1} for trilinolenin (18:3c).

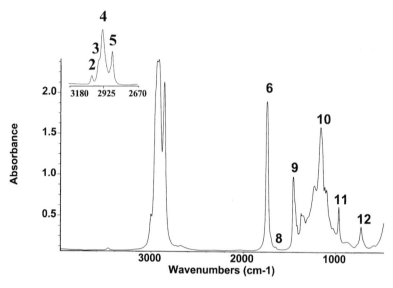

Figure 8.2 FTIR spectrum of a partially hydrogenated soybean oil in a 0.025 mm transmission cell. The inset shows the C—H stretching region of the spectrum recorded in a 0.010 mm transmission cell. The *trans* =C—H stretching absorption (band 1, Table 8.2) and the *trans* C=C stretching absorption (band 7, Table 8.2) are not visually discernible in the spectrum of this sample (*trans* content of the sample is ~40%).

Table 8.2 Selected characteristic infrared absorption bands of vegetable oils

Band	Band position (cm^{-1})	Assignment
Mid-IR[a]		
1	3025	v_{sym}(=C—H) (*trans* double bonds)
2	3008	v_{sym}(=C—H) (*cis* double bonds)
3	2953	v_{asym}(C—H) (CH_3 groups)
4	2925	v_{asym}(C—H) (CH_2 groups)
5	2854	v_{sym}(C—H) (CH_2 and CH_3 groups)
6	1746	v(C=O) (ester linkage)
7	1666	v(C=C) (*trans*)
8	1654	v(C=C) (*cis*)
9	1459	CH_2 scissoring deformation
10	1161	v(C—O)
11	966	Out-of-plane =C—H bending (*trans*)
12	722	CH_2 rocking
NIR[b]		
A	6200–5300	1st CH overtone region
B	4900–4500	Combination band regions
	7400–6700	
C	8900–8000	2nd CH overtone region

[a]See Figure 8.2.
[b]See Figure 8.3.

8.2.4 Solid-state spectra

The IR studies discussed thus far were conducted on either neat liquids or solutions. More complicated spectra are obtained in the solid state, and these spectra can yield additional valuable information. Sinclair *et al.* (1952a) reported that no qualitative differences were apparent between the spectra of saturated fatty acids of different chain lengths in solution; however, the spectra recorded from samples in the crystalline state showed major differences with changes in chain length. In particular, a progression of absorption bands of uniform spacing ($\sim$20 cm^{-1}) and intensity was observed between 1350 and 1180 cm^{-1}, with the number of bands in the progression increasing as a function of chain length (Jones *et al.*, 1952). These band progressions, and similar band progressions observed between 1070 and 710 cm^{-1}, are assigned to wagging and twisting-rocking vibrations of the methylene groups in the fatty acid chains and arise from different phase relationships between vibrations of adjacent methylene groups in the fatty acid chains. Keeney (1962) described an IR method for the determination of the average carbon chain length of saturated fatty acid esters from the band progressions. In subsequent work on the salts of fatty acids, Kirby *et al.* (1965) demonstrated that in the case of monounsaturated fatty acids the band progressions could be used to identify the position of the double bond in the chain. Thus, these band progressions proved highly useful in the elucidation of the structure of unknown fatty acids isolated from various plant sources. Solid-state spectra were also used to differentiate the polymorphic forms in which lipids crystallise (Chapman, 1965a).

8.2.5 NIR spectral characteristics of lipids

The near-infrared region (12 500–4000 cm^{-1}, 800–2500 nm) has been used in the analysis of fats and oils to a much lesser extent than the mid-IR region. The NIR spectra of fatty acids were first studied by Holman and Edmondson (1956). These authors recorded spectra between 900 and 3000 nm, which encompasses the region of first overtones of C—H stretching vibrations (6200–5300 cm^{-1} (1600–1900 nm)), the region of second overtones of C—H stretching vibrations (8900–8000 cm^{-1} (1100–1250 nm)), and the combination band regions (4900–4500 cm^{-1} (2000–2200 nm) and 7400–6700 cm^{-1} (1350–1500 nm)), which contain bands due to combinations involving the C—H stretching vibrations and other vibrational modes. These four major regions of interest in the analysis of lipids are shown in Figure 8.3, which presents the NIR spectrum of a partially hydrogenated soybean oil. Holman and Edmondson (1956) showed that *cis* double bonds gave rise to characteristic absorption bands

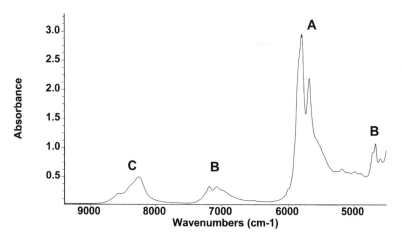

Figure 8.3   FT-NIR spectrum of a partially hydrogenated soybean oil in a 0.8 mm (o.d.) glass vial.

at 2190 and 2130 nm in the first combination band region as well as weak first overtones at 1680 nm and second overtones at 1180 nm. No bands due to *trans* double bonds were detected. These authors concluded that NIR spectroscopy could provide a suitable means for quantitation of *cis* unsaturation, which was not readily achieved by mid-IR spectroscopy because of the lack of a sufficiently intense absorption band characteristic of *cis* unsaturation that could be well resolved with the instrumental resolution available at that time. The development of a quantitative method for the determination of *cis* unsaturation by NIR spectroscopy was subsequently investigated by Holman *et al.* (1959) and Fenton and Crisler (1959). Another finding of early NIR work was the identification of bands at 1460 and 2070 nm as characteristic of hydroperoxides (Holman *et al.*, 1958). These bands, assigned to the overtone of the OO—H stretching absorption and to a combination absorption involving this vibrational mode, were better resolved from absorptions due to other hydroxyl-containing products (e.g., alcohols) than in the mid-IR spectra.

8.3   Fourier transform infrared spectroscopy

The work described thus far was conducted using dispersive IR spectrometers, so called because they employed a prism or a grating to disperse the radiation emitted by an infrared source into its component wavelengths. Beginning in the late 1960s, the principles of interferometry and Fourier transform spectroscopy were applied in the design of mid-IR spectrometers. The advent of these FTIR spectrometers, and the subsequent decrease in their cost, has almost eliminated the use of dispersive instruments and has increased the utility of mid-IR spectroscopy as an analytical tool.

The interferometer is the central component of an FTIR spectrometer. Interferometry provides a means of encoding both frequency and intensity information in the signal that reaches the detector, thereby eliminating the need for a dispersive element, and the Fourier transform provides the means of decoding this information. The advantages conferred by interferometry are the increased speed of data collection, making it possible to record the whole IR spectrum in one scan; the increased signal-to-noise ratio; the higher energy throughput; and the increased wavelength precision achieved through the use of an internal reference laser.

Figure 8.4 shows the trajectory of light through a Michelson interferometer, which is the type of interferometer employed in most FTIR spectrometers. In the Michelson interferometer, the beam is first divided into two equal components by a beamsplitter. One component of the original beam travels to a fixed mirror, while the other travels to a moving mirror. A laser beam, commonly from a helium–neon laser, is used to track the moving mirror. The two beam components are reflected back, and they recombine at the beamsplitter, producing a constructive/destructive interference pattern due to the varying differences in the distances travelled by each component. Half of the recombined IR beam is reflected back to the source and the other half reaches the sample holder, where the IR radiation is selectively absorbed by the sample. Sample handling devices will be discussed below.

After this, the beam arrives at the detector. FTIR spectrometers normally employ pyroelectric bolometers (e.g. deuterated triglycine sulfate (DTGS)) or photoconductive detectors. The detector response to the impinging beam is measured continuously. This analog signal is converted to a digital signal by an A/D converter, yielding an interferogram (Figure 8.5), which is a plot of the spectral data in the time domain. The data are then converted to the frequency domain by a fast Fourier transform (FFT) algorithm. The spectrum thus obtained, termed an emittance or single-beam spectrum, is a plot of the intensity reaching the detector against the frequency of the radiation and must be digitally ratioed against the emittance spectrum recorded with an open beam (i.e., no sample in the beam) to obtain the transmittance (T) spectrum of the sample. The latter spectrum may then be mathematically converted to an absorbance spectrum ($A = \log(1/T)$).

In NIR spectroscopy, dispersive and filter-based instruments are still widely used, as the advantages of interferometry are not as substantial in

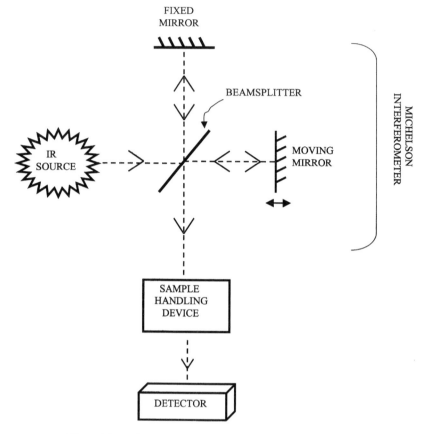

Figure 8.4 Trajectory of the IR beam in an FTIR spectrometer.

NIR spectroscopy as in mid-IR spectroscopy. However, the superior wavelength precision of FT-NIR instruments is an important advantage, as calibration drift due to wavelength stability over time, and the resulting need for frequent recalibration, is a major problem associated with the use of dispersive NIR spectrometers in routine analytical applications. The design principles of FTIR spectrometers operating in the mid-IR and NIR regions of the electromagnetic spectrum are identical, and, in fact, instruments that are capable of covering both regions are available. However, different sources and detectors are employed for the two regions. The source of mid-IR radiation can be a Globar (silicon carbide), a Nernst glower (rare-earth oxides), or a Nichrome coil. In FT-NIR spectroscopy, the source is usually a tungsten filament lamp. The materials employed as detector elements in the

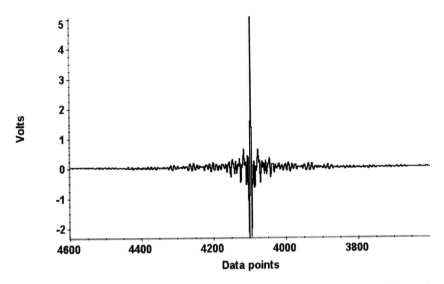

Figure 8.5   An interferogram recorded by an FTIR spectrometer. The sample analysed is canola oil.

mid-IR region spectroscopy are usually DTGS or mercury cadmium telluride (MCT) alloys; the latter are photoconductive and require cooling with liquid nitrogen. In FT-NIR spectroscopy, DTGS or lead sulfide is used for detection in the region between $10\,000$ and $4000\,\mathrm{cm}^{-1}$ (1–$2.5\,\mu\mathrm{m}$), and silicon is used for shorter wavelengths. An indium/gallium/arsenic composite can also be used.

8.4   Sample handling techniques

Several books contain detailed descriptions of the various sample handling techniques used in mid-IR (Coleman, 1993; Ferraro and Krishnan, 1990; Miller and Stace, 1979) and NIR spectroscopy (Williams and Norris, 1987; Osborne *et al.*, 1993). Only two will be discussed here: transmission measurements, which are employed in both mid-IR and NIR spectroscopy, and attenuated total reflectance, employed in mid-IR spectroscopy.

8.4.1   Transmission measurements

In mid-IR spectroscopy, a commonly used sample handling device is the transmission cell. This device consists of two polished windows, which are made of salt crystals (e.g. NaCl or KBr) and are separated by a Teflon spacer. The fluid sample is introduced between the windows, and the width of the spacer determines the path length. The IR beam passes through the sample and only the radiation that is not absorbed by the sample reaches the detector. Because of the low amount of energy produced by mid-IR sources and the high absorptivities of the absorption bands, very short path lengths are generally required. Figure 8.6 shows the transmittance spectrum of a vegetable oil in a transmission cell with a path length of 0.025 mm and illustrates that, even with this short path length, the percentage transmittance is very low in certain regions of the spectrum, such as the C—H absorption region between 3000 and 2850 cm^{-1}. In cases where these regions are not under study, 0.025 mm is a suitable path length for recording the spectra of neat oils. In cases where minor components, such as free fatty acids or oxidation products, are under investigation, the optimal path length depends on the concentration range of the compounds studied.

Sample handling for NIR transmission measurements is simpler than in the case of mid-IR spectroscopy because the lower absorptivities of the bands in NIR spectra and the greater amount of energy available from NIR sources allow for the use of longer path lengths (1–10 mm). In addition, spectra can be recorded from samples contained in quartz cuvettes or disposable glass vials because, unlike mid-IR radiation, NIR radiation is not strongly absorbed by these materials.

Many NIR spectrometers are configured for operation in the reflectance mode to allow for the analysis of opaque samples. With systems of this type, liquid samples can be analysed by the transflectance

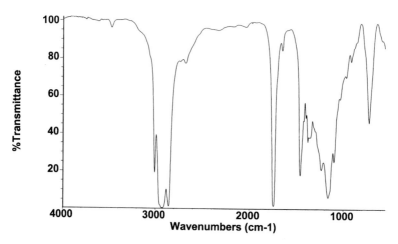

Figure 8.6   Transmittance spectrum of corn oil in a 0.025 mm transmission cell.

technique, which entails transmission of the NIR beam through the sample onto a reflecting surface and its reflection back through the sample to the detector, resulting in the effective doubling of the path length.

8.4.2 Attenuated total reflectance

The attenuated total reflectance (ATR) technique provides a means of overcoming many of the sample handling problems that have limited the full exploitation of mid-IR spectroscopy as an analytical technique in a number of areas, including food analysis (Sedman *et al.*, 1998a). ATR sample handling devices are based on the phenomenon of the total internal reflection of light (Harrick, 1967). When light travelling in a medium of high refractive index strikes an interface with a medium of lower refractive index at an angle equal to or greater than the critical angle, the light will totally internally reflect in the medium of higher refractive index. This gives rise to an evanescent wave in the medium of lower refractive index whose amplitude decays exponentially with distance from the interface. The attenuation of the evanescent wave due to the absorption of energy by species present in the medium of lower refractive index results in the extraction of energy from the totally internally reflected light beam. Thus, measurement of the energy of the totally internally reflected light as a function of frequency yields the absorption spectrum of these species.

In ATR sampling devices, the sample is placed in contact with an 'internal reflection element' (IRE), which is a crystal of an IR-transmitting material of high refractive index, such as zinc selenide or germanium. Light from the infrared source impinges on the IRE and reaches the interface between the IRE and the sample at such an angle that the beam totally internally reflects in the IRE. Thus, an evanescent wave emerges in the sample, the material of lower refractive index. Its amplitude decays exponentially with the distance from the interface. The depth of penetration, d_p, defined as the distance over which the wave decays to $1/e$ ($\sim$37%) of its amplitude at the surface, is given by

$$d_p = \lambda / \{2\pi n_1 [\sin^2(\theta) - (n_2/n_1)]^{1/2}\}$$

where λ is the wavelength of the radiation in the IRE, n_1 is the refractive index of the IRE material, n_2 is the refractive index of the sample, and θ is the angle at which the incident light strikes the interface. Because the depth of penetration increases with increasing wavelength, the spectral data must be mathematically corrected in order to make the spectra comparable to those obtained using a transmission sampling device.

The depth of penetration is very short: in the mid-IR region, it lies between 1 and 4 μm. When higher sensitivity is needed, the effective path length can be extended by using a multiple-bounce ATR device, in which the light is made to bounce through the IRE so that it undergoes multiple internal reflections (Figure 8.7). One of the most common ATR sampling accessories is the horizontal ATR (HATR, usually 1×8 cm), which can hold sample volumes between 0.2 and 0.5 ml. It is also possible to use a temperature-controlled HATR, which is particularly advantageous in the analysis of fats and oils.

Fibre-optic ATR probes are also available. They are useful in cases where the analysis must be performed at some distance from the spectrometer, or for the rapid sampling of bulk materials.

Much of the work reported in the literature on FTIR analysis of fats and oils has employed the ATR sampling technique. It has been used in the classification of oils based on their degree of unsaturation (Afran and Newbery, 1991; Safar et al., 1994); in the determination of the degree of *trans* unsaturation (Ali et al., 1996; Mossoba et al., 1996), iodine value and saponification number (van de Voort et al., 1992b), and free fatty acid content (Ismail et al., 1993); in the monitoring of a hydrogenation process (Dutton, 1974); and in the detection of adulteration of oils (Lai et al., 1994, 1995). In addition, heated HATR sampling accessories can be employed to monitor spectral changes as a function of time/temperature, thereby providing a powerful means for investigating temperature-dependent phenomena such as phase changes, thermal degradation, and lipid autoxidation. For example, when an oil sample is spread on the surface of a heated HATR, it undergoes accelerated oxidation due to the combination of elevated temperature and the large surface area of the sample that is exposed to air, allowing for real-time monitoring of the

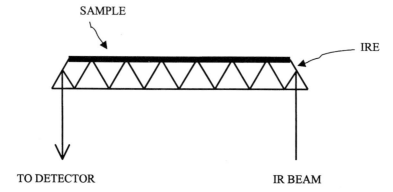

Figure 8.7 Trajectory of light through a multiple-bounce ATR sampling device.

formation of primary and secondary oxidation products by FTIR spectroscopy (van de Voort *et al.*, 1993). Sedman *et al.* (1996) demonstrated the potential utility of this approach for the evaluation of the performance of antioxidants.

The suitability of the ATR technique depends on the nature of the sample, and the analytical performance may not always match that of transmission IR techniques. The inherently short effective path length associated with the ATR technique is disadvantageous for applications requiring the detection of species present in low concentrations. Furthermore, because the evanescent wave interacts only with the molecules immediately at the interface, the sample must be homogeneous on the scale of the depth of penetration if the spectrum is to be representative of the composition of the whole sample. In the case of lipid analysis, extra care must be taken in the cleaning of the IRE surface to avoid cross-contamination due to the strong adhesion of lipid molecules to many of the common IRE materials (termed the 'memory effect'; van de Voort *et al.*, 1992). Despite these limitations, the ATR technique provides a simple and convenient mean of acquiring spectral information from a wide variety of samples, many of which are not readily amenable to IR analysis with the more common transmission sampling techniques.

8.5 Mathematical approaches to data analysis

IR spectroscopy is a secondary method of analysis, which requires a calibration with a set of standards of known composition, prepared gravimetrically or analysed by a primary chemical method, in order to establish the relationship between IR band intensities and the compositional variable(s) of interest. Various approaches are employed in the development of a quantitative analysis method. In the case of mid-IR spectroscopy, a Beer's law plot is an adequate basis for calibration in the case of a simple system, such as a single component dissolved in a non-interacting solvent. However, sophisticated multivariate analysis techniques are required for more complex systems and are almost always employed in NIR quantitative analysis owing to the complexity of NIR spectra. The data handling capabilities of modern computer-controlled spectrometers have allowed these latter techniques to be implemented in the instrument software and applied directly to spectral data, bringing a resurgence to quantitative IR spectroscopy during the past two decades.

As with other types of absorption spectroscopy (e.g., UV-visible spectroscopy), the basis of quantitative analysis in IR spectroscopy is the

Bouguer–Beer–Lambert law or Beer's law:

$$A_{\bar{v}} = e_{\bar{v}} bc$$

Here, $A_{\bar{v}}$ is the absorbance measured at wavenumber $\bar{v}$, $e_{\bar{v}}$ is the molar absorption coefficient of the absorbing species at this wavenumber, b is the path length of the IR cell, and c is the concentration of the absorbing species. Application of Beer's law for the determination of the amount of a compound present in a solution requires that $e_{\bar{v}}$ be determined by measuring the absorbance of a solution of known concentration. Of course, in order to attain better accuracy, it is preferable to prepare a series of solutions of different concentrations, spanning the concentration range of interest, and obtain $e_{\bar{v}}b$ from a plot of absorbance vs concentration by linear least-squares regression. This procedure averages out the errors due to instrumental noise, measurement errors, and other sources of random variation. In addition, it allows deviations from Beer's law in the concentration range of interest to be detected, such as may arise from hydrogen bonding, dimerisation, and other intermolecular interactions. Such interactions may then be modelled by the introduction of higher-order terms into the equation relating absorbance to concentration.

When more than one component is present in the solution, the above approach will generally not be satisfactory as it cannot account for any contributions of additional components to $A_{\bar{v}}$, nor can it model interactions between components. Therefore, a variety of multivariate calibration techniques have been applied in the analysis of multi-component systems (Martens and Naes, 1989), including the K-matrix and P-matrix methods and principal component regression (PCR) and partial-least-squares (PLS) regression, both of which are forms of factor analysis. Among these various approaches, PLS has emerged as the technique of choice in the FTIR analysis of complex multicomponent systems (Fuller *et al.*, 1988a,b; Haaland, 1988; Haaland and Thomas, 1988a,b; Heise *et al.*, 1994).

A key difference between PLS and the traditional approach to IR quantitative analysis, described above, is that a PLS calibration does not entail establishing direct relationships between concentration and absorbance measurements at specified frequencies (i.e. peak heights or peak areas). Instead, a PLS calibration model is developed by compressing the spectral data for the calibration standards into a set of mathematical 'spectra', known as the loading spectra or factors, which are linear combinations of the calibration spectra. The spectrum of each calibration standard is then decomposed into a weighted sum of the loading spectra, and the weights given to each loading spectrum, known as 'scores', are regressed against the concentration data for the standards. In the prediction step, the amounts of each loading spectrum employed in reconstructing the spectrum of the unknown, i.e. the 'scores', are then used to predict the concentration of the unknown.

A PLS calibration can, in principle, be based on the whole spectrum, although in practice the analysis is restricted to regions of the spectrum that exhibit variations with changes in the concentrations of the components of interest. As such, the use of PLS can provide significant improvements in precision relative to methods that use only a limited number of frequencies (Haaland and Thomas, 1988a). Furthermore, because PLS treats concentration rather than spectral intensity as the independent variable, PLS is able to compensate for unidentified sources of spectral interference, although all such interferences that may be present in the samples to be analysed must also be present in the calibration standards. The powerful data reduction capabilities of PLS can also be exploited to establish relationships between quality attributes or physicochemical properties and FTIR spectral data.

8.6 Recent applications in the analysis of fats and oils

The fats and oils industry routinely employs a number of 'official' analytical methods of the American Oil Chemists' Society (AOCS) and the Association of Official Analytical Chemists (AOAC) for process control, for quality control of its products and as a basis for setting product specifications. These methods are all well defined and widely used. However, many of them are time consuming, tedious, expensive and not easily amenable to automation. They also often require large amounts of solvents and reagents that are both hazardous and increasingly difficult to dispose of. The drawbacks associated with the traditional methods of analysis have prompted researchers working in IR spectroscopy to develop new methods of fats and oils analysis that circumvent these problems. In the following sections, we will briefly describe some of the established wet chemical methods used for the analysis of fats and oils, as the applications of IR spectroscopy that will be described later in this chapter pertain to these particular methods.

8.6.1 Chemical methods

8.6.1.1 Bulk characterisation
Fats and oils are composed predominantly of triglycerides, which can be characterised according to various structural features. Most important

are the chain lengths of the fatty acids attached to the glycerol backbone, the degree of unsaturation of the chains, and the geometric configurations of their double bonds. In addition, the solids content of fats is an important quality control parameter in the fats and oils industry because of its influence on the functional characteristics of margarines, short-enings and other fat blends.

Saponification number and iodine value. Two of the most widely performed analyses in the fats and oils industry (Sonntag, 1982) are the determination of saponification number (SN) and iodine value (IV), which are measures of mass-average molecular mass and total unsaturation, respectively. These analyses are titrimetric methods based on the reactions in equations (8.2) and (8.3), respectively:

$$-\text{CH}_2\text{OCOR} + \text{NaOH} \rightarrow -\text{CH}_2\text{OH} + \text{RCOONa} \qquad (8.2)$$

$$
\begin{array}{cc}
\text{RHC} = \text{CHR}' + \text{I}_2 \rightarrow \text{RHC} -\text{CHR}' & \\
| \quad | & (8.3) \\
\text{I} \quad \text{I} &
\end{array}
$$

with the saponification number being defined as the number of milligrams of potassium hydroxide required to saponify 1 g of the sample, and the iodine value being expressed as the number of centigrams of iodine absorbed per gram of sample (AOCS, 1989). Both of these analyses are tedious, time consuming, and expensive and involve the use of hazardous reagents.

Determination of trans *content.* The two most widely employed techniques for the analysis of *trans* content are gas chromatography (GC) and IR spectroscopy. The GC methods allow for the identification of individual *trans* as well as *cis* fatty acids; however, peak overlap can lead to an underestimation of C18:1*t* (Ratyanake *et al.*, 1990). Moreover, the GC methods require conversion of triglycerides to their fatty acid methyl esters and are not particularly suited to routine quality control applications in a process environment. The traditional dispersive IR method for the determination of total unconjugated *trans* isomers has been described in section 8.2.1, and recent modifications of this method that take advantage of the capabilities of FTIR instruments will be reviewed in section 8.6.2.1.

Solid fat index and solid fat content. In North America, the determination of solids content is performed by using dilatometry to obtain the solid fat index (SFI). SFI is an empirical measure of the change in the specific volume of the fat as a function of temperature (AOCS, 1989, Method Cd 10-57). In Europe, NMR spectroscopy is used to determine the solid fat content (SFC), and this method has recently been approved by the American Oil Chemists' Society (AOCS, 1989, Method Cd 16-81).

Both procedures involve measurements at a series of set temperatures and are fairly lengthy because tempering of the sample at each temperature is required in order to obtain reproducible values.

8.6.1.2 Oil quality indicators

Under aerobic conditions, fats and oils oxidise, forming hydroperoxides. Also, under hydrolytic conditions, the triglycerides undergo lipolysis, thereby releasing free fatty acids (FFAs). The concentrations of hydroperoxides and of FFAs, as well as of secondary oxidation products, are important indicators of oil quality, as these components strongly affect the organoleptic and functional properties of oils.

Peroxide value. Peroxide value (PV) is expressed in terms of milliequivalents of peroxides per kilogram of oil. Two methods for determination of PVs are accepted by the AOCS (AOCS, 1989, Methods Cd 8b-90 and Cd 8-53). Both methods are based on the stoichiometric conversion of KI to molecular iodine by hydroperoxides in an acidic environment and subsequent titration with standardised sodium thiosulfate to determined the amount of molecular iodine released:

$$ROOH + 2H^+ + 2I^- \rightarrow I_2 + ROH + H_2O \qquad (8.4)$$

$$I_2 + 2S_2O_3^{2-} \rightarrow S_4O_6^{2-} + 2I^- \qquad (8.5)$$

The two methods differ in the solvents that are used, with chloroform being discontinued because of environmental concerns.

Free fatty acids. One method for FFA determination is approved by the AOCS (AOCS, 1989, Method Ca 5a-40). It is based on the extraction of FFAs in ethanol and their titration with a defined concentration of base to a phenolphthalein endpoint.

Anisidine value. The breakdown of the hydroperoxides formed by the oxidation process to aldehydes, ketones, alcohols and hydrocarbons causes rancidity in fats and oils. Levels of aldehyde production can be calculated using the AOCS-approved anisidine value (AV) test (AOCS,

1989; Holm *et al.*, 1957), which is based on UV detection of the product formed by the reaction of *p*-anisidine with aldehydes. The extinction coefficient of the aldehyde/anisidine reaction products varies with the structure of the aldehyde. The AV test is particularly sensitive to α,β-unsaturated and α,β,γ,δ-unsaturated aldehydes. However, the extinction coefficients have been estimated for different classes of carbonyl compounds and so correlations are possible.

8.6.2   Mid-IR spectroscopy

8.6.2.1   Bulk characterisation
Iodine value and saponification number.   An ATR/FTIR method for the prediction of iodine value (IV) and saponification number (SN) was developed by van de Voort *et al.* (1992). A calibration set consisting of 18 pure triglyceride standards was employed in the development of partial-least-squares (PLS) calibration models for the prediction of the IV and SN of fats and oils. The use of these pure triglycerides as calibration standards has several advantages. First, the calibrations devised are 'universal' as they are applicable to all triglyceride-based oils and fats. Second, this calibration approach has the benefit of eliminating the need for chemical analyses of the calibration standards, as the reference values for the pure triglycerides are known from their molecular structure, and thus the accuracy of the IR method is not limited by the precision of a reference chemical method. Another key feature of this calibration approach is the use of PLS to establish correlations between the spectral data for these calibration standards and the corresponding reference values, as opposed to the traditional approach of attempting to identify a single absorption band whose height or area can be related to the measure of interest.

Van de Voort *et al.* (1995) subsequently extended this approach by developing PLS calibration models for the prediction of *cis* content, expressed as percentage triolein, and *trans* content, expressed as percentage trielaidin, by using an expanded set of triglyceride calibration standards. For this work, a heated transmission cell based sample handling accessory designed specifically for the analysis of oils and melted fats was constructed to facilitate automated analysis (van de Voort, 1994). In validation studies with more than 100 hydrogenated rapeseed and soybean samples, excellent internal consistency was obtained between the IV and *cis* and *trans* data predicted from the calibrations developed in this work (Sedman *et al.*, 1998b). In addition, for ~30 GC-analysed samples, the PLS IV predictions matched the IV calculated from the GC data within 1 IV unit. The SN calibration was validated with a set of 37 oil samples of different types (van de Voort *et al.*, 1992), covering a wide range of SN values (185–253), as the rapeseed and soybean validation samples did not span a sufficient range of SN values to allow an assessment of the predictive accuracy of this calibration. The validation data for the accuracy of the FTIR SN method relative to the chemical method (mean difference ~2.7 SN units, standard deviation of the differences ~2.0 SN units) were considered satisfactory, in view of the poor reproducibility of the chemical analyses (the standard deviation of the differences for duplicate analyses being 2.050 vs 0.333 for the FTIR method).

Determination of trans *content.* Over the years, a number of approaches aimed at increasing the accuracy of the IR method for the determination of *trans* content, described in Section 8.2.1, have been investigated (Firestone and Sheppard, 1992), and some of the improvements that have been suggested are incorporated in the revised AOCS method (AOCS, 1997, Method Cd 14-95). The newer method requires the conversion of all samples to methyl esters, regardless of *trans* content. As in the original method, samples are dissolved in CS_2 and their spectra recorded in a fixed path length (1 mm) transmission cell. Separate calibration equations are derived for the analysis of samples containing $\leq 10\%$ *trans* isomers and $> 10\%$ *trans* isomers. Other changes in the AOCS method concern the selection of baseline points in the measurement of the *trans* peak height; whereas fixed baseline points were formerly specified, the position at which the baseline is drawn in the modified method depends on the size of the *trans* peak.

Beyond updating the experimental protocol to reflect the data handling capabilities of modern IR spectrometers, the recent modifications to the AOCS method are largely directed towards improving the accuracy of IR *trans* analysis. A number of investigators have also made efforts to simplify the experimental procedure, particularly by analysing samples in their neat form and thereby eliminating the use of the volatile and noxious/toxic CS_2. For example, Sleeter and Matlock (1989) developed an FTIR procedure for measuring the *trans* content of oils, analysed as neat methyl esters using a 0.1 mm KBr transmission cell. This FTIR method was shown to provide better precision and a significant reduction in total analysis time (2.5 min/ sample) in comparison with the traditional AOCS method, as well as having the advantage of being amenable to automation (Sleeter and Matlock, 1989). These authors reported that more than 700 samples were run in the same cell over a period of 9 months without any need for recalibration, demonstrating the stability of the FTIR spectrometer (Sleeter and Matlock, 1989). Although this FTIR method greatly simplified the *trans* analysis, particularly by eliminating the need for CS_2, the analysis was still performed on methyl esters rather than neat fats and oils.

A means of eliminating the requirement for conversion to methyl esters is to employ the spectral ratioing capability of FTIR spectrometers to remove the contributions of triglyceride absorptions to the *trans* peak by ratioing the single-beam FTIR spectrum of the fat or oil being analysed against the single-beam spectrum of a similar reference oil that is free of *trans* groups (Mossoba and Firestone, 1996; Mossoba *et al.*, 1996; Sedman *et al.*, 1997). The ratioing method was adopted by the AOCS in 1996 as Recommended Practice Cd 14d-96 for the quantitation of isolated *trans* isomers at levels equal to or greater than 1% (AOCS, 1997; Firestone, 1996).

The main drawback of the ratioing method is the need to select a *trans*-free oil that is similar in composition to the samples to be analysed. When analysing partially hydrogenated soybean oils, Mossoba *et al.* (1996) found that the *trans* values obtained by using triolein as the reference oil were 2.6 percentage points higher than those obtained when a refined, bleached, and deodorised soybean oil served as the reference material. They attributed this difference primarily to the more similar composition of the soybean reference oil to that of the samples being analysed, making the ratioing out of the overlapping triglyceride absorptions more accurate. Thus, the limitations imposed on the ratioing method by the variability of the triglyceride absorptions among oils of different composition require more thorough examination, and the criteria for selection of an appropriate reference oil for a particular set of samples need to be defined.

Solid fat index (SFI). The underlying principle of SFI determination by FTIR spectroscopy is that the SFI profile of a fat is defined by its fatty acid composition and distribution, which in turn is characterised by the mid-IR spectrum of the melted fat, as the spectrum represents the superposition of all the contributions of the individual triglycerides making up the fat. By employing this approach, an FTIR SFI method was developed using 72 samples of partially hydrogenated soybean oil from 11 hydrogenation runs, obtained from a major US vegetable oil processor and preanalysed for SFI by the AOCS dilatometric method. The FTIR method was shown to have the potential to serve as a viable substitute for the traditional dilatometric method for the determination of solids content, with the advantage of a reduction in the analysis time from hours to minutes owing to the elimination of the tempering steps required in the dilatometric method (van de Voort *et al.*, 1996). A limitation of the FTIR approach to the determination of solids content is that the calibration models derived are only applicable to samples with similar characteristics to those of the standards used to derive the calibration.

8.6.2.2   Oil quality indicators
Determination of peroxide value.   Hydroperoxide functional groups can be quantitatively determined by mid-IR spectroscopy via measurement of their O—H stretching absorption, which is observed at $3550\,cm^{-1}$ in the spectra of fatty acid methyl ester hydroperoxides in CCl_4 solution (Fuzukumi and Kobayashi, 1972) but is shifted to $3444\,cm^{-1}$ and broadened in the spectra of oxidised oils owing to hydrogen bonding between the —OOH and the triglyceride ester C=O groups (van de Voort *et al.*, 1994). Although measurement of the hydroperoxide absorption band in the spectra of neat oils can serve as a basis for the determination of PV by FTIR spectroscopy, the development of an FTIR method is complicated by the need to account for spectral interferences due to other OH-containing species that may be present in oils, such as alcohols, mono- and diglycerides, free fatty acids, and water, all of which exhibit O—H stretching absorptions that overlap with the hydroperoxide band. Measurements of the height or area of the hydroperoxide absorption band are also affected by overlap with a triglyceride absorption band at $\sim3473\,cm^{-1}$, assigned to the first overtone of the ester carbonyl absorption at $1746\,cm^{-1}$, whose intensity varies among oils of different triglyceride compositions.

Van de Voort *et al.* (1994) explored a PLS calibration approach that could take all these effects into consideration. However, the inherent complexity of the OH stretching region in the spectra of edible oils and the need to extract a weak signal (owing to the low concentrations of hydroperoxides to be measured) from this complex spectrum made it difficult to develop a robust, 'universal' calibration (van de Voort *et al.*, 1994). Furthermore, the detection limit of this FTIR method, estimated to be 1.5 PV, was inadequate for the measurement of PV in freshly processed oils (PV < 0.5–1.0). For these reasons, the same group subsequently developed a simpler method that does not suffer from these limitations (Ma *et al.*, 1997). This method is based on the rapid reaction between hydroperoxides and excess triphenylphosphine (TPP), which leads to the formation of triphenylphosphine oxide (TPPO) in stoichiometric amounts (equation 8.6).

$$(8.6)$$

This reaction had previously been successfully employed by Nakamura and Maeda (Nakamura and Maeda, 1991) in a micro-assay for lipid hydroperoxides in biological samples, using a combination of HPLC and UV detection. The utility of this reaction for the determination of PV by FTIR spectroscopy is dependent on the ability to accurately quantitate TPPO in the presence of TPP. This proved to be readily achievable owing to the presence of a unique and sharp band at $542 \, \text{cm}^{-1}$ in the spectrum of TPPO, which is assigned to an X-substituent-sensitive phenyl vibration (Deacon and Green, 1968); the corresponding band in the spectrum of TPP is broad and shifted $\sim 40 \, \text{cm}^{-1}$ to lower frequency. A calibration equation relating PV to the height of the TPPO band at $542 \, \text{cm}^{-1}$ was derived by simple linear regression, and the calibration was validated by analysing both oxidised oils and oils spiked with t-butyl hydroperoxide. With a detection limit of 0.10 PV and a total analysis time of $\sim 2 \, \text{min}$, this simple FTIR method was judged by the authors to be highly suited for routine quality control applications in the fats and oils industry.

Determination of free fatty acid content. Oils containing free fatty acids (FFAs) exhibit a $v(C{=}O)$ absorption due to FFA dimers at $1711 \, \text{cm}^{-1}$. This band is difficult to measure accurately because it appears as a shoulder on the intense ester linkage carbonyl absorption band; at low FFA levels, the FFA band may not even be discernible. However, two methods of compensating for this band overlap in the determination of FFAs by FTIR spectroscopy have been proposed in the literature. Lanser *et al.* (1991) employed spectral deconvolution to sharpen the FFA band and diminish its overlap with the ester carbonyl absorption. The accuracy of this method in measuring the elevated levels of FFAs that are present in crude oils extracted from damaged soybeans was assessed with 19 samples; the FTIR-predicted values of FFA content were found to be within 0.5 percentage points of the values obtained by the AOCS titration method. In another study, Ismail *et al.* (1993) eliminated the spectral interference from the ester carbonyl absorption by ratioing the spectrum of the sample against the spectrum of an FFA-free oil of the same type. These authors demonstrated that FFA levels down to 0.05% could be detected after application of this spectral ratioing technique. However, accurate FFA analyses could not be obtained for oils that had undergone thermal or oxidative stress owing to the presence of various carbonyl-containing species (aldehydes and ketones) that spectrally interfere with the measurement of the FFA absorption band. For such samples, an indirect method was developed, involving extraction of the FFAs in 1% KOH–methanol and quantitation of the carboxylate anion by measurement of the $v(COO^-)$ absorption band at $1570 \, \text{cm}^{-1}$. Both the direct and the indirect method were shown to be comparable in precision and accuracy to the AOCS titration method.

Determination of anisidine value. Aldehydes exhibit strong bands in the 1730–1680 cm^{-1} region of the IR spectrum, owing to the high absorptivity of their C=O stretching vibrations, and the formation of aldehydes is readily observed in the FTIR spectra of oils undergoing oxidation, with individual peaks due to saturated aldehydes, α,β-unsaturated, and α,β,γ,δ-unsaturated being discernible (van de Voort *et al.*, 1993). On the basis of the ability to distinguish between these three classes of aldehydes, Dubois *et al.* (1996) formulated a synthetic calibration approach to the FTIR determination of AV involving individual quantitation of hexanal, *trans*-2-hexenal, and *trans,trans*-2,4-decadienal, these compounds having been selected to represent the three aldehyde classes.

Statistical analysis of the results indicated that the synthetic calibration approach yielded good predictive accuracy. Similar predictive accuracy was obtained from an alternative calibration approach based on the use of thermally stressed oils as calibration standards (Dubois *et al.*, 1996). Quantitative determination of AV by FTIR spectroscopy was shown to be feasible, and the synthetic calibration approach provided additional information on the aldehyde types present in a sample. This study provided the basis for the development of a rapid, automated FTIR method for AV analysis of thermally stressed fats and oils in their neat form without the use of chemical reagents, with possible application in the monitoring of the oxidative state of frying oils.

8.6.2.3 Authentication and detection of adulteration
The potential utility of FTIR spectroscopy for the authentication of vegetable oils has been examined in recent work by Lai *et al.* (1994), employing multivariate statistical methods, specifically, principal component analysis and discriminant analysis. These authors demonstrated that the spectral data for oils of different plant origin could be clustered by application of principal component analysis. However, they found that the differences between the spectra of some samples of different oil types were comparable in magnitude to the variations between replicate spectra of individual samples. It was therefore concluded that the development of a reliable method for classification of oils on the basis of their plant origin would require a larger data base of oil spectra, including replicate spectra of individual samples recorded over a period of time. These authors also demonstrated the ability of FTIR spectroscopy, in conjunction with discriminant analysis, to differentiate between extra virgin and refined olive oils, despite the strong similarities between the spectra of these two types of oil. They concluded that FTIR spectroscopy can potentially serve as a rapid and simple method for the detection, and possibly quantification, of adulteration of oils.

8.6.3 *NIR spectroscopy*

In recent years, several groups have used NIR spectroscopy in order to characterise and classify different types of oils and fats (Sato *et al.*, 1991; Bewig *et al.*, 1994; Sato, 1994; Wesley *et al.*, 1995, 1996). One of the major objectives of this work is to evaluate the potential utility of NIR spectroscopy for the detection of adulteration of oils. For example, Wesley *et al.* (1996) demonstrated the feasibility of identifying adulterants in extra virgin olive oil by NIR spectroscopy with the use of discriminant analysis. They also reported that the level of adulteration could be measured with an accuracy of $< 1\%$.

Ismail *et al.* (1996) investigated the suitability of FT-NIR spectroscopy as a technique for the bulk characterisation of oils through the measurement of iodine value and *cis* and *trans* content. For the development of calibration models, they utilised PLS and employed pure triglycerides as calibration standards, following the approach previously taken by these authors in the development of mid-IR methods (van de Voort *et al.*, 1992, 1995). The C—H second overtone region in the FT-NIR spectra of three of these triglyceride standards (tristearin (C18:0), triolein (C18:1*c*), and trielaidin (C18:1*t*)) is shown in Figure 8.8. From a visual comparison of these spectra, a band due to the *cis* double bond is readily discernible at $8563\,\mathrm{cm}^{-1}$ (1168 nm). This band and other NIR bands characteristic of *cis* isomers were identified as early as 1956 (see section 8.2.5). In this early work, no bands due to *trans* isomers were detectable, and it was frequently stated in the literature that *trans*

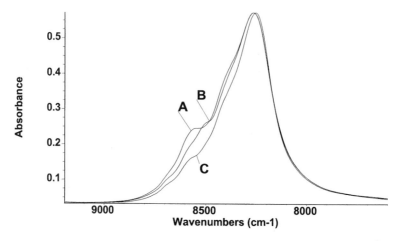

Figure 8.8   Region of second overtones of C—H stretching vibrations (8900–$8000\,\mathrm{cm}^{-1}$ (1100–1250 nm)) in the FT-NIR spectra of triolein (A), trielaidin (B), and tristearin (C).

unsaturation does not give rise to any characteristic absorption bands in the NIR spectra of lipids. The FT-NIR spectrum of C18:1t does, in fact, exhibit a shoulder at $8503 \, \text{cm}^{-1}$ (1176 nm) on the main band in this region, which would not have been resolvable by the instruments employed in the early work. This band is more clearly seen in the spectrum of trilinolelaidin (C18:2tt), which is overlaid on the spectrum of trilinolein (C18:2cc) in Figure 8.9. Thus, NIR spectroscopy should be suitable for the determination not only of *cis* unsaturation, as suggested in early work (section 8.2.5), but also of *trans* unsaturation and total unsaturation (IV). The PLS calibration models developed by Ismail *et al.* (1996) were employed to predict the *cis* content, *trans* content, and IV of a series of partially hydrogenated vegetable oils, and these predictions were compared to those obtained by mid-IR spectroscopy. The results showed an excellent correlation between the *cis* and IV predictions from the NIR and mid-IR calibrations; in the case of *trans* content, a plot of NIR versus mid-IR predictions was also linear but there was somewhat more scatter in the data, particularly for low *trans* values. On the basis of these preliminary findings, it was concluded that the determination of these three parameters by FT-NIR spectroscopy was feasible but that further refinement of the calibration model for the prediction of *trans* content was required in order to ascertain whether NIR spectroscopy can match the analytical performance of mid-IR spectroscopy in the determination of *trans* unsaturation.

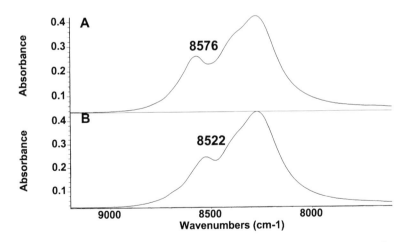

Figure 8.9 Region of second overtones of C—H stretching vibrations ($8900-8000 \, \text{cm}^{-1}$ (1100–1250 nm)) in the FT-NIR spectra of trilinolein (A) and trilinolelaidin (B), showing the characteristic absorption bands of the *cis* and *trans* double bonds at $8576 \, \text{cm}^{-1}$ (1166 nm) and $8522 \, \text{cm}^{-1}$ (1173 nm), respectively.

Other investigators employed NIR spectroscopy to measure the levels of oxidation products in oils. Boot and Speek (1994) performed NIR measurements in the transflectance mode on used frying oils and fats with a filter instrument. They found that NIR spectroscopy was well suited to the determination of free fatty acid content and dimer and polymer triglycerides. Dong *et al.* (1997) reported an FT-NIR method for the determination of PV, based on the stoichiometric reaction of TPP with hydroperoxides to form TPPO. In an analogous mid-IR method (Ma *et al.*, 1997), described in section 8.6.2.2, a simple univariate calibration allowed TPPO to be quantitated accurately in the presence of unreacted TPP. In the NIR spectra, the TPP and TPPO absorptions overlapped extensively, but accurate quantitation of TPPO was achieved through the use of PLS regression. The NIR method had comparable analytical performance to the mid-IR method but had the advantage of more convenient sample handling, as samples were analysed in 8 mm (o.d.) disposable glass vials whereas mid-IR measurements were conducted with a 0.10 mm transmission cell.

8.7   Future perspectives

Mid-IR spectroscopy has played a major role in fundamental research on lipid systems for the past five decades owing to the large amount of qualitative information that can be extracted from the IR spectra of lipids. It has also been an important tool in the routine analysis of hydrogenated fats and oils because it is one of the few techniques that provides a simple means of determining the level of *trans* unsaturation. However, apart from the widely utilised IR method for the determination of isolated *trans* isomers, quantitative analysis applications have been limited.

The advent of FTIR spectroscopy has revitalised quantitative IR spectroscopy owing to the superior performance and advanced data handling capabilities of FTIR systems. This has been paralleled by an even greater resurgence of NIR spectroscopy as a quantitative analytical tool. In recent years, a substantial amount of research has been focused on the possible applications of IR spectroscopy, and NIR spectroscopy to a lesser extent, in the fats and oils industry. This interest has been driven by increased demand in the industry for automated instrumental methods that can replace the traditional wet chemical and chromatographic methods of analysis. Although the improvements in efficiency and analytical performance that such instrumental methods can provide are important aspects, probably the major factor influencing the industry is a driving force away from the use of solvents and hazardous reagents because of environmental concerns and regulations and the costs of disposal.

In this context, FTIR and FT-NIR spectroscopy have the potential to rapidly become important analytical tools in the fats and oils industry, as analysis can be performed directly on oils and melted fats without any sample preparation or need for solvents. Thus, the development of FTIR methods for the rapid analysis of fats and oils has become a major focus of lipid-related IR research in recent years. Potential applications of FT-NIR spectroscopy in the analysis of fats and oils have been much less extensively investigated. However, this alternative has a number of practical advantages, including simpler sample handling and the availability of ruggedized instruments designed for industrial applications as well as of low-cost NIR-transmitting optical fibres that allow for remote analysis. It is anticipated that current and future research on IR and NIR analysis of fats and oils will lead not only to the implementation of new analytical methods in quality control laboratories but also to the application of these methods in at-line and on-line analysis and process monitoring in the fats and oils industry.

References

Afran, A. and Newbery, J.E. (1991) Analysis of the degree of unsaturation in edible oils by Fourier transform-infrared/attenuated total reflectance spectroscopy. *Spectrosc. Int.*, **3** 39-42.

Ahlers, N.H.E., Brett, R.A. and McTaggart, N.G. (1953) An infrared study of the *cis* and *trans* isomers of some C_{18} fatty acids. *J. Appl. Chem.*, **3** 433-443.

Ali, L.H., Angyal, G., Weaver, C.M., Rader, J.I. and Mossoba, M.M. (1996) Determination of total *trans* fatty acids in food: comparison of capillary-column gas chromatography and single-bounce horizontal attenuated total reflection infrared spectroscopy. *J. Am. Oil Chem. Soc.*, **73** 1699-1705.

AOCS (1989) *Official Methods and Recommended Practices of the American Oil Chemists' Society*, 4th edn, AOCS, Champaign, IL.

AOCS (1997) *Official Methods and Recommended Practices of the American Oil Chemists' Society*, 5th edn, AOCS, Champaign, IL.

Bewig, K.M., Clarke, A.D., Roberts, C. and Unklesbay, N. (1994) Discriminant analysis of vegetable oils by near-infrared reflectance spectroscopy. *J. Am. Oil Chem. Soc.*, **71** 195-200.

Boot, A.J. and Speek, A.J. (1994) Determination of the sum of dimer and polymer triglycerides and of acid value of used frying fats and oils by near-infrared reflectance spectroscopy. *J. AOAC Int.*, **77** 1184-1189.

Chapman, D. (1960) Infrared spectroscopic characterisation of glycerides. *J. Am. Oil Chem. Soc.*, **37** 73-77.

Chapman, D. (1965a) *The Structure of Lipids by Spectroscopic and X-ray Techniques*, Wiley, New York.

Chapman, D. (1965b) Infrared spectroscopy of lipids. *J. Am. Oil Chem. Soc.*, **42** 353-371.

Coblentz, W.W. (1905) *Investigations of Infra-Red Spectra*, Carnegie Institute of Washington, Publication No. 35.

Coleman, P.B. (1993) *Practical Sampling Techniques for Infrared Analysis*, CRC Press, Boca Raton, FL.

Deacon, G.B. and Green, J.H.S. (1968) Vibrational spectra of ligands and complexes—II. Infrared spectra (3650–375 cm^{-1}) of triphenylphosphine, triphenylphosphine oxide and their complexes. *Spectrochim. Acta*, **24A** 845-852.

Dong, J., Ma, K., van de Voort, F.R. and Ismail, A.A. (1997) Stoichiometric determination of hydroperoxides in oils by Fourier transform near infrared spectroscopy. *J. AOAC Int.*, **80**, 345-352.

Dubois, J., van de Voort, F.R., Sedman, J., Ismail, A.A. and Ramaswamy, H.R. (1996) Quantitative Fourier transform infrared analysis for anisidine value and aldehydes in thermally stressed oils. *J. Am. Oil Chem. Soc.*, **73** 787-794.

Dugan, L.R., Beadle, B.W. and Henick, A.S. (1949) An infrared absorption study of autoxidized methyl linoleate. *J. Am. Oil Chem. Soc.*, **26** 681-685.

Dutton, H.J. (1974) Analysis and monitoring of *trans* isomerization by IR attenuated total reflectance spectrophotometry. *J. Am. Oil Chem. Soc.*, **51** 407-409.

Fenton, A.J., Jr and Crisler, R.O. (1959) Determination of *cis* unsaturation in oils by near infrared spectroscopy. *J. Am. Oil Chem. Soc.*, **36** 620-623.

Ferraro, J.R. and Krishnan, K. (1990) *Practical FT-IR Spectroscopy, Industrial and Laboratory Chemical Analysis*, Academic Press, New York.

Feuge, R.O., Cousins, E.R., Fore, S.P., Dupre, E.F. and O'Connor, R.T. (1953) Modification of vegetable oils. XV. Formation of isomers during hydrogenation of methyl linoleate. *J. Am. Oil Chem. Soc.*, **30** 454-460.

Firestone, D. (1996) General referee reports; fats and oils. *J. AOAC Int.*, **79** 216-220.

Firestone, D. and De La Luz Villadelmar, M. (1961) Determination of isolated *trans* unsaturation by infrared spectrophotometry. *J. Assoc. Off. Anal. Chem.*, **44** 459-464.

Firestone, D. and LaBouliere, P. (1965) Determination of isolated *trans* isomers by infrared spectrophotometry. *J. Ass. Off. Anal. Chem.*, **48** 437-443.

Firestone, D. and Sheppard, A. (1992) Determination of *trans* fatty acids, in *Advances in Lipid Methodology* (ed. W.W. Christie), The Oily Press, Alloway, pp 273-322.

Fuller, M.P., Ritter, G.L. and Draper, C.S. (1988a) Partial least-squares quantitative analysis of infrared spectroscopic data. Part I: Algorithm implementation. *Appl. Spectrosc.*, **42** 217-227.

Fuller, M.P., Ritter, G.L. and Draper, C.S. (1988b) Partial least-squares quantitative analysis of infrared spectroscopic data. Part II: Application to detergent analysis. *Appl. Spectrosc.*, **42** 228-236.

Fukuzumi, K. and Kobayashi, E. (1972) Quantitative determination of methyl octadecadienoate hydroperoxides by infrared spectroscopy. *J. Am. Oil Chem. Soc.*, **49** 162-165.

Gibson, K.S. (1920) The infrared spectra of vegetable oils. *Cotton Oil Press*, **4** (5) 53.

Haaland, D.M. (1988) Quantitative infrared analysis of borophosphosilicate films using multivariate statistical methods. *Anal. Chem.*, **60** 1208-1217.

Haaland, D.M. and Thomas, E.V. (1988a) Partial least-squares methods for spectral analyses. 1. Relation to other quantitative calibration methods and the extraction of qualitative information. *Anal. Chem.*, **60** 1193-1202.

Haaland, D.M. and Thomas, E.V. (1988b) Partial least-squares methods for spectral analyses. 2. Application to simulated and glass spectral data. *Anal. Chem.*, **60** 1202-1208.

Harrick, N.J. (1967) *Internal Reflection Spectroscopy*, Wiley-Interscience, New York.

Heise, H.M., Marbach, R., Koschinsky, T. and Gries, F.A. (1994) Multicomponent assay for blood substrates in human plasma by mid-infrared spectroscopy and its evaluation for clinical analysis. *Appl. Spectrosc.*, **48** 85-95, and references therein.

Henick, A.S. (1951) Detection of deterioration products of autoxidising milk fat by infrared spectrophotometry. *Food Technol.*, **5** 145-147.

Holm, U., Ekbom, K. and Wode, G. (1957) Determination of the extent of oxidation of fats. *J. Am. Oil Chem. Soc.*, **34** 606-609.

Holman, R.T. and Edmondson, P.R. (1956) Near-infrared spectra of fatty acids and some related substances. *Anal. Chem.*, **28** 1533-1538.

Holman, R.T., Nickell, C., Privett, O.S. and Edmondson, P.R. (1958) Detection and measurement of hydroperoxides by near infrared spectrophotometry. *J. Am. Oil Chem. Soc.*, **35** 422-425.

Holman, R.T., Ener, S. and Edmondson, P.R. (1959) Detection and measurement of *cis* unsaturation in fatty acids. *Arch. Biochem. Biophys.*, **80** 72-79.

Honn, F.J., Bezman, I.I. and Daubert, B.F. (1949) Infrared absorption of hydroxy compounds in autoxidising linseed oil. *J. Am. Oil Chem. Soc.*, **1** 8127-8131.

Huang, A. and Firestone, D. (1971a) Determination of low level isolated *trans* isomers in vegetable oils and derived methyl esters by differential infrared spectrophotometry. *J. Assoc. Off. Anal. Chem.*, **54** 47-51.

Huang, A. and Firestone, D. (1971b) Comparison of two infrared methods for the determination of isolated *trans* unsaturation in fats, oils, and methyl ester derivatives. *J. Assoc. Off. Anal. Chem.*, **54** 1288-1292.

Ismail, A.A., van de Voort, F.R., Emo, G. and Sedman, J. (1993) Rapid quantitative determination of free fatty acids in fats and oils by Fourier transform infrared spectroscopy. *J. Am. Oil Chem. Soc.*, **70** 335-341.

Ismail, A.A., Charbonneau, C., Sedman, J. and van de Voort, F.R. (1996) Comparative analytical performance of mid-FTIR and FT-NIR spectroscopy in the determination of *cis* and *trans* content of fats and oils. *Book of Abstracts, Pittcon'96, Chicago, IL, March 3–8, 1996*, Abstract 012.

Kirby, E.M., Evans-Vader, M.J. and Brown, M.A. (1965) Determination of the length of polymethylene chains of salts of saturated and unsaturated fatty acids by infrared spectroscopy. *J. Am. Oil Chem. Soc.*, **42** 437-446.

Jackson, F.L. and Callen, J.E. (1951) Evaluation of the Twitchell isooleic method: comparison with the infrared *trans*-isooleic method. *J. Am. Oil Chem. Soc.*, **28** 61-65.

Jones, R.N., McKay, A.F. and Sinclair, R.G. (1952) Band progressions in the infrared spectra of fatty acids and related compounds. *J. Am. Chem. Soc.*, **74** 2570-2575.

Keeney, P.G. (1962) Estimating the average carbon chain length of saturated fatty acid esters by infrared spectroscopy. *J. Am. Oil Chem. Soc.*, **39** 304-306.

Knight, H.B., Eddy, C.R. and Swern, D. (1951) Reactions of fatty materials with oxygen. VIII. *Cis–trans* isomerisation during autoxidation of methyl oleate. *J. Am. Oil Chem. Soc.*, **28** 188-192.

Lai, Y.W., Kemsley, E.K. and Wilson, R.H. (1994) Potential of Fourier transform infrared spectroscopy for the authentication of vegetable oils. *J. Agric. Food Chem.*, **42** 1154-1159.

Lai, Y.W., Kemsley, E.K. and Wilson, R.H. (1995) Quantitative analysis of potential adulterants of extra virgin olive oil using infrared spectroscopy. *Food Chem.*, **53** 95-98.

Lanser, A.C., List, G.R., Holloway, R.K. and Mounts, T.L. (1991) FTIR estimation of free fatty acid content in crude oils extracted from damaged soybeans. *J. Am. Oil Chem. Soc.*, **68** 448-449.

Lemon, H.W. and Cross, C.K. (1949) The significance of an absorption band at $968\,\mathrm{cm}^{-1}$ in the infrared spectrum of methyl isolinoleate, *Can. J. Res.*, **27B** 610-615.

Lemon, H.W., Kirby, E.M. and Knapp, R.M. (1951) Autoxidation of methyl esters of peanut oil fatty acids. *Can. J. Technol.*, **29** 523-539.

Ma, K., van de Voort, F.R., Sedman, J. and Ismail, A.A. (1997) Stoichiometric determination of hydroperoxides in fats and oils by FTIR spectroscopy. *J. Am. Oil Chem. Soc.*, **74** 897-906.

Madison, B.L., DePalma, R.A. and D'Alonzo, R.P. (1982) Accurate determination of *trans* isomers in shortenings and edible oils by infrared spectrophotometry. *J. Am. Oil Chem. Soc.*, **59** 178-181.

Martens, H. and Naes, T. (1989) *Multivariate Calibration*, Wiley, Chichester.

Miller, R.G.J. and Stace, C. (1979) *Laboratory Methods in Infrared Spectroscopy*, Heyden and Sons, London.

Mossoba, M.M. and Firestone, D. (1996) New methods for fat analysis in foods. *Food Testing and Analysis*, **2** 24-32.

Mossoba, M., Yurawecz, M.P. and McDonald, R.E. (1996) Rapid determination of the *trans* content of neat hydrogenated oils by attenuated total reflection spectroscopy. *J. Am. Oil Chem. Soc.*, **73** 1003-1009.

Nakamura, T. and Maeda, H. (1991) A simple assay for lipid hydroperoxides based on triphenylphosphine oxidation and high-performance liquid chromatography. *Lipids*, **26** 765-768.

O'Connor, R.T. (1961) Recent progress in the applications of infrared absorption spectroscopy to lipid chemistry. *J. Am. Oil Chem. Soc.*, **38** 648-659.

O'Connor, R.T., Field, E.T. and Singleton, W.S. (1951) The infrared spectra of saturated fatty acids with even number of carbon atoms from caproic, C_6 (hexanoic) to stearic, C_{18} (octadecanoic), and of their methyl and ethyl esters. *J. Am. Oil Chem. Soc.*, **28** 154-160.

Osborne, B.G., Fearn, T. and Hindle, P.H. (1993) *Practical NIR Spectroscopy with Applications in Food and Beverage Analysis*, 2nd edn, Longman Scientific and Technical, Harlow, Essex.

Rasmussen, R.S., Brattain, R.R. and Zucco, P.S. (1947) Infrared absorption spectra of some octenes. *J. Chem. Phys.*, **15** 135-141.

Ratyanake, N.W.M., Hollywood, R., O'Grady, E. and Beare-Rogers, J.L. (1990) Determination of *cis*- and *trans*-octadecenoic acids by gas liquid chromatography–infrared spectrophotometry. *J. Am. Oil Chem. Soc.*, **67** 804-810.

Safar, M., Bertrand, D., Robert, P., Devaux, M.F. and Genot, C. (1994) Characterisation of edible oils, butters and margarines by Fourier transform infrared spectroscopy with attenuated total reflectance. *J. Am. Oil Chem. Soc.*, **71** 371-377.

Sato, T. (1994) Application of principal-component analysis on near-infrared spectroscopic data of vegetable oils for their classification. *J. Am. Oil Chem. Soc.*, **71** 293-298.

Sato, T., Kawano, S. and Iwamoto, M. (1991) Near infrared spectral patterns of fatty acid analysis from fats and oils. *J. Am. Oil Chem. Soc.*, **68** 827-833.

Sedman, J., Ismail, A.A., Nicodemo, A., Kubow, S. and van de Voort, F.R. (1996) Application of FTIR/ATR differential spectroscopy for the monitoring of oil oxidation and antioxidant efficiency, in *Natural Antioxidants* (ed. F. Shahidi), AOCS Press, Champaign, IL, pp 358-378.

Sedman, J., van de Voort, F.R. and Ismail, A.A. (1997) Upgrading the AOCS IR *trans* method for analysis of neat fats and oils by FTIR spectroscopy. *J. Am. Oil Chem. Soc.*, **74** 907-913.

Sedman, J., van de Voort, F.R. and Ismail, A.A. (1998a) Attenuated total reflectance spectroscopy: principles and applications in infrared analysis of food, in *Spectral Methods in Food Analysis* (ed. M. Mossoba), Marcel Dekker, New York, pp 397-425.

Sedman, J., van de Voort, F.R., Ismail, A.A. and Maes, P. (1998b) Industrial validation of FTIR *trans* and iodine value analyses. *J. Am. Oil Chem. Soc.*, **75** 33-39.

Shreve, O.D., Heether, M.R., Knight, H.B. and Swern, D. (1950a) Infrared absorption spectra [some long-chain fatty acids, esters and alcohols]. *Anal. Chem.*, **22** 1498-1501.

Shreve, O.D., Heether., M.R., Knight, H.B. and Swern, D. (1950b) Determination of *trans*-octadecenoic acids, esters and alcohols in mixtures. *Anal. Chem.*, **22** 1261-1264.

Sinclair, R.G., McKay, A.F. and Jones, R.N. (1952a) The infrared absorption spectra of saturated fatty acids and esters. *J. Am. Chem. Soc.*, **74** 2570-2575.

Sinclair, R.G., McKay, A.F., Myers, G.S. and Jones, R.N. (1952b) The infrared absorption spectra of unsaturated fatty acids and esters. *J. Am. Chem. Soc.*, **74** 2578-2586.

Sleeter, R.T. and Matlock, M.G. (1989) Automated quantitative analysis of isolated (nonconjugated) *trans* isomers using Fourier transform infrared spectroscopy incorporating improvements in the procedure. *J. Am. Oil Chem. Soc.*, **66** 121-127.

Sonntag, N.O.V. (1982) Analytical methods, in *Bailey's Industrial Oil and Fat Products*, vol. 2, 4th edn (ed. D. Swern), Wiley, New York.

Sreenivasan, B. and Holla, K.S. (1967) A rapid method for the estimation of *trans* unsaturation in hydrogenated oils and fats. *J. Am. Oil Chem. Soc.*, **44** 313-315.

Swern, D., Knight, H.B., Shreve, O.D. and Heether, M.R. (1950) Comparison of infrared spectrophotometric and lead salt–alcohol methods for determination of *trans* octadecenoic acids and esters. *J. Am. Oil Chem. Soc.*, **27** 17-21.

van de Voort, F.R. (1994) FTIR spectroscopy in edible oil analysis. *Inform*, **1994** (9) 1038-1042.

van de Voort, F.R., Sedman, J., Emo, G. and Ismail, A.A. (1992) Rapid and direct iodine value and saponification number determination of fats and oils by attenuated total reflectance/Fourier transform infrared spectroscopy. *J. Am. Oil Chem. Soc.*, **69** 1118-1123.

van de Voort, F.R., Ismail, A.A., Sedman, J. and Emo, G. (1993) Monitoring the oxidation of edible oils by FTIR spectroscopy. *J. Am. Oil Chem. Soc.*, **71** 243-253.

van de Voort, F.R., Ismail, A.A., Sedman, J., Dubois, J. and Nicodemo, A. (1994) The determination of peroxide value by Fourier transform infrared (FTIR) spectroscopy. *J. Am. Oil Chem. Soc.*, **71** 921-926.

van de Voort, F.R., Ismail, A.A. and Sedman, J. (1995) A rapid, automated method for the determination of *cis* and *trans* content of fats and oils by Fourier transform infrared spectroscopy. *J. Am. Oil Chem. Soc.*, **72** 873-880.

van de Voort, F.R., Memon, K.P., Sedman, J. and Ismail, A.A. (1996) Determination of solid fat index by Fourier transform infrared spectroscopy. *J. Am. Oil Chem. Soc.*, **73** 411-416.

Wesley, I.J., Barnes, R.J. and McGill, A.E.J. (1995) Measurement of adulterants of olive oils by near infrared spectroscopy. *J. Am. Oil Chem. Soc.*, **72** 289-297.

Wesley, I.J., Pacheco, F. and McGill, A.E.J. (1996) Identification of adulterants of olive oils. *J. Am. Oil Chem. Soc.*, **73** 515-518.

Williams, P. and Norris, K. (1987) *Near-infrared Technology in the Agricultural and Food Industries*, American Association of Cereal Chemists, St. Paul, MN.

9 Electron spin resonance studies of lipids

Christopher J. Rhodes and Timothy C. Dintinger

9.1 The principles of electron spin resonance

Much of the information on radicals derived from lipids has been gleaned from electron spin resonance (ESR) studies, as indeed is true of all radicals. Since we are principally ESR spectroscopists, we shall deal specifically with this aspect of the technique and its application to structural and mechanistic investigations of lipid systems, particularly in regard to biological membranes.

9.1.1 g-Value

The essence of the ESR method (Carrington and McLachlan, 1979) involves the irradiation of a sample containing unpaired electrons with microwave radiation, while simultaneously applying an external magnetic field (B). In the absence of the field, the two possible spin states that may be taken by unpaired electrons (denoted by the spin quantum numbers, $m_S = +\frac{1}{2}, -\frac{1}{2}$) have the same energy and are equally populated. On the application of B (Figure 9.1), this is no longer the case, and there is a net lowering of the energy of the state $m_S = -\frac{1}{2}$ and a corresponding increase in that of $m_S = +\frac{1}{2}$. The energy splitting (ΔE) between the $m_S = +\frac{1}{2}$ and the $m_S = -\frac{1}{2}$ states is a function of B and is given by equation (9.1):

$$\Delta E = h\nu = g\mu_B B, \qquad (9.1)$$

In this equation there is a dimensionless parameter (g), normally called the 'g-factor', which is structurally diagnostic—and is in fact exactly analogous to the chemical shift in NMR spectroscopy: increasing with the unpaired electron density on 'heavy-atoms', since it arises from spin–orbit coupling, where the mixing of excited states with electronic angular momentum into the ground state gives rise to a secondary field which couples with the magnetic moment of the electron owing to its spin. For a free electron, $g = 2.0023$, but shifts can occur to higher or lower values than this, depending on whether the unpaired electron orbital is coupled magnetically with a filled or with a vacant orbital, respectively. Since the term ΔE in equation (9.1) depends on the total field experienced, i.e. the sum of the external field (B) and the secondary field (above), for a given measured value of B, g is seen to vary.

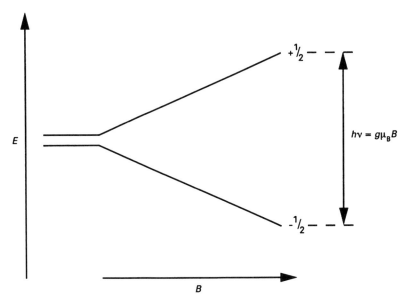

Figure 9.1 Schematic of energy levels for unpaired electron in an applied magnetic field B.

The practical consequence of the above may be illustrated by means of the radical $O_2^{-\bullet}$, in which the π^* manifold is degenerate, as shown in Figure 9.2. However, differential perturbation of the π_x^*, π_y^* levels by environmental effects such as hydrogen-bonding can give rise to a splitting of these energy states. When the radical is oriented with its $O—O(z)$ axis (**I**) parallel to the applied magnetic field B, electronic angular momentum is induced about that axis, and may be considered to arise from the motion of the unpaired electron in a circular path, flowing from the π_x^* to the π_y^* level, then back to π_x^*.

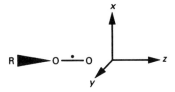

(**I**)

The ease of flow depends, therefore, on the splitting (ΔE), which provides a barrier to it. The shift (Δg) for a given orientation of the

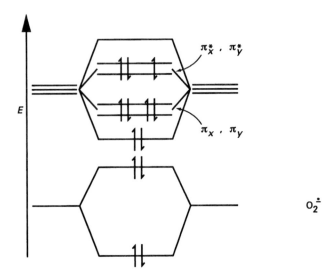

Figure 9.2 Diagram showing the degenerate π^*-manifold for the $O_2^{-\cdot}$ radical anion.

radical in the applied field B is often dominated by the contribution from a single atom, but depends on the contributions from all atoms (i) in the radical as may be expressed by equation (9.2).

$$\Delta g_z = \sum \rho_i \lambda_i / \Delta E_i \qquad (9.2)$$

E_i determines the sign of the shift (alluded to earlier) and is positive in the case of $O_2^{-\cdot}$, because the angular coupling of the SOMO is with a filled, lower energy, orbital, but is negative for coupling with a vacant, higher energy, orbital as in $O_2^{+\cdot}$. λ_i and ρ_i are, respectively, the spin–orbital coupling constant and the spin density for each atom (i).

The (isotropic) g-value observed from a rapidly reorienting radical, as in a liquid, differs from free-spin (2.0023) by the average of the shifts corresponding to each molecular Cartesian axis (x, y, z), (equation 9.3).

$$g(\text{isotropic}) = [(\Delta g_x + \Delta g_y + \Delta g_z)/3] + 2.0023 \qquad (9.3)$$

In the present example, the shift associated with the O—O axis is dominant ($B_{(z)}$), since ΔE is relatively small—it would be zero in the absence of an environmental perturbation—whereas it is larger for $B(y)(E\pi_x - E\pi_y)$, leading to a weaker coupling, and for $B(x)$ there is essentially no shift since circulation about $B(x)$ does not couple π_x^* with any other orbitals.

In view of their importance in lipid peroxidation, as is clear in later sections of this chapter, it is instructive to consider how a peroxyl radical, ROO·, might be formed by a *hypothetical* reaction between a proton (or other positively charged atomic centre) and the $O_2^{-\bullet}$ radical anion unit (Figure 9.3): if the bonding (with the σ-2p combination) is linear, as shown in (a), the electrostatic effect on the x,y levels in both the bonding and antibonding manifolds is the same. However, an angular bonding interaction, as shown in (b), will involve in addition the π_x levels, so that there is now a differential perturbation of the x, y levels in both bonding and antibonding orbital manifolds, and the important $\pi^*(x,y)$ degeneracy is lifted, leading to a reduced $g(z)$ shift: a typical value is ~ 2.035, reduced from that expected for complete degeneracy ($g = 4$); the other values are always close to $g_x = 2.003$, $g_y = 2.008$.

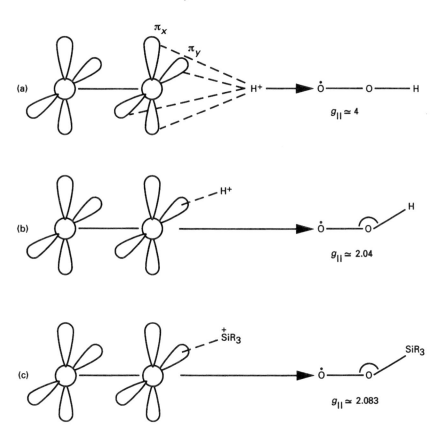

Figure 9.3   Diagram depicting the (hypothetical) formation of a peroxyl radical as a bonding perturbation of the orbitals in the $O_2^{-\bullet}$ radical anion, to explain the differing $g(z)$ shifts in these species.

9.1.2 Hyperfine coupling

Where it may be observed, hyperfine coupling—the analogue of nuclear spin–spin coupling in NMR spectroscopy, is diagnostically the most useful feature of ESR, since it reveals the nature of magnetic nuclei that are present in a particular paramagnetic species (free radical). In order to appreciate this effect, the simplest starting point is the hydrogen atom $H^{\cdot}$ which consists of a unique proton interacting with a single unpaired electron (Figure 9.4a). We need consider the $m_S = -\frac{1}{2}$ state of the electron since the ESR experiment measures the net absorption of energy due to the promotion of electrons from this to the $+\frac{1}{2}$ level. Accordingly, the associated proton may take either the $m_I = +\frac{1}{2}$ or the $-\frac{1}{2}$ state with respect to this, and, as may be depicted in Figure 9.4a, the magnetism due to the intrinsic proton spin may either augment $(+\frac{1}{2})$ or detract from $(-\frac{1}{2})$ the applied magnetic field (B). This means that each of the electron spin levels is further split into two, as shown. In the hydrogen atom, where the unpaired electron occupies the 1s orbital, owing to the finite probability density of the orbital at the nucleus the two magnetic entities (proton and electron) are able to influence one another by direct contact ('Fermi contact interaction'): the strength of this magnetic interaction is commonly quoted in units of gauss ($1G = 10^{-4}\,T$; equivalent to 2.8 MHz for a g-value close to free-spin, 2.0023) and for $H^{\cdot}$ amounts to $\sim$507 G. Most free radicals are more complex than $H^{\cdot}$ and contain more interacting magnetic nuclei; take, for example, the hypothetical radical $R_2CHCHR^{\cdot}$, where the groups R contain no magnetic nuclei, and there are two coupled protons that are inequivalent (i.e. with different coupling constants). This situation is just a development of that already discussed for $H^{\cdot}$, and there is now a sequential splitting by the two protons of the initial electron spin energy levels (Figure 9.4b). In first order, the quantum-mechanical selection rules forbid the reorientation of a nuclear spin ($\Delta m_I = 0$) simultaneously with that of the electron spin ($\Delta m_S = 1$); thus only the four transitions shown occur, and so four lines are observed in the ESR spectrum. If the two protons have equal couplings (Figure 9.4c), the situation is as above, but since the sequential splittings are the same, two of the transitions are coincident: thus three lines are observed with relative intensities $1:2:1$. The general rule is that if a group of n equivalent nuclei with spin I are coupled to the unpaired electron, $2nI + 1$ transitions and thus spectral lines will be observed. For the case of protons, or other nuclei with spin $I = \frac{1}{2}$, this reduces to $n+1$ lines for n equivalent nuclei, and the situation is therefore analogous to the familiar '$n + 1$' rule in NMR spectroscopy. Thus the simplest organic radical $CH_3^{\cdot}$ gives four lines.

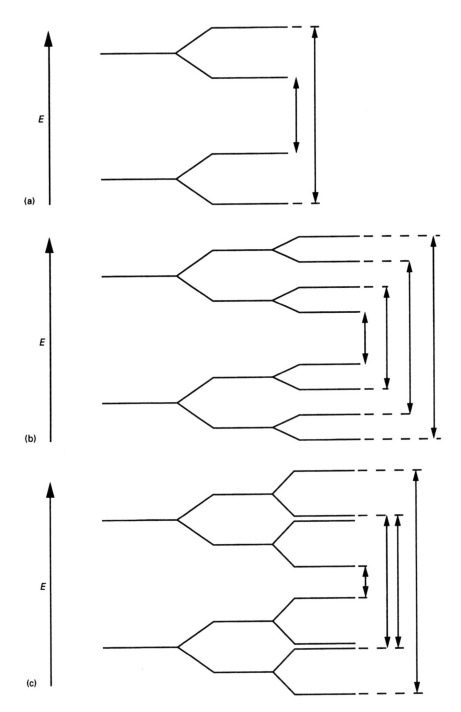

Figure 9.4 Magnetic energy levels for a radical containing: (a) a single proton; (b) two non-equivalent protons, (c) two equivalent protons.

9.1.3 Line intensities

As is the case in NMR spectroscopy, the relative intensities of the lines in a multiplet arising from the coupling of the odd electron with a set of n equivalent nuclei are given by the coefficients of the binomial expansion, and may conveniently be represented by 'Pascal's Triangle'. Hence the four lines in the ESR spectrum of $CH_3^{\cdot}$ are in the ratio 1:3:3:1.

9.1.4 Coupling to different sets of n-equivalent nuclei

In this situation, as in NMR, the total number of lines to be expected from coupling of the unpaired electron with different sets of nuclei n_x is given by $\prod (2n_x I + 1)$ or by $\prod (n_x + 1)$ for nuclei with spin $I = \frac{1}{2}$. To illustrate this, consider the case of the ethyl radical, $CH_3 CH_2^{\cdot}$. This has two sets of protons $n_1 = 3$ and $n_2 = 2$: thus the formula predicts $(3 + 1)(2 + 1) = 12$ lines; the coupling constants are $+27\,G$ for the methyl protons and $-22\,G$ for the methylene protons. The form of the ESR spectrum can be developed in terms of a stick diagram (Figure 9.5); the line intensities shown are arrived at by splitting each of the 1:3:3:1 intense lines into three more of relative intensities 1:2:1.

9.1.5 Couplings to nuclei with $I > \frac{1}{2}$

The most commonly encountered nucleus of this type is ^{14}N, with $I = 1$, so that it may take the values $m_1 = +1, 0, -1$, with each state exerting its own magnetic influence on the unpaired electron, leading to a splitting of each electron m_S level into three. Three lines are therefore observed in the ESR spectrum, with a common spacing that is equal to the ^{14}N coupling constant.

A radical often contains protons in addition to the nitrogen atom, so that its ESR spectrum arises from the combination of the coupling with both kinds of nucleus. The number of lines, as always, can be derived

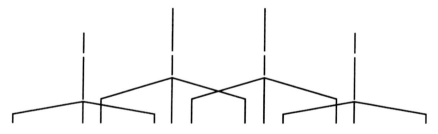

Figure 9.5 Stick diagram showing ESR pattern for the $CH_3 CH_2^{\cdot}$ radical.

from the formula $\prod(2n_x I + 1)$, but the important cases of radicals such as Ph(R)CHN(O⁺)Buᵗ and BuᵗN(O⁺)CH₂R (which are observed during spin-trapping experiments, as discussed in section 9.5) may be dealt with specifically by means of the coupling diagrams (Figures 9.6 and 9.7), where it is assumed that the proton coupling is smaller (Figure 9.6) or larger (Figure 9.7) than that of the ^{14}N nucleus.

9.1.5.1 Anisotropic hyperfine coupling

The foregoing refers explicitly to the simplified situation of radicals that are free to tumble rapidly in liquids (or very mobile solid matrices such as adamantane). In many instances, particularly with reactive σ-radicals such as phenyl, or almost all radical ions other than highly delocalised species usually derived from aromatic molecules, it is necessary to use matrix isolation methods. In the normal event, the tumbling rates of radicals isolated in solid matrices are relatively low, and the effects of the anisotropic magnetic dipolar nucleus–electron couplings are manifested in the ESR spectrum (see section 9.2.6). (Typically, the g-factor is also anisotropic, but for radicals centred on first-row elements this normally provides a relatively minor influence on the spectra). In situations where

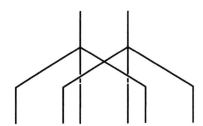

Figure 9.6 Stick diagram for radical PhRCHN(O⁺)Buᵗ, showing coupling to a single proton (^{1}H) and one ^{14}N nucleus; $a(^1\text{H}) < a(^{14}\text{N})$.

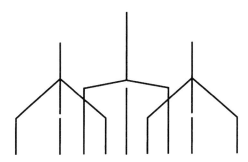

Figure 9.7 Stick diagram for radical BuᵗN(O⁺)CH₂R, showing coupling to two equivalent protons and one ^{14}N nucleus; $a(^1\text{H}) > a(^{14}\text{N})$.

the tumbling rate is intermediate between that typical for radicals in liquids and those in solids, line-broadening occurs from which motional information may be extracted (see section 9.6).

9.1.6   *Anisotropic coupling to the nucleus of the radical centre*

This is best illustrated with an example, that of the hypothetical amine radical cation $R_3N^{+\cdot}$. This is isoelectronic with the corresponding alkyl radical $R_3C^{\cdot}$, but has the advantage that the magnetic central nucleus (^{14}N) is almost 100% at natural abundance and so the influence of this is clearly seen (it is assumed that the groups R contain no strongly coupled magnetic nuclei and thus give no hyperfine splittings). In essence, there are two extremes (Figure 9.8): in one the radical is oriented so that the projection of the $N(2p_z)$ orbital is parallel to the applied magnetic field B

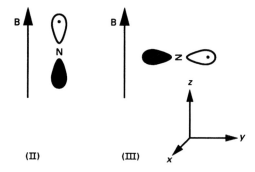

(along z) (**II**); in the other the orientation is perpendicular to B (**III**). It may be recognised that there are two equivalent axes (x,y) for which this second case applies. In the former case (**II**), the parallel coupling ($A_{\parallel}$) is the maximum possible, since the anisotropic (dipolar) coupling ($2B$) from the ^{14}N nucleus takes the same (positive) sign as the isotropic coupling (a), (equation 9.4).

$$A_{\parallel} = a + 2B \qquad (9.4)$$

In the perpendicular case the dipolar coupling is only half as great (B), and of opposite sign, so that there is a net cancellation of effects (equation 9.5), which for nitrogen-centred π-radicals leads to values for the perpendicular coupling ($A_{\perp}$) close to zero.

$$A_{\perp} = a - B \qquad (9.5)$$

If the radicals are formed in a single-crystal environment so that they are mainly all aligned with respect to the crystal axes, then rotation of the

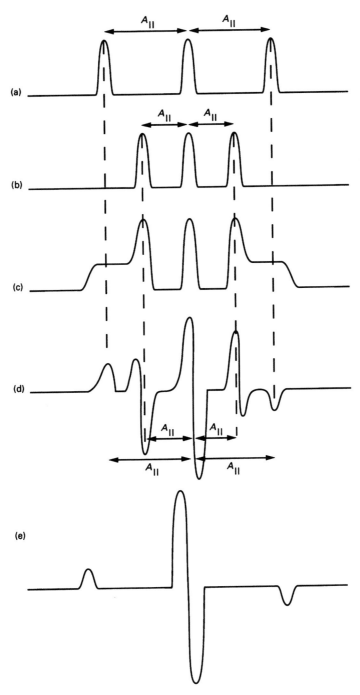

Figure 9.8 Illustration of anisotropic coupling to the ^{14}N nucleus in a radical $R_3N^{+\cdot}$ (see section 9.2.6).

crystal about these x,y,z axes leads to a simultaneous shift in the positions of the outermost two lines of the $I = 1$ triplet which will pass through the maximum (Figure 9.8a) and minimum (b) positions. On the other hand, in a randomly matrix-isolated system, as is more commonly encountered, there is no such overall alignment and so the resulting ESR spectrum is a composite of the spectra taken over all orientations, and the entire range of couplings intermediate between $A_{\parallel}$ and $A_{\perp}$ is encompassed, leading to the characteristic solid-state 'envelope' shape (c). When the first derivative of this pattern is measured (d), as is normal for ESR spectrometers, the parallel and perpendicular features stand out as shown, although for ^{14}N π-radicals the perpendicular peaks are often coincident with the parallel $|0>$ line (Figure 9.8e) because the perpendicular coupling $(A_{\perp})$ is close to zero and in the absence of splitting merely lends intensity to this central line, leading to a very intense central feature. However, the direct admixture of s-character in a ^{14}N σ-radical (Rhodes and Agirbas, 1990) usually renders these visible as $A_{\perp}$ is increased by the increased isotropic component (a).

The fact that $A_{\parallel}$ and $A_{\perp}$ often can be determined by direct inspection of the powder ESR spectrum is extremely important because, from these, estimates of the s and p orbital populations at the radical centre may be obtained by means of equations (9.6–9.9). The discussion given here for nitrogen-centred radicals holds true for most other elements, although corrections for orbital magnetic contributions are necessary for heavier elements and for transition metal complexes.

$$\% s = (a/a_0) \times 100 \qquad (9.6)$$

$$a = (A_{\parallel} + 2A_{\perp})/3 \qquad (9.7)$$

The isotropic coupling (a), while not measurable directly from the solid-state spectrum, can be estimated from equation (9.8), and the pure dipolar coupling $(2B)$ from equation 9.9.

$$\% p = (2B/2B_0) \times 100 \qquad (9.8)$$

$$2B = A_{\parallel} - a \qquad (9.9)$$

Since values have been tabulated in the literature, as calculated from appropriate atomic wavefunctions, for the coupling constants expected for unit occupancy of a particular s (a_0) or p $(2B_0)$ orbital, these may be used in equations (9.6) and (9.8) to map out the unpaired electron distribution over these orbitals.

9.2 Lipid peroxidation

Lipid peroxidation (Gutteridge and Halliwell, 1989) was first studied quantitatively by Saussure, as long ago as 1820, who observed that a layer of walnut oil on water absorbed 3 times its own volume of air in the first 8 months; this increased to 60 times in the next 10 days, then slackened off over the next 3 months, whereupon a total absorption of 145 times its own volume of oxygen had occurred; simultaneously, the oil had become rancid and viscous.

Subsequently, Berzelius proposed that such oxygen absorption might explain a variety of processes, including the fact that wool often ignited spontaneously when it had been lubricated with linseed oil; at that time, this was a notable scenario for catastrophe in textile mills.

In the 1940s, Farmer and others at the British Rubber Producers Association research laboratories deduced the sequence of reactions that is now understood to be the basis of lipid peroxidation: 'oxygen dependent deterioration' or 'rancidity' has long been recognised as a problem in the storage of fats and oils, and is even more so now, given the favour of 'polyunsaturated' margarines and cooking oils, and the widespread use of paints and other surface coatings, plastics and rubber, all of which can undergo oxidative degradation.

9.2.1 Lipids in membranes

Biological membranes (Gutteridge and Halliwell, 1989) consist mainly of lipids and proteins, the proportion of protein increases according to the level and complexity of the membrane function. Membrane lipids are generally amphipathic molecules, which means that they contain hydrophobic regions that exclude water and also polar groups that associate with water. In animal cell membranes, the major lipid components are phospholipids, based on glycerol, but other membranes, particularly plasma membranes, contain sphingolipids along with cholesterol; the most common phospholipid in animal cell membranes is lecithin (phosphatidylcholine).

The fatty acid side-chains of membrane lipids in animal cells have unbranched carbon chains and contain even numbers of carbon atoms, mostly in the range 14–24, and the $C{=}C$ double bonds are in the cis-geometry. The lipid composition of bacterial membranes depends largely on the species, but the lipid fraction is typically 10–30%. As the degree of unsaturation of a fatty acid increases, its melting point drops, and it becomes more susceptible to oxidative degradation; however, polyunsaturated fatty acid chains are present in many membrane phospholipids. The presence of saturated and polyunsaturated fatty acid side-chains in many membrane lipids confers fluidity on the membrane, so that the membrane interior gains the chemical nature and viscosity of a 'light oil'; oxidative damage to polyunsaturated chains tends to reduce membrane fluidity, known to be essential to the effective functioning of biological membranes, and so their action is impaired.

9.2.2   The peroxidation process

In a completely peroxide-free lipid system, a carbon-centred radical is formed by H-atom abstraction from a CH_2 group of a lipid chain; $\cdot OH$, widely implicated in biology, is readily capable of this (equation 9.10):

$$CH_2 + \cdot OH \rightarrow CH^\cdot + H_2O \qquad (9.10)$$

This may be demonstrated by the radiolysis of aqueous solutions, which generates $\cdot OH$. Both biological membranes, fatty acids and food lipids undergo peroxidation under these conditions and the process is inhibited by agents such as mannitol and formate, which are efficient $\cdot OH$ radical scavengers. Stimulation of lipid peroxidation in this way poses a problem in 'food irradiation', which aims to preserve foodstuffs.

The conjugate acid of superoxide ($HO_2^\cdot$), which is able to cross membranes easily, is a strong contender for an 'oxygen radical' that can initiate lipid peroxidation, although its definite role in this regard is as yet unproven.

Once the carbon-centred radical is formed, it will rapidly react with molecular oxygen to form a peroxyl radical (equation 9.11).

$$CH^\cdot + O_2 \rightarrow CHOO^\cdot \qquad (9.11)$$

The formation of lipid peroxyl radicals has been demonstrated by the ESR 'spin-trapping' technique (Perkins, 1980), which is described in section 9.5; this approach is often used as evidence that free-radical-mediated processes are involved in a given system, and some recent examples of this, pertinent to lipid peroxidation, are given subsequently.

It might be envisaged that peroxyl radical formation occurs competitively with other processes, such as radical combination or attack on other membrane components; the effective oxygen concentration might determine the relative effectiveness of these pathways, as might the 'anchoring' of the carbon radicals within the membrane structure.

The peroxyl radicals then enter the propagation phase, in which they abstract a hydrogen atom from another lipid chain, forming a lipid hydroperoxide (equation 9.12).

$$ROO^\cdot + CH_2 \rightarrow ROOH + CH^\cdot \qquad (9.12)$$

The 'new' radical can react with another O_2 molecule, and the entire process can be repeated, so constituting a chain-reaction.

An alternative step is the intramolecular attack of the peroxyl radical on a C=C bond of the same carbon chain, forming a cyclic peroxide (equation 9.13).

$$
\begin{array}{cc}
\underset{\diagup}{\text{O}-\text{O}^{\bullet}} & \underset{\diagup \quad \diagdown}{\text{O}-\text{O}} \\
\text{C}-\text{CH}_2-\text{C}{=}\text{C} \rightarrow & \text{C}-\text{CH}_2-\text{C}-\text{C}^{\bullet}
\end{array}
\qquad (9.13)
$$

Since the initial H-atom abstraction can occur at different points on the carbon chain, the products of lipid peroxidation are often formed as complex mixtures: for example, the peroxidation of arachidonic acid gives at least six different hydroperoxides along with cyclic peroxides and other products.

9.2.3   Protection against lipid peroxidation

Unsaturated fatty acids, dispersed in organic solvents or with detergents in aqueous solution, are readily peroxidised; phospholipids, similarly, are more rapidly peroxidised in simple micellar structures than in lipid bilayers. Ingold has proposed (Burton et al., 1985) that this is due to the relatively more polar character of lipid peroxyls than of unreacted hydrocarbon chains, which thus tend to avoid the weakly polar membrane interior, moving toward the bilayer surface, so reducing the initiation efficiency.

The presence of cholesterol also exerts an influence on the peroxidation sensitivity of surface membranes, probably by a combination of radical scavenging and modification of the internal structure of the membrane.

The susceptibility of a membrane to oxidative damage is also modified significantly by the presence of positively charged ions, which probably bind to the negatively charged head-groups of the phospholipid: this is manifested either by an increase or a decrease in the rate of peroxidation. For instance, a number of metal ions, e.g. Ca^{2+}, Co^{2+}, Cd^{2+}, Al^{3+}, Hg^{2+}, Pb^{2+}, have been shown to alter the rate of peroxidation of liposomes, erythrocytes and microsomal membranes: in general, peroxidation induced by the addition of iron ions is stimulated by their presence; the inhibiting influence of Zn^{2+}, Mn^{2+} or polyamines, found in certain cases, may be due to their ability to displace iron ions from phospholipid binding sites.

9.2.4   Antioxidants

Since lipid peroxidation is a 'bad' thing, particularly in the biological context where it destroys the essential integrity of membranes and so compromises their biological function, including the compartmentalisation effect so fundamental to the cellular system, ways of avoiding, or at least ameliorating, the process are of concern. We hear much, currently, of the role of antioxidants in the diet, which are thought to protect against the development of a variety of diseases, and even the process of ageing itself.

Antioxidants are generally regarded as 'radical scavengers', intercepting free radicals before they can attack sensitive cellular components, including the membranes themselves. Much could be written on this, but that would fall beyond the remit of the present review, so we now merely allude to some possible routes by which an 'antioxidant' might act to inhibit lipid peroxidation:

(1) Decreasing localised O_2 concentrations (e.g. sealing foods under vacuum or under nitrogen.
(2) Preventing the initiation stage of lipid peroxidation by scavenging reactive radicals like $^{\cdot}OH$.
(3) Complexing metal ions such that they are unable to produce initiating species such as $^{\cdot}OH$, ferryl, or $Fe^{2+}/Fe^{3+}/O_2$ and/or will not decompose lipid peroxides to peroxyl or alkoxyl radicals.
(4) Decomposing peroxides, without radical formation, to inactive products such as alcohols.
(5) Chain-breaking, i.e. scavenging intermediate radicals such as peroxyl and alkoxyl radicals, so preventing chain propagation via H-atom abstraction.

Antioxidants acting by mechanisms (2) and (3) can be called 'preventative antioxidants'. As an example of mechanism (4), glutathione peroxidase comes to mind. Chain-breaking antioxidants, often added to foodstuffs, are usually sterically hindered phenols or aromatic amines: these are also used widely to diminish the rate of peroxidation of surface coatings, e.g. paints.

In biological membranes, vitamin E is a highly effective antioxidant, thought to act by chain-breaking, since it is a phenol (equation 9.14).

$$\text{Vit-E-OH} + \text{ROO}^{\cdot} \rightarrow \text{Vit-E-O}^{\cdot} + \text{ROOH} \qquad (9.14)$$

It has been further argued that vitamin E also modifies the structure of membranes, since a protective effect is found even when vitamin E acetate is used (i.e. the reactive phenolic H-atom has been removed) (Gutteridge, 1978).

9.3   Lipid peroxyl radicals

9.3.1   Direct studies by ESR

As stated earlier, peroxyl radicals are involved in the destruction of cell membranes, and in the rancidification of fats and oils, by lipid peroxidation and there is the aim to prevent, or at least to ameliorate, the related undesired effects by means of deactivating the radicals that drive them. We have stressed that one solution is to administer 'antioxidants', some being phenolic, which can readily scavenge reactive radicals including peroxyl radicals. So motivated, Sevilla and his co-workers have used ESR to elucidate the mechanisms of reactions between carbon-centred radicals derived from lipids and phenolic antioxidants (*t*-butylhydroquinone (TBHQ), *n*-propyl gallate (PG), butylated hydroxyanisole (BHA), butylated hydroxytoluene (BHT) and vitamin E), using the lipid as a matrix at low temperatures (Zhu *et al.*, 1990): in all cases, the carbon-centred radicals were found to react at 135 K with molecular oxygen, forming peroxyl radicals. At 170 K, the peroxyl radicals were found to react with the added antioxidants, forming the corresponding phenoxyl radicals: those derived from tributyrin were found to decrease in their rate of reaction in the order BHT > TBHQ > E > PG > BHA, while those derived from triolein and trilinolein decrease in the order BHT > E > BHA > TBHQ. These differences are accounted for in terms of the way in which the antioxidants are dispersed in the different matrices. The fraction of unsaturated lipid peroxyl radicals which abstract from the antioxidant is lower than that in saturated lipids, and in the case of tributyrin the initial carbon-centred radicals can abstract directly from the antioxidant, while those more stable, conjugated, radicals derived from unsaturated lipids do so only to a negligible extent. The decay of all five antioxidant phenoxyl radicals in tributyrin follows second-order kinetics with an apparently common activation energy of $96 \, \text{kJ} \, \text{mol}^{-1}$. The same group previously determined the reactivity of radicals formed from cholesterol and its analogues by γ-irradiation (Sevilla *et al.*, 1986). In oxygen-free samples of cholesterol itself are produced a tertiary side-chain radical (**IV**) and an allylic radical (**V**); the structure of the latter is confirmed by experiments using two cholesterol analogues, namely 7-hydroxycholesterol (**VI**) and the selectively deuterated 7-deutero-7-hydroxycholesterol (**VII**), both of which produce the allylic radical after radiolysis by loss of the hydroxyl group. When oxygen is present in the samples, evidence is provided for the formation of two distinct peroxyl radicals, and these are suggested to possess differing reactivities resulting from their differing motional freedoms; it is further suggested that the products detected

IV

V

VI

VII

following radiation-induced autoxidation at or close to ambient temperature are consistent with the differing reactivities of these intermediary peroxyl radicals.

The autoxidation phenomena has also been explored for the lipids tributyrin, triarachidin, triolein, triliolein, trilinolenin and linolenic acid (Yanez *et al.*, 1987). It was found that the *unsaturated* lipids generally contained sufficient hydroperoxides that their direct UV photolysis as neat samples at 77 K resulted in intense ESR signals; in contrast, it was necessary either to add *t*-butyl hydroperoxide to the samples of *saturated* lipids prior to photolysis, or instead to use γ-irradiation of the pure material, the latter being found to give a more even distribution of radicals in the sample. Diffusion of molecular oxygen was found to occur in each case at a particular temperature, which is characteristic of the particular lipid, when it reacts with those carbon-centred radicals produced initially, and forms the peroxyl radicals. Most interestingly, for unsaturated lipids the peroxyl radicals were found to decay with further annealing, with the concomitant production of the lipid allylic or pentadienyl radicals, respectively, depending on the number of double bonds in the chains. We see, then, the essential free radical chemistry involved in the peroxidation process: a hydrogen atom is abstracted from a weak C—H bond; the resulting stabilised radical then adds molecular oxygen to form a peroxyl radical, which then abstracts a weakly bound hydrogen atom. In experiments with linolenic acid, subsequent introduction of oxygen resulted in the formation of further peroxyl radicals at low temperatures but the signal from the pentadienyl radical was found to prevail at higher temperatures; an explanation is proposed in terms of the relative rates of oxygen migration and hydrogen-atom abstraction processes.

An investigation of the kinetics of the autoxidation of triglycerides was published three years later, again using ESR (Zhu and Sevilla, 1990). It was found that, following the initial production of carbon-centred radicals the following distinct reaction stages then occur: formation of peroxyl radicals by addition of O_2 molecules, and so the depletion of the oxygen content of the sample; conversion of the lipid peroxyl radicals into allylic or pentadienylic radicals, by hydrogen-atom abstraction; and finally, combination of these carbon-centred radicals. From the kinetics of these stages it is concluded that the peroxidation step is controlled by O_2 diffusion, which has an apparent activation energy of $24 \, kJ \, mol^{-1}$ in unsaturated lipids. The subsequent H-atom abstraction step (autoxidation cycle) depends on the lipid structure, and partly on the relative C—H bond strength for the atom being abstracted, since the activation energies are 9 ± 2, 34 ± 8, $88\pm11 \, kJ \, mol^{-1}$, for trilinolenin, trilinolein and triolein, yielding radicals conjugated presumably with two C=C double bonds in the former two cases, but with only one C=C bond in the latter. The cause of the difference in activation energy between trilinolenin and trilinolein is not clear. In the final stage, these carbon-centred radicals combine at temperatures approaching the softening point of the particular lipid matrix, with a common activation energy of $\sim 40 \, \text{kJ mol}^{-1}$. For saturated lipids, the peroxyl radical signal decays with second-order kinetics, indicating a bimolecular radical combination mechanism. In contrast, the unsaturated lipid peroxyl decay, which is unimolecular, can be explained on the basis that the strength of the ROO—H bond is less than that of a C—H bond in a *saturated* molecule, thus preventing its mode of decay by H-atom abstraction.

9.3.2   Motional behaviour of peroxyl radicals in solid matrices

In all chemical and biological processes, it is molecular reorientation and diffusion that permits the constituent molecules to interact with each other, and thus to undergo molecular reactions; efforts, therefore, to understand these motional phenomena are manifest and profound. The study of these aspects of reactive intermediates is generally very difficult, with the greatest success being met with free radicals, particularly by means of ESR and related kinds of spectroscopy. Indeed, stable free radicals (usually nitroxides, R_2N—$O^{\bullet}$, see section 9.6) may be deliberately incorporated as functional groups into polymers, proteins (enzymes) and in membranes so that ESR may monitor their motional behaviour and therefore that of the local molecular environment, at the behest of particular events.

Peroxyl radicals may be similarly utilised, as is of direct relevance to those oxidation processes that they mediate; the gross changes in the ESR spectra that signify both type and rate of molecular reorientation are shown in Figure 9.9, in which (a) represents the condition of very slow reorientation, in which three g-values are measured (see above), while (b) shows the case in which the peroxyl radical reorients rapidly, and equally, about all three (x, y, z) axes. Although not a lipid system, a good example of this is found in a paper (Mach et al., 1989) that reports the detection of the tropenylperoxyl radical, formed in an adamantane matrix. In consequence of its relatively small size in relation to the cavities present in solid adamantane (which is a 'rotor-solid' in which radicals have frequently been isolated to obtain sharp, liquid-like ESR spectra), rapid tumbling occurs at 130 K, and the ESR spectrum is isotropic (Figure 9.9b), while one of type (a) is recorded at 77 K, because any reorientational motion is very slow on the ESR spectra time scale. The ESR parameters measured at 77 K ($g_1 = 2.035$, $g_2 = 2.009$, $g_3 = 2.002$; $a_1 = 8.4$, $a_2 = 9.6$, $a_3 = 5.3$ G) are seen to average to 2.0153 and 7.78 G,

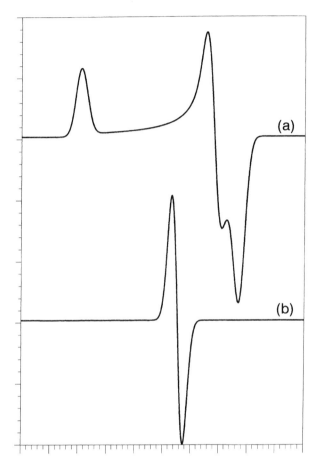

Figure 9.9   ESR spectra of a radical $RO_2^{\cdot}$ that is reorienting (a) very slowly and (b) very fast on the spectral time scale.

which are similar to those of 2.0157 and 7.5 G measured at 130 K: the drop in a at the higher temperature is probably real, and reflects an increased torsional oscillation of the O—O group relative to the unique CH unit. In a less simple system, preferential reorientation about one axis may occur, so that the ESR spectrum is only partially averaged, but in a manner that reveals which axis this is; with reference to the axis system in (**I**), Figure 9.10 indicates the effect of such motion about each axis shown.

When polycrystalline Ph_3CCO_2H or Ph_3CCl was irradiated with UV light (Melamud and Silver, 1974) at room temperature, the $Ph_3C^{\cdot}$ radical was formed, as signified by a single line at $g = 2.0024$. On admission of O_2 gas to the evacuated sample, a nearly symmetrical signal was produced,

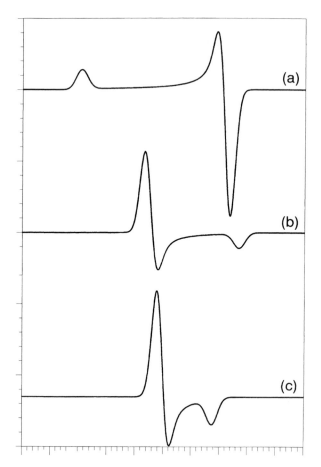

Figure 9.10 Partial averaging of ESR spectra of RO$_2$· radicals by preferential reorientation of the peroxy unit about: (a) z-axis, (b) x-axis, (c) y-axis.

but at $g = 2.014$. To distinguish between mechanisms in which either the whole molecule is tumbling rapidly or the peroxy group is rotating with respect to the remainder of the molecule, the [17]O-enriched Ph$_3$CO$_2$· radical was studied; this showed conclusively that the latter process was dominant.

Sevilla *et al.* have formed peroxyl radicals from lipids (fatty acid esters) present in their clathrates with urea (Sevilla *et al.*, 1989), by exposure of the γ-irradiated material to O$_2$. Orientation and temperature dependence studies show that, at room temperature, the peroxyl radicals rotate or oscillate, as indeed do the precursor carbon-centred radicals, in the urea channels about their long axis: the rotational jump mechanism found for the peroxyl radicals usually follows the trigonal symmetry of the urea host channels, which are distorted at low temperatures. At room temperature the entrapped molecules rotate freely about their long axis, and in consequence, orientation of the samples at the 'magic-angle' (54.7°) results in isotropic couplings. ^{17}O coupling data are also reported, and are typical for hydrocarbon peroxyl radicals. In previous, related, work this group also investigated the motion of lipid peroxyl radicals in the neat lipids, which were triarachidin and linoleic acid (Becker *et al.*, 1987). The spectra were interpreted, as above, in terms of models used previously for peroxyl radicals formed in hydrocarbon polymers: for triarachidin, it was found that simulations based on the modified Bloch equations suggest a 90° jump about the chain axis with the plane of the peroxy group perpendicular to that axis. For linoleic acid, simulations were made assuming that the principal axis system of the *g*-tensor is first rotated by 42° about the *y*-axis which approximates to a rotation about the plane of the COO˙ group. Good agreement with the experimental spectra was found based on a chain axis jump of 180°, but an alternative model which involves rotation about the C—O bond by 90° jumps is also considered. The activation energy for rotation in triarachidin is 4 times that in linoleic acid ($\sim 3.4\,\text{kJ}\,\text{mol}^{-1}$).

9.4 Spin-trapping

9.4.1 Background to method

As free radicals are usually highly reactive species, it is often difficult to obtain them in sufficiently high concentrations for direct detection by ESR to be possible: this is true particularly in reacting systems that are not deliberately designed to form radicals in large amounts, but which it is desired to investigate for radical intermediacy. One very important approach to this problem is called 'spin-trapping', which is a general procedure for converting transient radicals into more stable radicals that can accumulate in the system to detectable concentrations (Perkins, 1980). The spin-traps most commonly used are *C*-nitroso compounds and nitrones, although some success has been met with thiones, most notably in the trapping of organometallic radicals. General reactions for the trapping of a transient radical (R˙) are shown in equations (9.15) to (9.17).

$$R^{\cdot} + R'NO \rightarrow R(R')N\!-\!O^{\cdot} \qquad (9.15)$$

$$R^{\cdot} + PhCH\!=\!N^{+}(O^{-})Bu^{t} \rightarrow PhCH(R)\!-\!N(O^{\cdot})Bu^{t} \qquad (9.16)$$

$$R'_2C\!=\!S + R^{\cdot} \rightarrow R'_2C^{\cdot}\!-\!SR \qquad (9.17)$$

Adducts of nitrones tend to be more stable than are those of nitroso compounds but, unfortunately, the trapped radical is located at a position in which its nuclei are too remote from the site of high spin density for their couplings to be resolved. Thus identification of the radical is difficult, and for the case of adducts with the general formula $PhCH(R)N(O^{\cdot})Bu^{t}$ assignment of the nature of $(R^{\cdot})$ normally rests upon small variations in the couplings to the ^{14}N nucleus and the unique proton: it is then necessary to produce the adduct by independent means to 'fingerprint' these coupling constants. In the case of C-nitroso adducts, the trapped radical $(R^{\cdot})$ is bound directly to the nitrogen atom, so that any α-protons in $(R^{\cdot})$ are now β-to the spin density at the nitrogen atom and are strongly coupled by hyperconjugation. Thus, identification of the trapped radical is much more certain. The magnitude of such β-proton couplings varies according to a $B\cos^{2}\theta$ dependence, and so is determined by conformational preferences of the trapped group (R) in the particular spin adduct.

The range of application of spin-trapping is enormous, covering as it does very many chemical and biological systems. The latter, however, dominate by far and the field of biological spin-trapping is one of increasing growth, as the search for radical involvement in living processes expands; since lipid peroxidation is regarded as ubiquitous in many disease states, there is a vast and burgeoning literature which involves, *inter alia*, spin-trapping as a means to determine its importance in particular processes.

9.4.2 Some recent examples of spin trapping in lipid peroxidation

We have discussed generally oxidative damage to lipids, in which an initial carbon-centred radical is formed by H-atom abstraction from an 'allylic' CH_2 group. However, a study by Muller *et al.*, (1997) has provided some indication that radical attack on the glyceryl moiety of a phospholipid, producing the corresponding 2-glyceryl radical, might precipitate lipid damage via rapid β-cleavage of the adjacent C—O bond, releasing an acyl anion. Moreover, the rate of cleavage is greater in non-polar solvents, at least in model systems, so it may be particularly effective in membranes.

In addition to the ubiquitous implication of the $^{\cdot}OH$ radical in all manner of disease states, 'peroxynitrite' follows from the almost equally ubiquitous 'nitric oxide' (NO), which can form peroxynitrite/peroxynitrous acid by reaction with superoxide (equation 9.18)

$$NO + O_2^{-\cdot} \rightarrow {}^{-}OONO/HOONO \qquad (9.18)$$

Peroxynitrite may be considered as a toxic form of NO, and is formed either inadvertently or deliberately in the cellular defence mechanism. Being strongly oxidising, it is believed capable of initiating lipid peroxidation, and is the subject of many investigations.

9.4.2.1 Thiyl radicals in lipid peroxidation

In one study, the decomposition of peroxynitrite (HOONO) in the presence of the spin-trap 5,5-dimethyl-1-pyrroline-N-oxide (DMPO) generated 5,5-dimethyl-2-pyrrolidone-1-oxyl, without generation of the ·OH adduct of DMPO; the peroxynitrite decomposition was enhanced by formate, but no formate derived radicals were generated. Glutathione, cysteine, penicillamine and ascorbate reacted with peroxynitrite to generate the corresponding thiyl and ascorbyl radicals. The results show that the decomposition of peroxynitrite does not generate significant amounts of ·OH radicals and that single-electron reduction of peroxynitrite by ascorbate may be one of the important detoxification pathways (Shi *et al.*, 1994a). Of course, as in all detoxification mechanisms that produce secondary radicals, the subsequent fate of these radicals is a matter of concern, e.g. for thiyl radicals, which are highly reactive, especially in non-polar media (Rhodes, 1998; Rhodes *et al.*, 1998). We note that thiyl radicals are undetectable by ESR spectroscopy in solution, but a method involving positive muons has been used to study them with great success (Rhodes *et al.*, 1997).

Oxidation of thiols, present in cells to defend against reactive radicals, often results in the formation of thiyl radicals, which, although generally undetectable in solution by ESR directly (Rhodes *et al.*, 1997) may be spin-trapped. A novel phosphorylated spin-trap, 5-diethoxyphosphoryl-5-methyl-1-pyrroline-N-oxide, which may be considered an analogue of DMPO, has been used to investigate the reactions of sulfur-centred radicals produced by the oxidation of thiols with peroxynitrite (Karoui *et al.*, 1996a). In all cases, the predominant species trapped is the corresponding thiyl radical, from glutathione and N-acetyl-DL-penicillamine, but the sulfite radical anion from sulfite ions. All these radicals react with ammonium formate to form the $CO_2^{-·}$ radical anion. It is concluded that the direct reaction of peroxynitrite with thiols and with sulfite forms thiyl radicals and the sulfite radical anion by a mechanism independent of ·OH radicals; pathological implications of thiyl radical formation, and subsequent oxyradical-mediated chain reactions, are discussed. It is suggested that oxygen activation by thiyl radicals formed during peroxynitrite-mediated oxidation of glutathione (GSH) may limit its effectiveness against peroxynitrite-mediated toxicity in cells.

In a re-examination of the formation and reactions of radicals formed by the peroxynitrite-mediated oxidation of thiols and sodium bisulfite, it was found that thiyl radicals were indeed formed from GSH, L-cysteine and N-acetyl-D,L-penicillamine, but the sulfite radical anion was formed from bisulfite, as detected by spin-trapping. Additionally the formation of the OH-DMPO adduct was completely inhibited by low molecular mass superoxide dismutase (SOD) mimics. This suggests that the OH-DMPO adduct originates from the decay of the superoxide radical adduct of DMPO. In the presence of these SOD mimics, the DMPO-sulfur radical adducts were more persistent, implying that superoxide is partly responsible for their decay. Again, the origin of these sulfur radicals by an ˙OH radical-independent mechanism is concluded (Karoui *et al.*, 1996b).

Using a low-frequency (1.1 GHz; L-band) ESR spectrometer, the haemoglobin thiyl radical was detected in living rats. The radical was spin-trapped using DMPO, in blood samples, following the intragastric administration of phenylhydrazine. Pretreatment of the rats with ascorbate and diethylmalonate decreased the signal intensity; incubation of diethylmalonate with rat blood containing preformed thiyl radical adduct showed no effect on the signal, while incubation with ascorbate caused a reduction in its intensity, so providing direct evidence that the haemoglobin thiyl radical is formed *in vivo* and may be studied as such, free from artefacts that can occur when using *ex vivo* methods (Jiang *et al.*, 1996).

9.4.2.2 Haloalkyl radical-initiated peroxidation

Following immediately from the previous section on thiyl radicals is a study of the role of thiyl radicals in lipid peroxidation by CCl_4-derived radicals (Stoyanovsky and Cederbaum, 1996). Elevation of cellular calcium levels has been shown to occur after exposure to hepatotoxins such as CCl_4, and has been associated with inhibition of the Ca^{2+}, Mg^{2+}-ATPase, which pumps calcium into the endoplastic reticulum. Elevated Ca^{2+} may also result from activation of calcium-releasing channels. In the presence of NADPH, CCl_4 produced a concentration-dependent release of calcium from liver microsomes, but after a time lag. The lag period was shorter with microsomes from pyrazole-treated rats in which the enzyme CYP2E1 (a cytochrome P450) is induced. The calcium release process appears to be very sensitive to activation by CCl_4, and it is indicated that CCl_4 metabolism is required for the activation of its release. Production of $CCl_3{}^{\cdot}$ was shown by spin-trapping experiments, which further demonstrated that calcium release was prevented by the presence of spin-traps, confirming that $CCl_3{}^{\cdot}$ and possibly other reactive radicals are indeed required. It is notable that lipid peroxidation was not observed at the levels of CCl_4 used. CCl_4-induced calcium release could be partly reversed by lipophilic thiols such as mercaptoethanol or cysteamine, whereas GSH was ineffective; so the activity is present exclusively in the membrane.

There is concern over the toxic—especially carcinogenic—effects of chromium compounds. One paper (Tezuka *et al.*, 1991) reports the formation of CCl$_3$· radicals in the livers of mice that had been administered Cr(III); the lipid peroxidation in liver microsomes induced *in vitro* by CCl$_4$ in the presence of NADPH was decreased by the preadministration of Cr(III) to the mice. The activity of NADPH-cytochrome *c* reductase, which presumably catalyses the formation of CCl$_3$· from CCl$_4$ in liver microsomes, was depressed by Cr(III) administration and remained lower for at least 2 hours after dosing with CCl$_4$ than in the control group. Spin-trapping in the liver homogenate demonstrated that CCl$_3$· production was reduced by Cr(III) preadministration in a manner similar to that by DL-α-tocopherol. It is believed that Cr(III) actually scavenges CCl$_3$· radicals in the liver cells. However, another group (Hanna *et al.*, 1993) reported a study of the protective role of Cr(III) and also Zn(II) and metallothionein against CCl$_4$ toxicity *in vivo* (Hanna *et al.*, 1993). Their conclusion is that there is no difference in either the amount or in the rate of formation of CCl$_3$· metabolites in the presence of these agents compared with the control. Another study used the formation of CCl$_3$· spin-adducts as a measure of the conversion of CCl$_4$ to CCl$_3$· radicals *in vivo*; this primary bioactivation step was found to occur at similar rates in female rats aged 5, 14 and 28 months, and so the attenuation of CCl$_4$-induced hepatotoxicity is not explained on the basis of decreased bioactivation to reactive species (Rikans *et al.*, 1994).

9.4.2.3 *Studies of lipid peroxides*

As discussed earlier, a crucial event in lipid peroxidation is abstraction of a hydrogen atom by a lipid peroxyl radical, forming a hydroperoxide: this is a propagation step since a carbon-centred radical is also produced, which can add dioxygen to form another peroxyl radical. Since the products of lipid peroxidation arise partly from subsequent reactions of lipid hydroperoxides, an incentive is provided for their study, as in the following examples.

Cr(IV) is known to mediate free radical production from thiols, and in a study of cysteine and penicillamine, with hydrogen peroxide and model lipid peroxides, using spin-trapping, thiyl radicals were detected (Shi *et al.*, 1994b) With cysteine, the radical became detectable at a relative cysteine:Cr(IV) concentration of ∼5, reached its highest level at ∼30 and thereafter declined; similar results were obtained with penicillamine. Incubation of Cr(IV), cysteine or penicillamine and H$_2$O$_2$ led to ·OH radical generation, as was verified by quantitative competition experiments using ethanol. The mechanism is considered to be a Cr(IV)-mediated Fenton-type reaction. When model lipid hydroperoxides such as t-butyl hydroperoxide and cumene hydroperoxide were used instead of H_2O_2, hydroperoxide radicals were produced. It is suggested that since thiols such as cysteine exist in cells at relatively high concentrations, Cr(IV)-mediated free radical generation in the presence of thiols may participate in the mechanisms of Cr(IV)-induced toxicity and carcinogenesis.

On the toxicity of thiols, an independent study has shown that thiyl radicals generated from cysteine and glutathione by nitrogen dioxide enhance the lipid peroxidation of liposome composed of 1-palmitoyl-2-arachidonylphosphatidylcholine. ESR spin-trapping using DMPO confirms the presence of thiyl radicals, while additional analysis shows the effective induction of both lipid peroxidation and DNA strand breaks (Kikugawa et al., 1994).

The relative toxicities of methyl linoleate-9,10-ozonide (MLO) and cumene hydroperoxide (CumOOH) have been compared (Hempenius et al., 1992). Both agents caused a dose-dependent decrease in the phagocytosing activity of alveolar macrophages isolated from rat lungs; MLO was found to be 3 times more toxic than CumOOH. Supplementation of the macrophages with vitamin C caused a decrease in their sensitivity towards MLO but an increase in their sensitivity towards CumOOH, suggesting that different mechanisms underlie their toxic action. This was supported by data obtained on GSH and vitamin E depletion: in both cases, depletion of the antioxidant was more extensive on exposure to CumOOH; additionally, following GSH depletion, the sensitivity of the macrophages towards CumOOH was more increased than towards MLO. Further, MLO was unable to enhance peroxide formation from methyl linoleate, whereas CumOOH initiated its peroxidation. The results of ESR spin-trap experiments further support that MLO-induced toxicity is independent of lipid peroxidation.

From all of this, it is concluded that both mechanisms known to be of importance in peroxide-induced cell toxicity, i.e. depletion of cellular GSH levels and/or lipid peroxidation, are not the main processes leading to MLO toxicity in vivo.

Endothelial cells have been shown to generate primary oxygen-centred radicals (Kramer et al., 1995) ($^{\bullet}$OH, $O_2^{-\bullet}$) during post-anoxic reoxygenation, but little evidence is available concerning subsequent initiation of lipid peroxidative injury. ESR spin-trapping, using phenyl-t-butyl nitrone (PBN), was employed to monitor lipid-derived free radicals formed by cultured bovine aortic endothelial cell suspensions subjected to anoxia (N_2 gas for 45 min) followed by reoxygenation (95% O_2, 5% CO_2 for 15 min). In some experiments, SOD ($10\,\mu g\,ml^{-1}$) was introduced immediately prior to reoxygenation to assess the effects of this primary free radical scavenger on lipid radical production. At various times, aliquots were removed and PBN was introduced to either the cell suspension or the corresponding cell-free filtrate, prior to extraction with toluene and ESR analysis. A lipid derived alkoxyl radical adduct of PBN was detected during reoxygenation using both procedures, with maximal production at 4–5 min followed by a second maximum at 10 min; SOD effectively reduced the RO˙ production. HPLC analysis of hydroperoxide production showed comparable maxima at 4–5 min and 10 min implicating the RO˙ radicals in the peroxidation mechanism.

The mechanism of free radical production from hydroperoxides by cytochrome c has been investigated (Barr and Mason, 1995). When t-butyl hydroperoxide and cumene hydroperoxide were treated with the enzyme in the presence of DMPO, the presence of methyl, peroxyl and alkoxyl radicals was demonstrated. From detailed analysis of the relative concentrations of these species, it was concluded that the alkoxyl radical was that initially produced from the hydroperoxide, presumably by homolytic cleavage of the O—O bond by ferric cytochrome c; this contrasts with a previous study that proposed a heterolytic mechanism for the reaction of cytochrome c with hydroperoxides. Methyl radicals were produced by β-scission of the alkoxyl radicals, while the peroxyl radicals were shown to be secondary products arising from the reaction of dioxygen with methyl radicals (MeOO˙). In separate experiments, visible absorption spectroscopy revealed that the haem centre of the cytochrome c was destroyed during the reaction; both the haem destruction and the production of radical adducts were inhibited by addition of cyanide, presumably by the formation of a cyanohaem complex.

From the same laboratory comes another study, this time of the action of cytochrome P450 on cumene hydroperoxide (Barr et al., 1996). Cumene hydroperoxide-derived peroxyl, alkoxyl and carbon-centred radicals were formed and spin-trapped during the course of the reaction; by means of 2-methyl-2-nitrosopropane as the spin-trap, the carbon-centred radicals were identified as methyl, hydroxymethyl and a secondary carbon-centred radical ($R_2CH˙$). The reaction did not require NADPH-cytochrome P450 reductase or NADPH, and the proposed mechanism is as above for cytochrome c.

The reaction of cytochrome P450 with linoleic acid hydroperoxide was also studied and compared with chemical systems where $Fe(II)SO_4$ or $Fe(III)Cl_3$ was substituted for cytochrome P450 (Rota et al., 1997). In the P450 system, DMPO adducts of ˙OH, $O_2^{-˙}$, peroxyl, methyl and acyl radicals were detected; the same radicals were found in the $Fe(II)SO_4$ system, but only the ˙OH and carbon-centred radical adducts were detected using $Fe(III)Cl_3$. It is proposed that polyunsaturated fatty acid hydroperoxides are initially reduced to form alkoxyl radicals, which then undergo intramolecular rearrangement to form epoxyalkyl radicals: the epoxyalkyl radical then reacts with dioxygen to form a peroxyl radical that decomposes with the elimination of $O_2^{-\bullet}$.

9.4.2.4   Lipid peroxidation by solid particles: silica, asbestos, coal-dust

There is current concern over the toxicity of various minerals caused by their inhalation, leading to lung diseases, including specific types of cancer. Their precise mode of action remains uncertain; however, as shown in the examples below, there is evidence from spin-trapping studies that these materials can initiate free radical formation, and so might precipitate lipid peroxidation of the pulmonary cell membranes.

It is known that iron is present in crocidolite asbestos and that it can reduce O_2 and participate in Fenton-type reactions (Gulumian et al., 1993a). Because of the importance of these reactions in crocidolite-induced toxicity, studies have been made on three different types of crocidolite fibres to determine the factors that control the activity of iron in catalysing the two reactions. Results show that the total concentration of iron in crocidolite fibres is not an appropriate parameter for characterising the activity of this mineral, which seems rather to be controlled by the valency and the location of the iron in the lattice, and also by its availability for mobilization from these minerals.

Following this is a study of crocidolite that has been 'detoxified' (Gulumian et al., 1993b). The fibres were treated with ferric oxide salts to form a metal-micelle polymer surface coating that prevented physiological reactions with the mineral. This detoxified crocidolite was tested for its ability to produces 'OH radicals from H_2O_2. It was found that the intensity of the DMPO-OH radical adduct signal was indeed reduced from that obtained from the native crocidolite fibres. Similar experiments showed that the ability of the detoxified crocidolite to reduce oxygen was also decreased compared with the native mineral. The availability of ferrous iron present in the two crocidolite fibres to catalyse the above reactions was investigated with the chelating agent ferrozine. The results indicate that ferrozine was able to remove fewer ferrous ions from detoxified crocidolite than the native form; moreover, Mössbauer spectroscopy shows that the detoxification process results in both bulk and surface changes in the coordination chemistry of the detoxified sample. The detoxification process also introduces a surface coating comprising ferric ions that shields near-surface ferrous ions and consequently reduces the Fenton-type reactivity of the fibres. On the subject of 'detoxification' of crocidolite, the same group (Gulumian et al., 1997) report the treatment of crocidolite fibres with microwave radiation at different temperatures: this reduced the Fe^{2+}/Fe^{3+} ratio, according to Mössbauer measurements, and produced a concomitant decrease in the ability of the fibres to peroxidize linoleic acid.

At least one *in vivo* study has been made of the toxicity of asbestos using a spin-trapping technique (Ghio *et al.*, 1998). Rats (180 days old) were instilled intratracheally with either 500 µg of crocidolite or saline; 24 hours later, histological examination of the lungs revealed a neutrophilic inflammatory response. ESR examination of the chloroform extract from lungs exposed to asbestos showed a spectrum consistent with a carbon-centred radical adduct while those spectra from lungs instilled with saline revealed a far weaker spectrum. The adducts are nearly identical with ethyl and pentyl radical adducts, providing evidence of *in vivo* lipid peroxidation resulting from asbestos exposure.

Another report shows that the 'OH-generating potential of coal dust correlates positively with the surface iron content of the coal dust (Dalal *et al.*, 1995). Two other papers describe results of the inhibition of quartz-induced lipid peroxidation. In one (Shi *et al.*, 1995), tetrandrine, an alkaloid used in China to treat the lesions of silicosis, is tested for its antioxidant activity: it is found that tetrandrine reacts efficiently with 'OH radicals generated by the reaction of freshly fractured quartz particles with an aqueous medium, and also scavenged $O_2^{-\bullet}$ radicals produced from xanthine/xanthine oxidase. A significant inhibition of linoleic acid peroxidation by freshly fractured quartz particles was also found. Taurine-based compounds were similarly investigated (Shi *et al.*, 1997): it was discovered that hypotaurine, but not taurine, caused a significant reduction in silica-induced peroxidation, again using linoleic acid as a model lipid.

9.5   Spin-labelling

9.5.1   *Background to method*

Essentially, this technique (Berliner, 1976) involves the incorporation of a stable radical, usually a nitroxide, into a medium as a probe of molecular mobility, and rests upon the averaging of anisotropic *g*- and hyperfine tensors, as illustrated in Figure 9.11. In Figure 9.11a is represented the limit of slow motion, where the full anisotropy in both *g*- and *A*-tensors is present; the other extreme is shown in (c), which is the limit of very fast reorientation so that the anisotropies of the tensors are completely averaged, and the spectrum shows three lines of equal intensity, spaced by the isotropic ^{14}N coupling constant and centred on the isotropic *g*-factor. The intermediate situation (b) is called the region of 'intermediate motion'; the anisotropies are not completely averaged, but their presence results in an asymmetric broadening of the lines as shown, being greatest

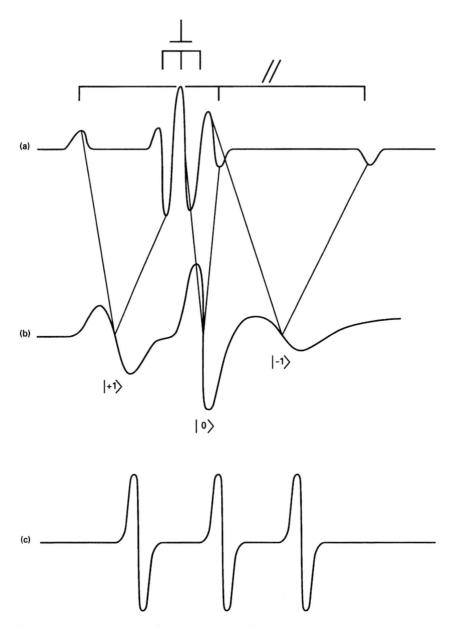

Figure 9.11   ESR spectrum of a spin-label ($R_2NO^{\bullet}$): (a) the limit of slow motion, (b) line broadening in the region of intermediate motion, (c) the limit of fast motion.

for the high-field line. By a detailed analysis of these lineshapes, it is possible to derive the rate of motion of the probe and thus the local degree of motion of its environment.

An extension of this is to chemically bond the probe to a particular site, to assess details of the anisotropy of the motion (for example, a long, straight molecular chain will generally reorient preferentially along the chain axis). This has found its greatest use in biological systems, where the probe may be attached, for example, to an enzyme protein or to a lipid in a cell membrane, in order to measure the molecular response of a biomolecule to a particular interacting substrate. The analysis of spin-label spectra under conditions of such (slow) anisotropic motion is a difficult theoretical task, but it is rendered more practical by developments in computational techniques, particularly the availability of 'super-computers'.

9.5.2 Spin-labelling studies of membranes

The literature on spin-labelling studies of membranes is enormous and covers many diverse aspects of membrane function and biochemistry, but we hope to have illustrated some major recent advances in the understanding of membrane properties using spin-labelling techniques. As an introduction to this field, we call attention to an excellent review of magnetic resonance studies of membranes, which includes both spin-labelling and NMR work (Knowles and Marsh, 1991).

9.5.3 Some recent examples

A model previously developed for describing the dynamics of flexible alkyl chains has been adapted to the analysis of spin-label spectra obtained from an oriented phospholipid membrane. In this model, rotation around each C—C bond of the labelled alkyl chain is characterized by two inequivalent minima, with one end of the chain fixed to mimic the phospholipid headgroup, and with the dynamic effects of the nitroxide label included explicitly. This model is integrated with that for the overall rotation of the phospholipid in the mean orientational potential of the aligned membrane, and it is incorporated into the stochastic Liouville equation that describes the ESR lineshape in the presence of these dynamic processes. The analysis is simplified by introducing the fact that the relatively rapid internal modes of motion can be treated by motional narrowing theory and a time scale separation can be made with respect to the much slower overall motions of the phospholipid. A series of ESR spectra from the spin-label 16-PC in the lipid dimyristoylphosphatidylcholine were obtained in the range of temperature 35–65°C in the L-alpha phase for various orientations of the normal to the bilayer plane relative to the magnetic field. Very good agreement with experiment is obtained from this model by using least-squares fitting (Cassol et al., 1997).

Another paper by this group (Patyal et al., 1997) reports a study of lipid–gramicidin interactions using two-dimensional Fourier transform

ESR. It is shown that 2D-FT-ESR spectra provide substantially enhanced spectral resolution to changes in the dynamics and ordering of the bulk lipids—as compared with continuous wave (cw) ESR—that result from addition of gramicidin to membrane vesicles of phosphatidylcholine in excess water containing 16-PC as the lipid spin-label. Both the rotational and translational diffusion rates of the bulk lipid are substantially decreased by addition of gramicidin (GA), whereas the ordering is only slightly increased, for a 1:5 ratio of GA to lipid.

No significant evidence is found in the 2D-FT-ESR spectra for a second immobilised component, which is seen in cw ESR measurements, and simulations of the FT spectra suggest that this component, usually ascribed to 'immobilized' lipid is inconsistent with its being characterized by increased ordering, but is more consistent with a component with a significantly reduced diffusion rate. This is because the 2D-FT-ESR spectra exhibit a selectivity, favouring components with longer homogeneous transverse relaxation times.

2D-FT and cw ESR spectra at X-band frequencies were recorded over a broad range of temperatures covering the solid and melt states of a liquid crystalline polymer. The cw spectra were analysed using conventional motional models. The nematic phase was macroscopically aligned in the magnetic field, whereas the solid state showed microscopic order but macroscopic disorder; an end-label on the polymer showed smaller ordering and larger reorientational rates than those of the cholestane spin-probe dissolved in the same polymer, since the former can reorient by local internal chain modes. It was demonstrated that the 2D-FT-ESR experiments provide greatly enhanced resolution to the order and dynamics of the end-label, especially when performed as 2D-ELDOR (electron-electron double resonance) experiments as a function of the mixing time. Instead of the conventional model of Brownian reorientation, a model of a slowly relaxing local structure that enables differentiation between the local internal modes experienced by the end-label and the collective reorganization of the polymer molecules around the end-label yielded much improved fits to the experiments in the nematic phase (Xu et al., 1996).

In the functioning of a biological membrane, lipid–protein interactions play a major role. It has been found that the stoichiometries of lipid–protein interactions, obtained from spin-label ESR experiments with integral membrane proteins, deviate from values predicted from simple geometric models for the intramembranous parameter that are based on the predicted numbers of transmembrane helices. These deviations provide evidence for oligomerization of the protein in the membrane and/or the existence of more complex arrangements of the transmembrane segments (Marsh, 1997).

The specific binding of hen egg white avidin to phosphatidylcholine lipid membranes containing spin-labelled N-biotinylphosphatidyethanolamines was investigated by ESR (Swamy and Marsh, 1997). Spin-labelled derivatives were prepared with the nitroxide group at positions C-5, C-8, C-10, C-12 or C-14 of the lipid chain, and binding of avidin caused a strong and selective restriction of lipid mobility at all positions of labelling. Overall, the results indicate that the biotinylphosphatidylethanolamines are partially withdrawn from the membrane, with a vertical displacement of 7–8 Å, on complexation with avidin.

Conventional ESR of spin-labelled lipids and saturation-transfer ESR of spin-labelled proteins are used to study lipid–protein interactions and the mobility of integral proteins, respectively, both in biological membranes and in reconstituted lipid–protein systems. Conventional ESR reveals two spin-labelled lipid populations, the mobility of one being hindered by direct interaction with the integral membrane proteins (Marsh, 1996). The proportion of the latter component increases with increasing protein content and with increasing selectivity of the lipid species for the protein. Lipid exchange rates at the protein interface obtained by spectral simulation are found to be consistent with fast-exchange found by ^{2}H NMR on similar systems and to reflect the lipid selectivity observed by ESR. Protein-reactive covalent labels were used to study the rotational diffusion and aggregation states of membrane proteins via saturation-transfer ESR. The integral protein rotation is uniaxial and the anisotropic motion is analysed to obtain the principal component of the diffusion tensor: this is sensitive to the cross-sectional dimensions of the protein in the membrane, and hence to its state of assembly. A variety of novel experiments based on the power saturation properties of the spin-labelled components were also used to determine lipid exchange rate, protein translational diffusion rates, and the location and penetration of proteins in membranes.

Finally, we mention the use of very high-frequency (94.3 GHz) ESR in a determination of the partitioning and dynamics of 2,2,6,6-tetramethyl-1-piperidinoxyl nitroxide radicals in large unilamellar liposomes. The nitroxide was completely resolved in both lipid and aqueous phases on account of the large resolution of the differing isotropic g-factors in the two media (Smirnov et al., 1995).

References

Barr, D.P. and Mason, R.P. (1995) Mechanism of radical production from the reaction of cytochrome c with organic hydroperoxides—an ESR spin-trapping investigation. J. Biol. Chem., **270** 12709-12716.

Barr, D.P., Martin, M.V., Guengerich, F.P. and Mason, R.P. (1996) Reaction of cytochrome P450 with cumene hydroperoxide—ESR spin-trapping evidence for the homolytic scission of the peroxide O—O bond. *Chem. Res. Toxicol.*, **9** 318-325.

Becker, D., Yanez, J., Sevilla, M.D., Alonso-Amigo, M.G. and Schlick, S. (1987) An electron spin resonance investigation of the motion of lipid peroxyl radicals. *J. Phy. Chem.*, **91** 492-496.

Berliner, L.J. (ed.) (1976) *Spin-Labelling: Theory and Applications*, Academic Press, New York.

Burton, G.W., Foster, D.O., Perley. B., Slater, T.F., Smith. I.C.P. and Ingold, K.U. (1985) Biological antioxidants. *Phil. Trans. R. Soc. London B*, **311** 565-578.

Carrington, A. and McLachlan, A.D. (1979) *Introduction to Magnetic Resonance*, Chapman and Hall, London.

Cassol, R., Ge, M.T., Ferriani, A. and Freed, J.H. (1997) Chain dynamics and the simulation of ESR spectra from oriented phospholipid membranes. *J. Phys. Chem.*, **101** 8782-8789.

Dalal, N.S., Newman, J., Pack, D., Leonard, S. and Vallyathan, V. (1995) Hydroxyl radical generation by coal-mine dust—possible implication to coal-workers pneumoconiosis. *Free Radic. Biol. Med.*, **18** 11-20.

Ghio, A.J., Kadiiska, M.B., Xiang, Q.H. and Mason, R.P. (1998) In-vivo evidence of free radical formation after asbestos instillation: an ESR spin-trapping investigation. *Free Radic. Biol. Med.*, **24** 11-17.

Gulumian, M., Bhoolia, D.J., Theodorou, P., Rollin, H.B., Pollak, H. and Vanwyk, J.A. (1993a) Parameters which determine the activity of transition-metal iron in crocidolite asbestos. *S. Afr. J. Sci.*, **89** 405-409.

Gulumian, M., Vanwyk, J.A., Hearne, G.R., Kilk, B. and PollaK, H. (1993b) ESR and Mössbauer studies on detoxified crocidolite—mechanism of reduced toxicity. *J. Inorg. Biochem.*, **50** 133-143.

Gulumian, M., Nkosibomvu, Z.L., Channa, K. and Pollak, H. (1997) Can microwave radiation at high temperatures reduce the toxicity of fibrous crocidolite asbestos? *Environ. Health Perspect.*, **105** 1041-1044.

Gutteridge, J.M.C. (1978) Effect of cholesterol and cholesteryl acetate on peroxidation of ox-brain phospholipid membranes. *Res. Commun. Chem. Pathol. Pharmacol.*, **22** 563-571.

Gutteridge, J.M.C. and Halliwell, B. (1989) *Free Radicals in Biology and Medicine*, Clarendon Press, Oxford.

Hanna, P.M., Kadiiska, M.B., Jordan, S.J. and Mason, R.P. (1993) Role of metallothionein in zinc(II) and chromium(III) mediated tolerance to carbon tetrachloride hepatotoxicity—evidence against a trichloromethyl radical-scavenging mechanism. *Chem. Res. Toxicol.*, **6** 711-717.

Hempenius, R.A., Rietjens, I.M.C.M., Grooten, H.N.A. and Devries, J. (1992) Comparative study on the toxicity of methyl linoleate-9,10-ozonide and cumene hydroperoxide to alveolar macrophages. *Toxicology*, **73** 23-34.

Jiang, J.J., Liu, K.J., Jordan, S.J., Swartz, H.M. and Mason, R.P. (1996) Detection of free-radical metabolite formation using in-vivo EPR spectroscopy—evidence of rat hemoglobin thiyl radical formation following administration of phenylhydrazine. *Arch. Biochem. Biophys.*, **330** 266-270.

Karoui, H., Hogg, N., Frejaville, C., Tordo, P. and Kalyanaraman, B. (1996a) Characterization of sulfur-centred radical intermediates formed during the oxidation of thiols and sulfite by peroxynitrite. *J. Biol. Chem.*, **271** 6000-6009.

Karoui, H., Hogg, N., Joseph, J. and Kalyanaraman, B. (1996b) Effect of superoxide-dismutase mimics on radical adduct formation during the reaction between peroxynitrite and thiols. *Arch. Biochem. Biophys.*, **330** 115-124.

Kikugawa, K., Hiramoto, K., Okamoto, Y. and Hasegawa, Y.K. (1994) Enhancement of nitrogen dioxide-induced lipid peroxidation and DNA strand breaking by cysteine and glutathione. *Free Radic. Res.*, **21** 399-408.

Kramer, J.H., Dickens, B.F., Misic, V. and Weglicki, W.B. (1995) Phospholipid hydroperoxides are precursors of lipid alkoxyl radicals produced from anoxia reoxygenated endothelial cells. *J. Mol. Cell. Cardiol.*, **27** 371-381.

Knowles, P.F. and Marsh, D. (1991) Magnetic resonance of membranes. *Biochem. J.*, **274** 625-641.

Mach, K., Novakova, J., Hanus, V. and Raynor, J.B. (1989) Gas-phase radicals in co-condensed adamantane matrix: proton splitting in the ESR spectrum of the cycloheptatrienylperoxyl radical. *Tetrahedron*, **45** 843-848.

Marsh, D. (1996) Membrane assembly studied by spin-label ESR. *Braz. J. Med. Biol. Res.*, **29** 863-871.

Marsh, D. (1997) Stoichiometry of lipid–protein interaction and integral membrane protein structure. *Eur. Phys. J. Biophys. Lett.*, **26** 203-208.

Melamud, E. and Silver, B.L. (1974) Triphenylmethylperoxyl. *J. Magn. Reson.*, **14** 112-116.

Muller, S.N., Batra, R., Senn, M., Geise, B., Kisel. M. and Shadyro, O. (1997) Chemistry of 2-glyceryl radicals: indications for a new mechanism of lipid damage. *J. Am. Chem. Soc.*, **119** 2795-2803.

Perkins, M.J. (1980) Spin trapping. *Adv. Phys. Org. Chem.*, **17** 1-64.

Patyal, B.R., Crepeau, R.H. and Freed, J.H. (1997) Lipid-gramicidin interactions using two-dimensional Fourier-transform ESR. *Biophys. J.*, **73** 2201-2220.

Rhodes, C.J. (1999) Radical behaviour. *Chem. Britain* (in press).

Rhodes, C.J. and Agirbas, H. (1990) Electron paramagnetic resonance of imine radical cations in low-temperature solid matrices. *J. Chem. Soc., Faraday Trans.*, 3303-3308.

Rhodes, C.J., Hinds, C.S. and Reid, I.D. (1997) Muon spectroscopy applied to biological systems: a study of thiyl radicals. *Free Radic. Res.*, **27** 347-352.

Rhodes, C.J., Hinds, C.S. and Reid, I.D. (1998) Transverse field studies of thiyl radicals. *PSI Annual Report (1997)* (in press).

Rikans, L.E., Hornbrook, K.R. and Cai, Y. (1994) Carbon tetrachloride hepatotoxicity as a function of age in female Fischer-244 rats. *Mech. Ageing Dev.*, **76** 89-99.

Rota, C., Barr, D.P., Martin, M.V., Guengerich, F.P., Tomasi, A. and Mason, R.P. (1997) Detection of free radicals produced from the reaction of cytochrome P-450 with linolein acid hydroperoxide. *Biochem. J.*, **328** 565-571.

Sevilla, C.L., Becker, D. and Sevilla, M.D. (1986) An electron spin resonance investigation of radical intermediates in cholesterol and related compounds: relation to solid-state oxidation. *J. Phys. Chem.*, **90** 2963-2968.

Sevilla, M.D., Champagne, M. and Becker, D. (1989) Study of lipid peroxyl radicals in urea clathrate crystals: oxygen-17 couplings and rotational averaging. *J. Phys. Chem.*, **93** 2653-2658.

Shi, X.L., Rojanasakul, Y., Gannet, P., Liu, K.J., Mao, Y., Daniel, L.N., Ahmed, N. and Saffiotti, U. (1994a) Generation of thiyl and ascorbyl radicals in the reaction of peroxynitrite with thiols and ascorbate at physiological pH. *J. Inorg. Biochem.*, **56** 77-86.

Shi, X.L., Dong, Z.G., Dalal, N.S. and Gannett, P.M. (1994b) Chromate-mediated free radical generation from cysteine, penicillamine, hydrogen peroxide and lipid hydroperoxides. *Biochim. Biophys. Acta Mol. Basis Dis.*, **1** 65-72.

Shi, X.L., Mao, Y., Saffiotti, U., Wang, L.Y., Rojanasakul, Y., Leonard, S.S. and Vallyathan, V. (1995) Antioxidant activity of tetrandrine and its inhibition of quartz-induced lipid peroxidation. *J. Toxicol. Environ. Health*, **46** 233-248.

Shi, X.L., Flynn, D.C., Porter, D.W., Leonard, S.S., Vallyathan, V. and Castranova, V. (1997) Efficiency of taurine based compounds as hydroxyl radical scavengers in silica induced peroxidation. *Ann. Clin. Lab. Sci.*, **27** 365-374.

Smirnov, A.I., Smirnova, T.I. and Morse, P.D. (1995) Very high-frequency EPR of 2,2,6,6-tetramethyl-1-piperidinoxyl in 1,2-dipalmitoyl-*sn*-glycero-3-phosphatidylcholine liposomes —partitioning and molecular dynamics. *Biophys. J.*, **68** 2350-2360.

Stoyanovsky, D.A. and Cederbaum, A.I. (1996) Thiol oxidation and cytochrome P450-dependent metabolism of CCl$_4$ triggers Ca^{2+} release from liver-microsomes. *Biochemistry*, **35** 15839-15845.

Swamy, M.J. and Marsh, D. (1997) Spin-label studies on the anchoring and lipid–protein interactions of avidin with N-biotinylphosphatidylethanolamines in lipid bilayer membranes. *Biochemistry*, **36** 7403-7407.

Tezuka, M., Ishii, S. and Okada, S. (1991) Chromium (III) decreases carbon tetrachloride-originated trichloromethyl radical in mice. *J. Inorg. Biochem.*, **44** 261-265.

Xu, D.J., Crepeau, R.H., Ober, C.K. and Frred, J.H. (1996) Molecular-dynamics of a liquid-crystalline polymer studied by 2-dimensional Fourier-transform and c.w. ESR. *J. Phys. Chem.*, **100** 15873-15885.

Yanez, J., Sevilla, C.L., Becker, D. and Sevilla, M.D. (1987) Low-temperature autoxidation in unsaturated lipids: an electron spin resonance study. *J. Phys. Chem.*, **91** 487-491.

Zhu, J., Johnson, W.J., Sevilla, C.L., Herrington, J.W. and Sevilla, M.D. (1990) Reactions of lipid peroxyl radicals with antioxidants. *J. Phys. Chem.*, **94** 7185-7190.

Zhu, J. and Sevilla, M.D. (1990) Kinetic analysis of free-radical reactions in the low-temperature autoxidation of triglycerides. *J. Phys. Chem.*, **94** 1447-1452.

10 UV/visible light spectroscopy of lipids

Andrew J. Young and Richard J. Hamilton

10.1 Introduction

By 1949, UV spectroscopy had been applied routinely to fats and oils for a decade. A review published by Hilditch in 1949 was notable for the fact that the only instrumental technique employed was UV spectroscopy. To produce the spectra shown in Figure 10.1, the wavelength was set manually and the absorption was read off. Such was the familiarity with the technique that Hilditch was able to note the adulteration of Stillingia oil (*Sapium sebiferum*) by tung oil (*Aleurites fordii*), which contains eleostearic acid. Similarly, van der Hulst (1935), and Dann and Moore (1933) had published spectra of 'pure' ethyl oleate and of isomerised sodium linoleate during the 1930s.

Although UV/visible light spectroscopy is not generally employed for lipids, as the majority do not display distinctive absorption bands above 220 nm, some compounds (generally those possessing a conjugated system) lend themselves to this technique. Because of their long conjugated double bond systems (the chromophore), the carotenoids absorb in the visible or, in some cases, in the UV region of the spectrum. Absorption spectra can therefore be determined routinely as one of the important criteria used for the partial identification of a carotenoid but also for quantitative analysis (see Britton (1995) for a recent review). Valuable information concerning the identity of a compound and its geometric configuration can be obtained by examining both the position of the absorption maxima (λ_{max}) and the shape or fine structure of the spectrum.

10.2 Theory: origins of the spectrum

10.2.1 The electromagnetic spectrum

The ultraviolet region of the spectrum runs from the lower visible region (the blue end of the spectrum) at 400 nm, to below 100 nm (see Chapter 1). Visible light has wavelengths in the region 400–700 nm, and higher wavelengths are referred to as infrared light. Wavelengths are often also expressed in angstroms: $1 \text{ Å} = 10^{-10} \text{ m}$ (0.1 nm). Thus, a wavelength 500 nm is also expressed as 5000 Å. Spectroscopists also use the term wavenumber (cm^{-1}), which represents the number of waves that occupy a

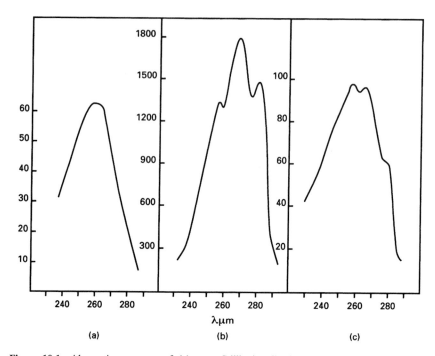

Figure 10.1 Absorption spectra of (a) pure Stillingia oil, (b) α-eloeostearic acid, and (c) commercial Stillingia oil. (After Hilditch, 1947.)

length of 1 cm. Thus, for light of wavelength 500 nm the wavenumber $= 2 \times 10^4 \, cm^{-1}$. The use of wavenumber or frequency is often preferred by spectroscopists because these are proportional to the energy of a photon of the radiation, whereas the wavelength is inversely proportional to energy. High energy thus corresponds to high wavenumber but low wavelength.

10.2.2   *Energy levels in atoms and molecules*

Absorption in the near-UV results from the interaction of electromagnetic radiation with unsaturated groups in a molecule or with atoms carrying unshared pairs of electrons.

The term *chromophore* is used to refer to the conjugated, unsaturated part of a molecule (containing delocalised π electrons) that is responsible for the absorption of light in the near UV or visible regions of the spectrum.

In organic molecules, when an sp^3 hybridised orbital of carbon overlaps with the 1s orbital of a hydrogen atom, a sigma bond is formed.

In methane, for example, four σ orbitals are formed, which contain the bonding electrons of methane. For each bonding σ orbital there is a corresponding antibonding orbital σ*, which has a much higher energy. When UV radiation falls on methane, one electron is promoted from the σ to the σ* orbital. Such σ→σ* excitation is observed at 120 nm.

The relative positions of the energy levels of orbitals in organic molecules are shown in Figure 10.2. The non-bonding electron pair in an atom such as iodine or nitrogen present in the molecule can be promoted to the antibonding σ* orbital, which is normally unoccupied. This $n \to \sigma^*$ energy gap is less than for σ→σ*. The relationship between energy and wavelength is given by $E = h\nu$ where E is energy, h is Planck's constant and ν is the frequency. The frequency $\nu = c/\lambda$, where c is the velocity of light $(3 \times 10^{10} \, \text{cm s}^{-1})$. From these two equations we can thus derive $E = h \times c/\lambda$ (i.e. as energy decreases, the wavelength increases). So the transition $n \to \sigma^*$ is associated with higher wavelengths than σ→σ* and is at 259 nm for iodomethane. The π bond in an ethylenic group, for instance, is formed by the overlap of an unhybridised p lobe and an sp^2 hybridised carbon atom. This π bond has a corresponding antibonding orbital π*. The transition π→π* is near 175 nm for a monosubstituted ethene (see Chapter 1).

Some typical absorption maxima are given in Table 10.1.

Because the absorption leads to broad bands in the UV, little structural information can be determined for lipids. However, these same broad bands permit quantitative measurements concerning sample concentration.

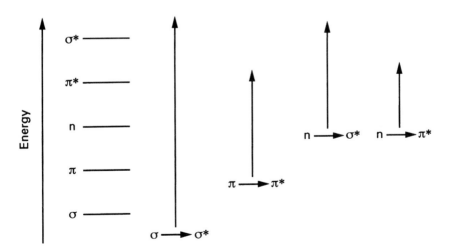

Figure 10.2 The relative positions of energy levels and electronic transitions in organic chemistry.

Table 10.1 Absorption maxima for simple chromophores

Chromophore	λ_{max} (nm)	$\log_{10} \varepsilon_{max}$
$\pi \rightarrow \pi^*$		
$RCH\!=\!CH_2$	175	4.1
$RCH\!=\!CH\!-\!R$ *cis*	176	4.1
$RCH\!=\!CHR$ *trans*	187	3.9
$RC\!\equiv\!CH$	187	2.65
$n \rightarrow \pi^*$		
$CH_3\,C\overset{\displaystyle \overset{\cdot\cdot}{O}}{\underset{H}{\diagup}}$	294	1.08
$CH_3\,C\overset{\displaystyle \overset{\cdot\cdot}{O}}{\underset{CH_3}{-}}$	279	1.14
$n \rightarrow \sigma^*$		
$CH_3\!-\!\overset{\cdot\cdot}{\underset{\cdot\cdot}{C}}l\!:$	173	2.30
$CH_3\!-\!\overset{\cdot\cdot}{B}r\!:$	204	2.30
$CH_3\!-\!\overset{\cdot\cdot}{\underset{\cdot\cdot}{I}}\!:$	259	3.56
$CH_3\!-\!\overset{\cdot\cdot}{N}H_2$	215	2.78
$CH_3\!-\!\overset{\cdot\cdot}{\underset{\cdot\cdot}{O}}H$	184	2.18

The position of the band—its wavelength—can permit significant estimates of structural features to be made.

10.2.3 Conjugation

In molecules that have two double bonds that are separated from each other, each will behave as described above. However, if these double bonds are adjacent, the effect is to bring these C=C bonds into conjugation and to bring the π orbital energy levels closer together (that is, the energy required to excite an electron from a π to a π* orbital is reduced). The electronic structure of the simplest conjugated molecule, butadiene, will be used as a model structure. The molecular orbitals of butadiene are shown in Figure 10.3. The two ethylenes in conjugation in butadiene result in the formation of four orbitals: two bonding orbitals π+π and π−π and two antibonding orbitals π*+π* and π*−π*. The electronic transition occurs as π−π→π*+π*. Since this is clearly a smaller energy gap than the π→π* of ethylene, it appears at a longer wavelength (Figure 10.3). As a result, butadiene has an absorption maximum of 217 nm.

For a highly conjugated system (such as β-carotene) where the π electrons are highly delocalized, the excited state is of relatively low energy. As the length of the chromophore increases, it is easier to promote one of the bonding π electrons to an excited state—the energy of the excited state and of the transition are reduced. For a conjugated

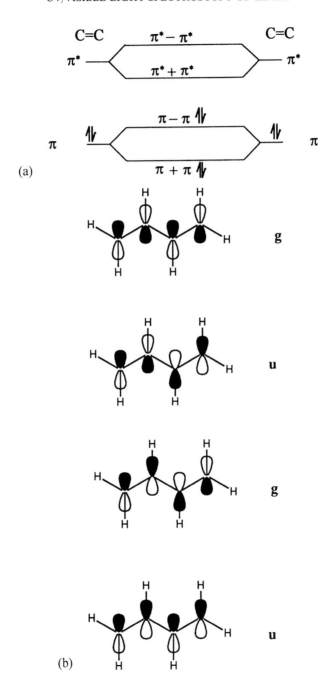

Figure 10.3 (a) The energy levels and (b) the π-electron molecular orbitals (showing symmetry labels to the right) of butadiene.

system, such as a carotenoid molecule, the key feature in terms of their light absorption properties is the delocalisation of the π electrons along the polyene backbone. It is this that gives these molecules their characteristic electronic spectra but also affects their photochemical properties. In the minimum energy (ground state) configuration (termed 1^1A_g) all the bonding molecular orbitals are doubly occupied. The promotion of one electron from the highest-energy occupied molecular orbital (HOMO) to the lowest-energy unoccupied molecular orbital (LUMO) generates the lowest excited state (1^1B_u). Other electron promotions can lead to the formation of higher energy configurations. The strong $1^1A_g \rightarrow 1^1B_u$ transition is responsible for the pronounced absorption in the UV and visible regions of the spectrum that is characteristic of all polyenes and carotenoids.

10.2.4   Position of the absorption maxima

The absorption band of lipids and molecules such as carotenoids can be affected by a number of factors, including alterations to their molecular environment. Such changes can be classified as follows: bathochromic shift (shift of absorption maximum to a longer wavelength); hypsochromic shift (shift of absorption maximum to a shorter wavelength); hyperchromic effect (an increase in absorbance); hypochromic effect (a decrease in absorbance). The relationship between the structure of the molecule and the effects of changes to its molecular environment are discussed below.

Most carotenoids have three absorption maxima (λ_{max}), the positions of which are characteristic of the carotenoid chromophore, although the solvent does have some influence over this. Absorption spectra are usually determined in petrol or hexane for the carotenes and in ethanol for the more polar xanthophylls. The values recorded in these solvents are almost identical, although other solvents may greatly influence the λ_{max} of the carotenoid (see below).

The λ_{max} values of a carotenoid are largely a function of the length of the conjugated double bond chromophore, and the λ_{max} values increase as the number of conjugated double bonds increases. This is clearly seen in a series of acyclic carotenoid molecules that form the biosynthetic route for carotenoid synthesis in biological systems, shown in Figure 10.4; as the chromophore increases from 3 conjugated double bonds in phytoene to 11 in lycopene, a clear relationship with the λ_{max} of the compound is evident. Double bonds that are not in conjugation do not contribute to the chromophore (as in the isolated double bond of the ε-end-group of α-carotene).

Extension of the conjugated double bond system into a ring (particularly the C-5,6 double bond of the β-ionone ring) does extend

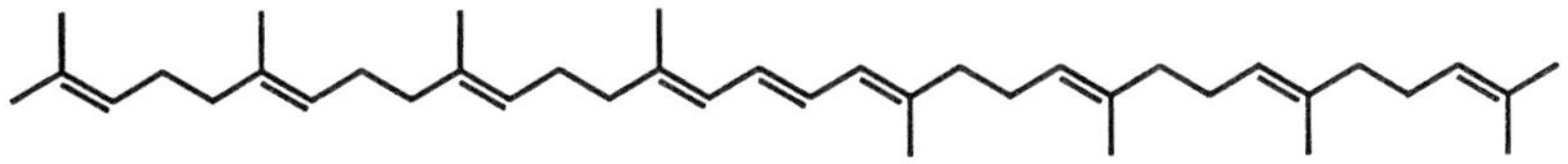

Phytoene ($n = 3$; $\lambda_{max} = 286$ nm)

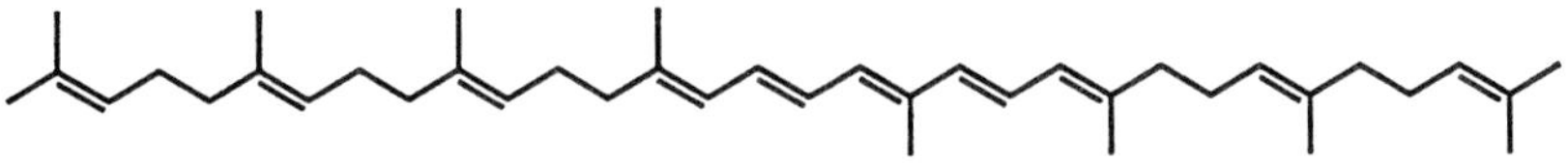

Phytofluene ($n = 5$; $\lambda_{max} = 348$ nm)

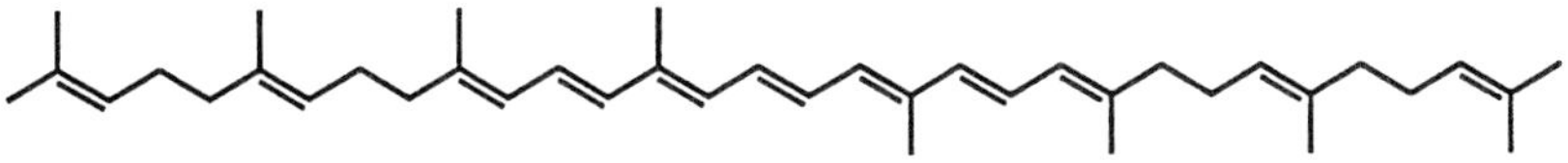

ζ-Carotene ($n = 7$; $\lambda_{max} = 400$ nm)

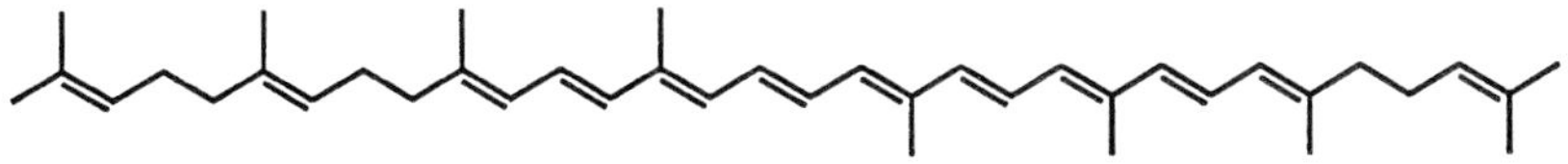

Neurosporene ($n = 9$; $\lambda_{max} = 440$ nm)

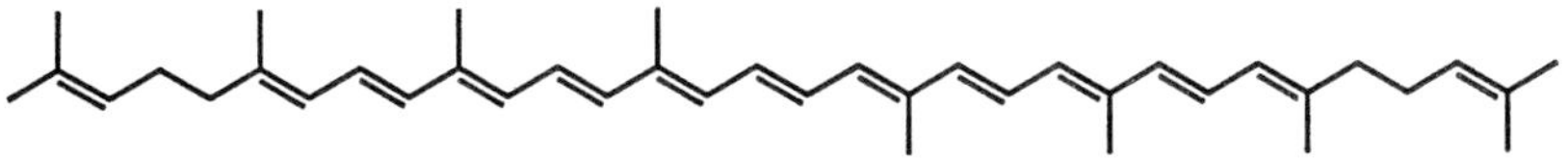

Lycopene ($n = 11$; $\lambda_{max} = 470$ nm)

Figure 10.4 Absorption maxima (λ_{max}) for a series of acyclic carolenoids with increasing conjugated double bond system.

the chromophore but, because the ring double bond is not coplanar with the main polyene chain, the absorption maxima occur at shorter wavelengths than those of the acyclic carotenoid with the same number of conjugated double bonds. Thus, although they are all conjugated undecaenes, the acyclic, monocyclic and dicyclic isomers lycopene, γ-carotene and β-carotene have λ_{max} at 444, 470 and 502 nm, at 437, 462, 494 nm and at 425, 450, 478 nm, respectively.

Carbonyl groups, in conjugation with the polyene system, also extend the chromophore. A conjugated keto group in a ring increases the λ_{max} by 10–20 nm, so that echinenone and canthaxanthin, the 4-oxo and 4,4′-dioxo derivatives of β-carotene, have λ_{max} at 461 and 478 nm, respectively.

The presence of hydroxy and methoxy groups and non-conjugated oxogroups generally does not affect the chromophore or the λ_{max} of the carotenoid, so that all such substituted carotenoids show λ_{max} values that are virtually identical with those of their parent hydrocarbon molecule. Thus, β-carotene and its hydroxy derivatives β-cryptoxanthin and zeaxanthin have virtually identical spectra, with λ_{max} at 425, 450 and 478 nm (in ethanol). Carotenoid acyl esters also possess the same λ_{max} as the parent carotenoid.

The solvent shift behaviour of carotenoids and other polyenes is a well known phenomenon. The absorption band shifts to lower energies in relation to the refractive index of the solvent (η). A shift of $\sim 10^4 \, cm^{-1}$ to lower energy is observed for a unit change in $(\eta^2 - 1)/(\eta^2 + 2)$ (Kohler, 1995). Thus, λ_{max} values recorded in acetone are greater by ~ 4 nm, those recorded in chloroform or benzene greater by 10–12 nm, and those in carbon disulfide by as much as 35–40 nm when compared to those obtained in n-hexane. When spectra are determined on-line by HPLC, it must be remembered that the values for λ_{max} in the eluting solvent frequently do not correspond to the published values that were recorded in a different pure solvent.

10.2.5   Spectral fine structure

The effects of vibrational fine structure in the UV/visible spectra of molecules will not be discussed here in any detail and readers are instead referred to standard texts (e.g. Hollas, 1992). Acyclic carotenoids such as lycopene typically display a three-peak structure for their main absorption band. This arises from transitions from the lowest vibrational level of the ground state to the lower vibrational level of the excited state.

The fine structure of the absorption spectrum of an organic molecule can be diagnostic. Spectral fine structure may be expressed as a numerical function %III/II; i.e. the ratio of peak heights III/II (see Figure 10.5). Values for %III/II may range from 0 in conjugated ketocarotenoids such as canthaxanthin, to 25 for β-carotene, 65 for an acyclic carotenoid such as lycopene, and up to 118 for the C_{20} acyclic dialdehyde crocetindial.

With the carotenoids, the degree of fine structure is generally dependent on the extent of planarity that the chromophore can achieve. Thus, acyclic carotenoids, in which the conjugated double bond system can adopt an almost planar conformation (e.g. lycopene), are characterised by having an absorption spectrum that has sharp maxima and minima (= persistence). The degree of fine structure decreases a little when the chromophore exceeds nine double bonds. Cyclic carotenoids in which the conjugation does not extend into the rings behave as linear polyenes and have similar well-defined spectral fine structure. When the conjugation is broken (e.g. by the presence of C5-C6 epoxide groups), this reduces the λ_{max} but also increases the degree of fine structure.

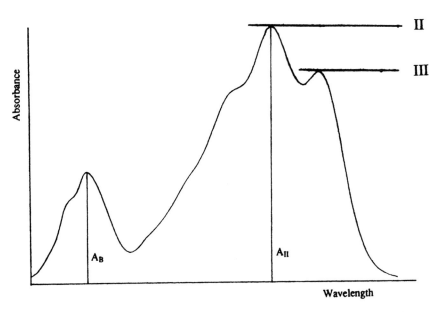

Figure 10.5   The relative intensity of a *cis*-peak ($\%A_B/A_{II}$) and of spectral fine structure ($\%III/II$) of a carotenoid.

When the conjugation extends into a β-ring, steric strain or hindrance, and the resulting twisting of the molecule to adopt a conformation in which the ring double bond is not coplanar with the π electron system of the polyene chain, give rise to spectra that show a lower degree of fine structure. Spectral fine structure therefore decreases in the order lycopene (ψ,ψ-acyclic) > γ-carotene(β,ψ-monocyclic, one ring double bond conjugated) > α-carotene (β,ε-dicyclic, one ring double bond conjugated) > β-carotene (β,β-dicyclic, two ring double bonds in conjugation). The effect is even more pronounced in carotenoids that have carbonyl groups in conjugation with the polyene chain; thus the spectra of ketocarotenoids such as canthaxanthin have only a single, rounded, almost symmetrical absorption peak. A slight degree of fine structure can be observed if the spectrum is determined in a non-polar solvent such as petrol or hexane rather than ethanol.

In general, the solvent used to determine the absorption spectra of carotenoids can have an influence on the degree of fine structure observed, as well as λ_{max} (see above), and therefore at least two solvents should normally be used for the determination of absorption spectra. Other factors that affect the molecular environment of the carotenoid may significantly alter the absorption spectrum. Carotenoids in aqueous solutions may form aggregates (of the J- and H-types), the absorption characteristics of which are very different from the monomeric forms of the compounds (in terms of both absorption maxima and spectral fine structure). Thus, the λ_{max} of the dihydroxy carotenoid zeaxanthin is shifted from ~450 nm in 100% ethanol to ~380 nm in 60% (v/v) ethanol–water. In addition, all spectral fine structure is lost (in this case J-type aggregates are formed) (Ruban *et al.*, 1993). Carotenoids may also be found in biological systems complexed with lipid and/or protein. In some cases the *in situ* absorption characteristics of the compound are relatively unaltered. In others, the effects of caroteno–protein complex formation can be dramatic. Perhaps the best example of this can be seen for the caroteno–protein complexes of marine invertebrates such as the lobster or starfish. The carapace of the lobster changes from blue to red on cooking, owing to the liberation of the carotenoid astaxanthin (λ_{max} ~480 nm) from the α-crustacyanin–protein complex (λ_{max} ~630 nm). Spectral fine structure is little altered (Britton *et al.*, 1997).

10.3   Geometric isomers

The absorption spectra of carotenoids that contain one or more *cis* double bonds in the chromophore show several characteristic differences from the spectrum of the all-*trans* compound (Zechmeister, 1962). For the *cis* isomers, the λ_{max} values are generally 1–5 nm lower, the spectral fine structure is decreased, and a new absorption peak, commonly referred to as the *cis*-peak appears at a characteristic wavelength in the UV region 142 (±2) nm below the longest wavelength peak in the main visible absorption region. In addition the intensity of the main absorption maxima is reduced as the *cis*-peak appears. These effects, especially the intensity of the *cis*-peak, are greatest when the *cis* double bond is located at or near the centre of the chromophore. This is illustrated in the series of *cis* isomers of β-carotene shown in Figure 10.6. With the acyclic carotenoids that have a very long chromophore, an overtone peak is usually present in the *cis*-peak region of the spectrum of the all-*trans* isomer, though this peak is usually of lower intensity than in the spectra of the *cis* isomers. The relative intensity of the *cis*-peak can be expressed as a function of the middle absorption band: $\%A_B/A_{II}$ (Figure 10.5).

10.4   Practical considerations and diagnostic tests

10.4.1   Oils and fats

Fatty acids and glycerides do not show any absorption in the visible region of the spectrum (400–700 nm). However, natural fats and oils

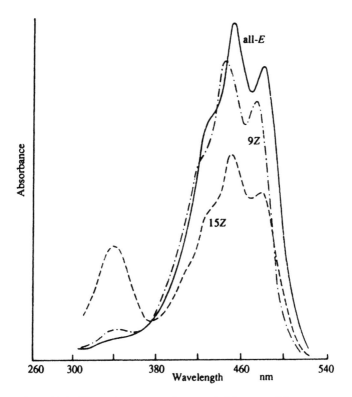

Figure 10.6 Absorption spectra of geometrical isomers of β-carotene.

contain pigments that do absorb light in this region (see Chapter 12). Gossypol has a strong absorption band at 337 nm, chlorophyll *a* at ~660 nm (together with the Soret absorption at ~431 nm) and chlorophyll *b* at ~641 nm (Soret at ~453 nm). Note that the position of the actual maxima is dependent upon the solvent.

As discussed, compounds with conjugated double bond systems (or aromatic rings) are not common among naturally occurring lipids. Despite being the most common type of detector for HPLC, UV detection is rarely employed. Derivatisation can be used to permit detection using such detectors. Thus, fatty acids can be converted into aromatic esters, glycolipids can be benzoylated and diacylglycerols derived from phospholipids can be esterified with aromatic acid derivatives (Sewell, 1992). For example, derivatisation of fatty acids to form *p*-bromophenyl esters permits their separation by HPLC with on-line UV detection at 288 nm (Durst *et al.*, 1975). Further practical applications and examples are given by Sewell (1992).

Using the only instrumental techniques available at that time for sterol structural determination, Woodward and Fieser developed a set of rules that enabled close approximations to be calculated for actual band absorption maxima. An example of the application of these rules is given in Table 10.2 and an explanation of their use can be found in Silverstein *et al.* (1991).

Saturated fatty acids (e.g. caprylic and myristic) absorb at 205 nm and 185 nm owing to the $C=O$ of the carboxyl group, while oleic and elaidic acid have corresponding bands at 200 nm and 153 nm. It should be noted that, although these peaks are not very intense, especially when compared to compounds with conjugated chromophores, they have been used to detect lipids eluting from HPLC columns before the advent of the so-called mass detector (or evaporative light-scattering detector).

The UV spectra of conjugated di and tri-unsaturated acids are especially characteristic:

$18:2^{9t,11t}$ $\lambda_{max} = 231$ nm with inflections at 227 nm and 239 nm
$18:3^{9t,11t,13t}$ $\lambda_{max} = 260, 269, 281$ nm
$18:4^{9t,11t,13t,15t}$ $\lambda_{max} = 288, 302, 316$ nm

It was found that when oils containing linolenic and linoleic acids were treated with alkali, the double bond systems were moved into conjugation (Morton *et al.*, 1931; Hilditch, 1949). The resulting peaks at 233 nm and 268 nm corresponded to diene and triene acids. The spectroscopic method for the analysis of these compounds was first developed by Mitchell *et al.* (1943). The method first measures the absorbance of the untreated oil at 233, 262, 268 and 274 nm so that any small amount of diene or triene acids can be allowed for. Approximately 100 mg of oil is then heated at 180°C under nitrogen in the presence of 6–11% (w/v) KOH in glycol for 25 min. The solution is cooled and transferred to a graduated flask and made up to 100 ml. The absorbance can then be measured. The method is an empirical one and it is necessary to work to the standardised conditions.

This procedure does not apply to modified drying oils, fats containing *trans* fatty acids as a result of hydrogenation, fish oils and oils containing

Table 10.2 Predicted and observed absorption maxima

	Predicted λ_{max} (nm)	Observed λ_{max} (nm)
Cholesta-2,3-diene	273	276
Cholesta-3,4-diene	237	235
Cholesta-5,7-dien-3-ol	263	262
Cholesta-ten-3-one	227	231
Cholesta-3,5-dien-7-one	275	278

pigments whose absorption changes as a result of the alkali treatment. Occasional problems were encountered with the quantitative analysis of linoleic and linolenic acids using this technique and GC was adopted as the accepted method.

More recently, these same UV bands at 223 and 268 nm have been used to monitor the rancidity of oils and fats and of fatty foods (Prior and Loliger, 1995). Autoxidation of linolenic acids yields 9- and 13-hydroperoxides that have conjugated double bonds. One of the simplest tests for rancidity is to measure the spectrum from 200 to 300 nm, which monitors the formation of the hydroperoxides and other oxidised lipids. Equally, the use of normal-phase HPLC with a UV detector permits the levels of the four isomers that result from the autoxidation of linolenic acid to be determined (Hsieh and Kinsella, 1986).

10.4.2 Carotenoids

A rapid and simple test for the presence of 5,6-epoxide groups in a carotenoid (as in violaxanthin, for example) can be carried out in a cuvette. The spectrum of the carotenoid is determined in ethanol and redetermined following the addition of 1 drop of $0.1\,mol\,dm^{-3}$ HCl. A hypsochromic shift of about 20 nm for the mono-epoxide and 40 nm for the di-epoxide is observed owing to the furanoid rearrangement to 5,8-epoxide groups. Little overall change in the fine structure of the spectrum is observed.

Carotenoids containing conjugated carbonyl groups (e.g. astaxanthin) usually exhibit a spectrum (in ethanol) with little or no fine structure. Again, the presence of such groups can be rapidly determined in a spectrophotometer cuvette. The carotenoid is dissolved in ethanol and the absorption spectrum recorded. A small amount (1–2 mg) of $NaBH_4$ is added and thoroughly mixed and the spectrum is redetermined after 30 s, 5 min and 30 min. This will show whether aldehyde groups (these are usually reduced within 30 s to 1 min) or ketogroups (5–30 min) are present in conjugation with the main polyene chain; reduction of the carbonyl groups to alcohol groups results in an increase in spectral fine structure and a corresponding reduction in the λ_{max}.

Methods have been developed for the spectroscopic determination of carotenoids in a range of biological and food/feed formulations. Perhaps the most important of these concerns the analysis of β-carotene and astaxanthin. These two carotenoids more than any others are of considerable commercial interest; the former is used as an antioxidant and food colorant, the latter is used as a pigment in aquaculture and is responsible (with canthaxanthin) for the coloration of farmed salmon and trout. Chromatographic and spectroscopic methods for the analysis of these carotenoids have been published (Schierle *et al.*, 1995; Schüep and Schierle, 1995) and the quantitation of carotenoids by spectroscopic means is described below.

One of the many problems encountered during their analysis is the initial extraction of these compounds from the biological or food/feed matrix. Saponification can be used, but losses of β-carotene are commonly reported and carotenoids such as astaxanthin (which posses a 3-hydroxy, 4-keto end group) are highly sensitive to alkali treatment and even some chromatographic absorbents (e.g. alumina). Equally, their sensitivity to light, oxygen and high temperatures means that great care should be taken in their handling during extraction and isolation procedures. Enzyme digestion of fish feed containing synthetic carotenoid formulations (Carophyll Pink and Carophyll Red, F. Hoffmann-La Roche) using an alkaline protease (e.g. Maxatase, Genencor Int.) is a standard technique to ensure complete recovery of astaxanthin and canthaxanthin, respectively, prior to quantitative determination by HPLC or UV/visible spectroscopy (Schierle and Härdi, 1994).

10.5 Quantitative determination

10.5.1 Spectrophotometry

Quantitative determination of carotenoids is generally carried out using the specific absorption coefficient $A_{1\,cm}^{1\%}$ (= specific extinction coefficient $E_{1\,cm}^{1\%}$), although the molar absorption or extinction coefficient ε_{mol} (the absorbance of a $1\,mol\,dm^{-3}$ solution, pathlength 1 dm) is sometimes used. Values for $A_{1\,cm}^{1\%}$ and ε_{mol} can be determined by accurately weighing 1–2 mg of carotenoid that is completely free from impurity or contamination and dissolving this in the appropriate solvent. This can, in itself, be fraught with difficulty, especially with crystalline compounds for which it may be difficult to ensure complete dispersion in the solvent. In this case a small volume of dichloromethane (< 3% v/v) may be added to aid solution of the carotenoid.

$$\varepsilon = A_{1\,cm}^{1\%} \times \text{molecular mass} \qquad (10.1)$$

The $A_{1\,cm}^{1\%}$ is defined as the absorbance of a 1% (w/v) solution in a 1 cm path cuvette at a defined wavelength. An arbitrary value of $A_{1\,cm}^{1\%} = 2500$ (i.e. a $1\,\mu g\,ml^{-1}$ solution would give an absorbance of 0.25) is usually used when no specific values are available for an individual carotenoid or for a mixture of pigments. The amount of carotenoid present (x mg) in y ml of solvent is determined as

$$x = \frac{A_y \times 1000 \times y}{A_{1\,cm}^{1\%} \times 100} \tag{10.2}$$

The specific absorption coefficient is related to the molecular mass of a compound, so that $A_{1\,cm}^{1\%}$ (zeaxanthin) $= A_{1\,cm}^{1\%}$ (β-carotene) $\times (536/568)$.

Absorption spectra can be determined for $\sim$1–2 µg of carotenoid, although on-line detection using HPLC may allow a good-quality spectrum to be obtained from as little as a few nanograms of compound (see below). As with any other compound, accuracy is greatly improved if the absorbance of the sample is be adjusted to give absorbance in the range 0.3–0.7.

Tabulated values for $A_{1\,cm}^{1\%}$ of carotenoids have been published elsewhere (Britton, 1995). Table 10.3 presents λ_{max} values and extinction coefficients for compounds of industrial or commercial interest.

10.5.2   Quantitative determination by HPLC

The choice of the mobile phase for HPLC analysis of lipids is critical if UV absorption is to be used for detection. The UV cut-off (the wavelength at which the transmission is $< 10\%$) is a critical property as many solvents routinely used for lipid analysis (e.g. chloroform) absorb in the same wavelength region as lipids (typically 200–210 nm). Other solvents may contain impurities (plasticisers, etc.) or antioxidants that can result in significant levels of absorption in the UV region.

HPLC provides the most sensitive, accurate and reproducible method for quantitative analysis of carotenoids. The relative amounts of each compound can be determined, provided the peak area can be calculated at its λ_{max}. If monitoring is possible at only a single wavelength, corrections must be made for the difference between the absorbance at that wavelength and at λ_{max} for each component. External calibration can be achieved by injecting known amounts (determined spectro-photometrically) of pure carotenoid, determining peak areas, and creating a calibration graph (usually over the range 10–2000 pmol) which allows the amount of each carotenoid to be estimated with great precision at the nanogram level. An alternative strategy is to use an internal standard that is added to the sample itself (e.g. decapreno-β-carotene). The amount of each component in the chromatogram is then estimated by comparing its peak area with that of the standard, again correcting for the difference between absorbance at λ_{max} and that at the monitoring wavelength, if necessary. The main criteria for the selection of the

Table 10.3 Structures, extinction coefficients and absorption maxima (λ_{max}) of the major commercially used carotenoids

Carotenoid	Structure	$A^{1\%}_{1\,cm}$	ε_{mol} (dm^2 mol^{-1})	λ_{max} (nm)	Solvent
Astaxanthin		2100	125100	470	Hexane
Canthaxanthin		2200 / 2092	124100 / 118000	466 / 484	Petroleum ether / Benzene
Capsorubin		2200	132000	460, 489, 523	Benzene
β-Carotene		2592 / 2620	138900 / 140400	425, 450, 477 / 450, 476	Petroleum ether / Ethanol
Ethyl-8′-apo-β-carotene-8′-oate		2640	109800	457	Petroleum ether
Lutein		2550 / 2236	144800 / 12700	422, 445, 474 / 432, 458, 487	Ethanol / Benzene
Lycopene		2450 / 3370	184900 / 180600	444, 470, 502 / 455, 487, 522	Petroleum ether / Benzene
Zeaxanthin		2348 / 2340	133400 / 132900	424, 449, 476 / 430, 452, 479	Petroleum ether / Acetone

standard are that it should not be present in the sample under
investigation, that it should clearly separate from all components in that
extract, that its stability and λ_{max} should be similar to those of the natural
components, and that its $A_{1\,cm}^{1\%}$ should be known accurately. Alternatively, a compound present in the sample under study may be used.
Following analysis of the extract by HPLC (and determination of the
peak areas), a known amount of pure standard compound is added to the
sample and a second HPLC analysis is performed. The difference in peak
area between the two analyses will allow the determination the amounts
of all substances that are present in the extract.

It should be noted that, in general, the extinction coefficients for
compounds in their HPLC eluting solvent (which may, of course, be a
complex mixture) are not available. It is therefore essential that the
chromatographic system is fully calibrated through the use of internal or
external standards. Some adjustment may also be necessary if significant
amounts of *cis* isomers are present.

10.6   Instrumentation

UV/visible detectors are the most commonly used detectors for HPLC
analysis, particularly for the analysis of compounds with a conjugated
double bond system. Naturally occurring lipids generally lack such
conjugated systems although some fatty acids found in seed oils and lipid
hydroperoxides may contain conjugated double bonds. Derivatisation of
fatty acids to form aromatic esters, for example, can be used to permit
UV detection (see below). Typically therefore UV/visible detection is
limited to compounds such as the carotenoids and tocopherols.

In the last few years, diode-array detection has become a routine
laboratory technique for analysing compounds on the basis of their
absorption spectra. Certainly, its impact in the area of carotenoid
research has been significant in the last few years. Figure 10.7 shows a
diagramatic representation of the 1100 series diode-array detector
produced by Hewlett-Packard. In this device, an achromatic lens system
focuses polychromatic light into the flow cell (typically 13 µl, although
smaller, high-pressure flow cells ($\sim$1.7 µl) may be used for SFC or LC-
MS). After passing through the sample, the beam then passes through a
programmable slit (1–16 nm), disperses on the surface of a diffraction
grating, and falls on the photodiode array. Typically, light ranging from
190 to 950 nm can be measured within a few milliseconds by the
electronics associated with the array. Each diode measures a narrow band
of wavelengths of the spectrum, effectively permitting simultaneous data
collection. In the Hewlett-Packard device, the array consists of 1024

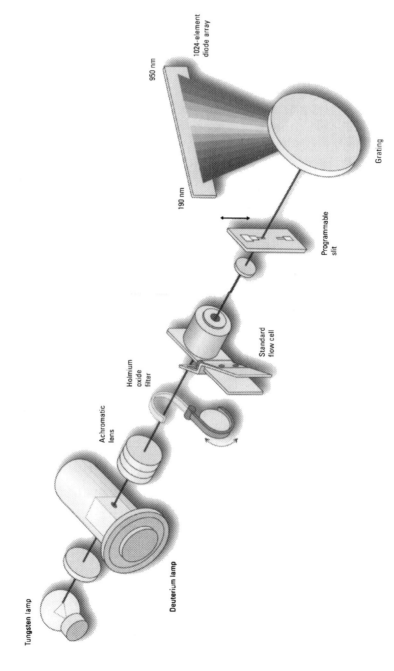

Figure 10.7 Diagramatic representation of a HPLC diode-array detector (Hewlett-Packard 1100 series).

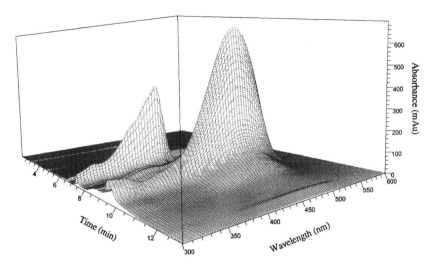

Figure 10.8 3D plot from a diode array detector showing on-line spectral resolution over the wavelength range 300–600 nm of a mixture of carotenoids and chlorophyll dervitatives separated by normal-phase HPLC.

diodes, with a spectral resolution of 1 nm. Dual lamps (tungsten and deuterium) are used to ensure that high sensitivity is maintained over the whole wavelength range.

The advantages of the diode-array detector are considerable for compounds that exhibit characteristic absorption spectra and some of these are outlined below, although it is still the case that most diode-array systems are less sensitive than their single-wavelength or multi-wavelength counterparts. One of the main strengths of these detectors is the ability to display a 3D or so-called 'isoabsorbance' plot (Figure 10.8). Such displays are of considerable assistance in the optimisation of chromatographic separations, especially when a mixture of compounds with widely differing λ_{max} is present. For carotenoids, the presence of *cis* isomers can be quickly confirmed as the *cis*-peak may be readily observed at a lower wavelength than the main absorption maxima (see above).

References

Britton, G. (1995) in *Carotenoids*, vol. 1B, *Spectroscopy* (eds G. Britton, S. Liaaen-Jensen and H. Pfander), Birkhaüser, Basel, pp 13-62.

Britton, G., Weesie, R.J., Askin, D., Warburton, J.D., Gallardo-Guerrero, L., Jansen, F.J., de Groot, H.J.M., Ludtenburg, J., Cornard, J.-P. and Merlin, J.-C. (1997) *Pure Appl. Chem.*, **69** 2075-2084.

Dann, W.J. and Moore, T. (1933) *Biochem. J.*, **27** 116-119.

Durst, H.D., Milano, M., Kikta, E.J., Connelly, S.A. and Grushka, E. (1975) *Anal. Chem.*, **47** 1797.

Hilditch, T.P. (1949) *J. Oil Colour Chem. Assoc.*, **XXXII** 5-23.

Hollas, J.M. (1992) *Modern Spectroscopy*, 2nd edn, Wiley, New York.

Hsieh, R.J. and Kinsella, J.E. (1986) *J. Food Sci.*, **51** 940.

Kohler, B.E. (1995) in *Carotenoids*, vol. 1B, *Spectroscopy* (eds G. Britton, S. Liaaen-Jensen and H. Pfander), Birkhaüser, Basel, pp 1-12.

Mitchell, J.H., Kraybell, H.R. and Zscheile, F.P. (1943) *Ind. Eng. Chem.*, **15** 1.

Morton, R.A., Helbron, I.M. and Thompson, A. (1931) *Biochem. J.*, **25** 20.

Prior, E. and Loliger, J.D. (1995) in *Rancidity in Foods* 3rd edn. (eds J. Allen and R.J. Hamilton), Chapman and Hall, London.

Ruban, A.V., Horton, P and Young, A.J. (1993) *J. Photochem. Photobiol.*, **21** 229-234.

Schüep, W. and Schierle, J. (1995) in *Carotenoids*, vol. 1A, *Isolation and Analysis* (eds G. Britton, S. Liaaen-Jensen and H. Pfander), Birkhaüser, Basel, pp 273-276.

Schierle, J. and Härdi, W. (1994) in *Analytical Methods for Vitamins and Carotenoids in Feed* (eds P. Hofman, H.E. Keller, J. Schierle and W. Schüep), F. Hoffmann-La Roche, Basel, supplement.

Schierle, J., Härdi, W., Faccin, N., Bühler, I. and Schüep, W. (1995) in *Carotenoids*, vol. 1A, *Isolation and Analysis* (eds G. Britton, S. Liaaen-Jensen and H. Pfander), Birkhaüser, Basel, pp 265-272.

Sewell, P.A. (1992) in *Lipid Analysis* (eds R.J. Hamilton and S. Hamilton), Oxford University Press, Oxford, pp 153-204.

Silverstein, R.M., Bassler, G.C. and Morrill, T.C. (1991) *Spectrometric Identification of Organic Compounds*, 5th edn, Wiley, New York.

van der Hulst, L.J.N. (1935) *Rec. Trav. Chim.*, **54** 639-6443, 644-650.

Zechmeister, L. (1962) *Cis-Trans Isomeric Carotenoids, Vitamins A and Arylpolyenes*, Springer-Verlag, Wien.

11 X-ray diffraction of lipids

Peter Laggner

11.1 Introduction: what can a fat and oil chemist and technologist today expect from X-ray diffraction?

Every chemist or chemical technologist is aware of the enormous impact X-ray diffraction has had throughout the 20th century, and particularly over the last few decades, in the development of molecular science and technology. Our present understanding of nucleic acids and proteins—the key structures of modern biotechnology—and of synthetic macro- or supramolecules as elements for new materials, are just the more spectacular examples of where X-ray diffraction has laid foundations and opened new horizons. At the same time, however, the more practically oriented scientists and technologists still have certain reservations against its use, based upon two traditional misconceptions: that X-ray diffraction works only with crystals; and that X-ray diffraction is slow and tedious.

Neither is true. Firstly, noncrystalline (gel or powder), partly ordered (liquid crystalline), or even completely random (colloidal dispersion) systems, can be most beneficially analysed by X-ray diffraction. Secondly, today's X-ray techniques are neither slower nor any more tedious than other laboratory techniques: technological developments in modern X-ray instrumentation—refined optics, powerful detectors, as well as advanced computers and software—have rendered the general experimental practice simple, not dependent on the presence of the talented magician in the laboratory, and fast. Laboratory X-ray measurements (except for the most demanding single-crystal analyses) are presently a matter of a few minutes to an hour, and at high-flux synchrotron X-ray sources, which, despite all the user-friendliness offered by these large facilities, are still rather elaborate to deal with, exposure times of milliseconds and below can be reached, so that real-time observations of dynamic processes are becoming possible.

Table 11.1 summarizes the various techniques presently available and of potential use in the field of fats and lipids. This chapter will briefly illustrate the essential features of this methodological spectrum and give the reader a practice- and application-oriented overview. For greater depth in the theoretical and experimental treatment, the reader is referred to the ample literature of textbooks, monographs, and reviews. This chapter is written for readers who know something about fats and lipids

Table 11.1 Techniques and applications of X-ray diffraction with lipids

Technique	Type of information	Remarks
Single-crystal analysis	High-resolution atomic coordinates, three-dimensional molecular structure	Only possible with well-developed single crystals
Powder diffraction	Type and dimensions of supramolecular arrangement (bilayers, rods, etc.), definition of hydrocarbon chain packing	Used to define T-, p-, c- phase diagrams; to study interactions with additives; suitable for *time-resolved studies* with synchrotron radiation
Diffuse small-angle scattering	Size and shape of lipid micelles or vesicles at low resolution; soluble lipid–protein complexes	Useful where neither ideal crystals nor liquid crystals can be obtained, e.g. with lipoproteins, or soluble mixed micelles
Surface diffraction	Electron density profile of surface monolayer at low resolution; in-plane arrangement of hydrocarbon chains	Unique technique for studying the structure of surface films; extremely low signal: needs synchrotron radiation

but are non-experts in X-ray diffraction: it should be useful to fat and oil chemists and technologists who need molecular structure information and are in search of the right approach. It should also convince those who still adhere to the above reservations to consider X-ray techniques as normal analytical tools side by side with spectroscopic techniques. The techniques are introduced briefly, and their potential is demonstrated on selected examples. This selection is certainly biased by the author's unavoidable preferences, even more so as this is not intended to be a review in the classical sense.

What then, in a nutshell, can fat and oil chemists and technologists today expect from X-ray diffraction? In R&D, X-ray crystallography together with neutron and electron diffraction is still the best source for high-resolution structural data—if suitable crystals can be grown. One might argue that NMR spectroscopy could do the same, and perhaps more easily. This is true, but in general only for molecules in solution, and they might well be different in conformation from the aggregated state. The combination of X-ray crystallography, advanced NMR techniques, and molecular modelling is certainly the most powerful approach to study the structure of lipids. A lucid account of this has been given in the review by Pascher *et al.* (1992).

X-Ray powder diffraction of polycrystalline material can also be used to obtain important structural information, and this is by far the most suitable approach in studies on mixed lipid systems, where microphase separation or limited co-crystallisation prevent the growth of single crystals suitable for X-ray analysis. Experimentally, this is a very simple, rapid, and comparatively cheap technique, so that its potential can be realized in any standard laboratory situation. Its strength lies primarily in the sensitivity to structural changes induced by additives, such as drugs, vitamins, or lipids of different composition, and by physical parameters such as pressure and temperature. Through their sensitivity, the methods of powder diffraction are potent tools not only in R&D, but also in the analytical laboratory for product quality assessment, and for industrial production control.

A relatively recent addition to the arsenal of X-ray diffraction techniques comes from the development of powerful synchrotron radiation sources. These are electron (or positron) accelerators that produce X-ray beams of ten thousand-fold and higher flux density than provided by conventional laboratory X-ray generators. This opens the possibility of performing fast time-resolved structure studies, reaching down to the millisecond and microsecond time domains. Owing to the very fact that these are large installations—ring tunnels in the order of hundreds of metres in circumference—they are not routine instruments. However, as complementary tools for studying the dynamics of, for example, structural transitions, be it in a technological or in a biological context, they are a revolutionary addition to the methodological spectrum of X-ray diffraction.

The micellar state of lipids, in which molecules aggregate to a limited size and shape, is also amenable to X-ray diffraction studies by a special technique called small-angle X-ray scattering, abbreviated SAXS. This is particularly useful for studies on soaps and detergents in water, for microemulsions, or for biological lipid–protein complexes such as lipoproteins.

11.2 The most important methods and some representative examples

11.2.1 Single-crystal diffraction

This is the method with the ultimate structural resolution, as it has the potential for unambiguously evaluating the precise atomic positions, except for hydrogen atoms, which are of such a low electron density that they are practically unseen by X-rays. Its theoretical background is excellently covered by elementary textbooks and monographs (Guinier, 1963; Warren, 1969; Sherwood, 1976), and any further repetition here would be redundant.

A typical result illustrating the information content in the area of lipids, is shown in Figure 11.1. Such information, although obtained from a single crystal, which must be considered a rather artificial state of lipids, is of the greatest importance in discussing the results of other, frequently simpler, techniques that provide data of lower resolution but under different conditions.

The greatest difficulty and limitation in the application of single-crystal diffraction in this field is certainly the growth of crystals suitable for high-resolution diffraction. Lipids will mostly crystallise in thin plates with rather limited extension in the direction of their long molecular axes, and consequently the resolution in this direction will be often limited; also these crystals are normally very deformable and difficult to handle. This technique, therefore, has to be considered a specialist domain, hardly suitable for routine analytical purposes. In the literature dating back as far as the early 1920s (Müller, 1923), a solid basis of information about the crystal structure of alkanes, alcohols, fatty acids, soaps and glycerides exists. Much less information is available on membrane lipids, such as phospho- and sphingolipids, but over the last 20 years several important structures have been reported. A comprehensive coverage of the crystallographic lipid structures can be found in the volume on the physical chemistry of lipids by Small (1986).

11.2.2 Small-angle and wide-angle powder diffraction

Fats and lipids, when not in their molten state, are normally polycrystalline materials: they consist of randomly oriented crystalline domains, separated by more or less extensive regions of disordered, amorphous regions. The crystallites are varying in their sizes and can reach up to submillimetre dimensions. An X-ray beam with a larger cross-section than that of the crystallites will simultaneously hit all orientations of the crystallites and, therefore, at any given time all possible Bragg reflections will be excited. The resulting diffraction pattern, therefore, is a set of rings, centred around the primary beam, as shown schematically in Figure 11.2. Small-angle powder diffraction, owing to its ease of application and the fact that crystallographic data are obtained without having to grow large, single crystals, is the most widely used diffraction technique with fats and lipids. The type of information that can be obtained is illustrated in Figure 11.3 by the composition–temperature phase diagram of two closely related phospholipid species (Lohner *et al.*, 1987). For details of the powder diffraction technique, see Jenkins and Snyder (1986).

There is no fundamental reason for distinguishing conceptually between small-angle and wide-angle powder diffraction, except that the

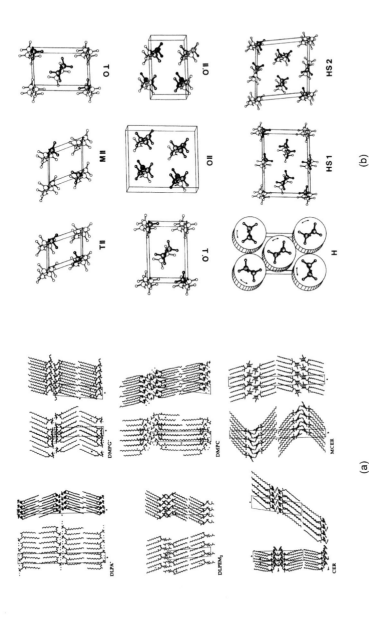

Figure 11.1 (a) Packing arrangements of double-chain lipids in crystals. The molecular packing of the different lipids is shown in two projections along the two short unit cell axes, usually along the *a* (left) and *b* axes (right). For some of the structures, with two conformationally different molecules in the unit cell, the packing arrangements give overlapping projections. In these cases the two molecules are shown only within the frame of the unit cell. To the left and to the right of the unit cell, however, only one type of the two molecules (A and B or D and L) is shown. (b) Chain packing modes in crystalline lipids. The lateral packing arrangements of the hydrocarbon chains in simple and hybrid subcells are shown in a view along the chain axes. (From Pascher *et al.*, 1992, with permission.)

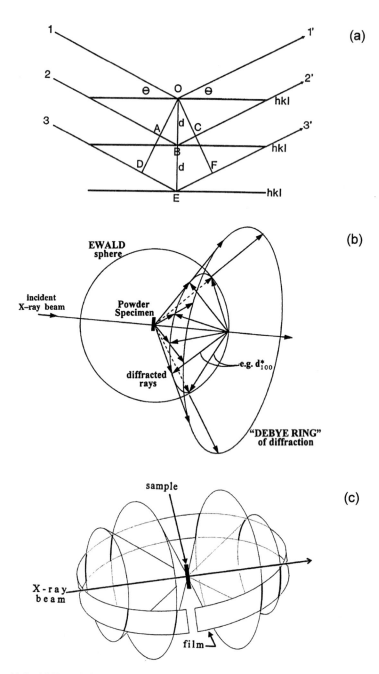

Figure 11.2 (a) Bragg's law is easily seen to arise from an optical analogy to crystallographic planes reflecting X-rays. (b) The intersection of $\mathbf{d}^*_{100}$ vectors from a powder with the Ewald sphere. (c) The origin of Debye diffraction rings. (From Small, 1986, with permission.)

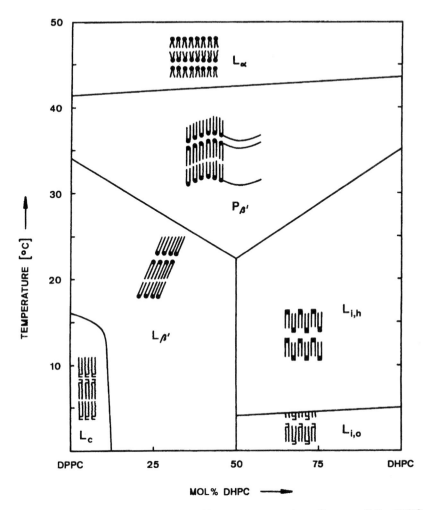

Figure 11.3 Schematic, unrefined composition–temperature phase diagram of the DPPC–DHPC system in presence of excess water.

resolution ranges are quite different. This is shown in Figure 11.4, where the real space dimensions in nanometres are plotted against the scattering angle. It becomes immediately obvious that the instrumental resolution limits, given by the finite size and divergence of any practical primary beam and the pixel size of the detector, will be different in small-angle and wide-angle diffraction. The more stringent conditions, experimentally, are those in the small-angle region, and therefore this has to be in the foreground of attention in designing the experiment. It is easier to extend a small-angle camera to large angles, than vice versa.

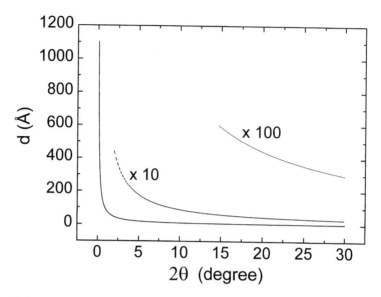

Figure 11.4 Plot of Bragg's spacing d as a function of scattering angle 2θ, for the typical wavelength of 1.54 Å ($C_u k_\alpha$).

Why is it important to measure small-angle and wide-angle diffraction from lipids? Remember that lipid molecules are rather anisotropic, and that they frequently are amphiphilic, which leads to a characteristic packing with distinct short and long spacings: the short spacings originate from the side-by-side packing lattice of the hydrocarbon chains, and are in the order of 3–5 Å, as for example in polyethylene, while the long spacings arise from the periodicity in end-to-end packing, and are in the order of the molecular length, i.e. anything between 20 and 60 Å. This situation is shown schematically in Figure 11.5.

It is clear, therefore, that a structural description of a lipid phase has to contain both the long- and short-spacing data. A particular point has to be made about the necessity of performing small-angle and wide-angle diffraction experiments simultaneously on the same sample. Lipids and fats, as soft condensed matter in general, are very complex systems, not only in their static structures but also with respect to their kinetics of supramolecular structure formation. Hysteresis phenomena or super-cooling can gravely complicate the task of defining the underlying structures and the boundaries in a phase diagram.

As an illustrative example, the phase transition of phosphatidylcholines with two saturated hydrocarbon chains, e.g. DPPC (1,2-dipalmitoyl-*sn*-glycero-3-phosphocholine), will be considered here (Janiak *et al.*, 1976; Komura and Furukawa, 1988; Bóta *et al.*, 1997). This transition proceeds

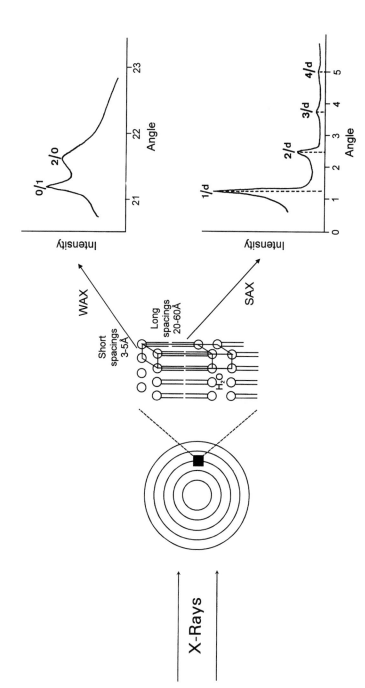

Figure 11.5 Scheme of small- and wide-angle scattering from a multilamellar liposome. The small-angle pattern records the long-period order, and the wide-angle pattern records the chain packing.

between a corrugated 'ripple' structure ($P_{\beta'}$) with partly ordered hydrocarbon chains in nearest-neighbour (hexagonal) packing and a low-temperature lamellar phase structure ($L_{\beta'}$) in which the bilayers are planar and the chains are packed in an orthorhombic lattice, tilted with respect to the bilayer normal (Figure 11.6). The transition occurs over a calorimetric temperature interval of about 3°C, centred at about 33°C. The kinetics of $L_{\beta'}$ formation upon quenching from the $P_{\beta'}$ phase are quite complex: the transition velocity has a minimum close to the transition temperature ('critical slowing down'), and the thermodynamic driving force, i.e. the difference in free energy between the unstable $P_{\beta'}$ phase and the stable $L_{\beta'}$ phase is very low (see sketch in Figure 11.7). Lowering the temperature first leads to a strong increase in the transition velocity, which reaches a maximum about 5°C below the equilibrium transition temperature, and then the transition velocity decreases again. The range of transition times spanned by this behaviour is a few minutes to some ten hours. Furthermore, there seems to be a kinetic difference between the packing of the hydrocarbon chains and the development of the long-range order, i.e. the proper development of the ordered lamellar lattice. In principle this behaviour has been discussed in terms of two competing factors: first the thermodynamic driving force increasing from zero at the transition temperature to ever increasing values as the system is cooled to lower temperatures, and secondly the viscosity of the system, which also increases with cooling and works against a rapid transition process. Both factors taken together lead to the type of kinetic behaviour found here. Strictly speaking, this also has the consequence that the real, low-temperature $L_{\beta'}$ structure can only be approached asymptotically in a limited span of time. From all this, it follows that observations of the chain packing, as reflected by the wide-angle pattern, and of supramolecular, lamellar lattice formation, as reflected by the small-angle pattern, have to be done in one experiment, at the same time.

The importance of this simultaneous small-angle and wide-angle scattering approach is further illustrated by another example: the detection of the so-called sub-main transition (Pressl et al., 1997a) in phospholipid–water systems, in the presence of alkali salts, which for a long time escaped recognition because it is a minor effect close in temperature to the major structural rearrangement of the main or chain-melting transition of phospholipids. The structural changes can be recognised both as a slight decrease, by about 1 Å, of the long spacing and in the wide-angle pattern by a fractional decrease in the ordered chain-packing and a concomitant increase in the diffuse liquid chain band around 4.5 Å (Figure 11.8). The whole process consists, according to this study, of a partial melting of hydrocarbon chains in the so-called ripple phase ($P_{\beta'}$), thus conferring higher bilayer flexibility to the system, which

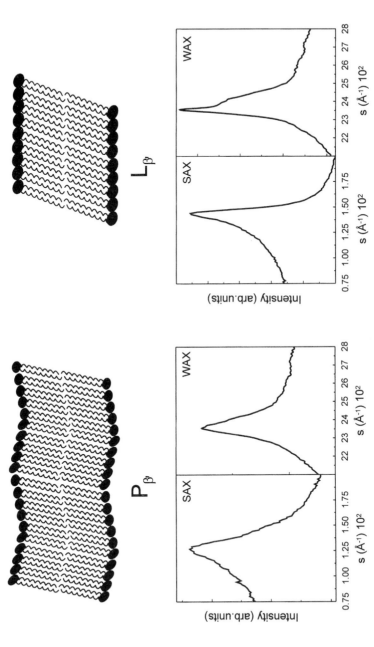

Figure 11.6 (Upper) Schematic of the $L_{\beta'}$ and $P_{\beta'}$ phases, respectively, of saturated phosphatidylcholines. (Lower) Typical SAX and WAX patterns of the first order.

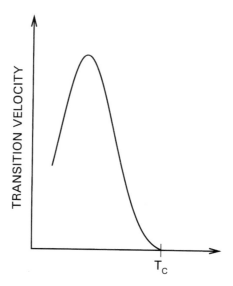

Figure 11.7 Schematic course of the transition velocity after quenching to different temperatures below the phase transition $P_{\beta'} \rightarrow L_{\beta'}$.

in turn reduces the ripple amplitudes. It has to be emphasised that, with separate small-angle and wide-angle experiments this process would have remained undetected, because the isolated effects would have been too small to be taken seriously, while in combination the results gain reliability and structural significance.

There are numerous other examples in which small-angle and wide-angle scattering measured simultaneously have substantially contributed to advances in the understanding of lipid structures, and it is to be hoped that this becomes routine in lipid structure studies.

11.2.3 Diffuse small-angle scattering

Diffuse small-angle X-ray scattering (SAXS) always occurs from materials that contain non-crystalline density fluctuations at a colloidal length scale, i.e. the nanometre scale, typically between 1 and 100 nm. For comprehensive treatments of the method and its theory, which is just a variant of the other methods described above, reference is made to Glatter and Kratky (1982), Baltá-Calleja and Vonk (1989), Feigin and Svergun (1987) and Kratky and Laggner (1992). The lipid chemist will encounter such systems most frequently in the forms of gels, (micro-) emulsions and micellar dispersions. The prime characteristic is the continuous nature of the curves and absence of sharp diffraction peaks,

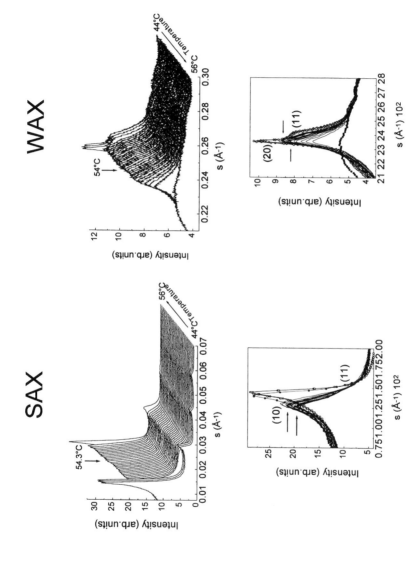

Figure 11.8   Small- and wide-angle diffractograms of the pre- and main-transition of DSPC, showing the sub-main transition. (From Pressl *et al.*, 1997a, with permission.)

which is the direct consequence of the absence of crystallinity. In general, also, the scattering is much weaker than in ordered, crystalline systems, and therefore the precise measurement requires particular precautions, which are different from those for crystallography or powder diffraction. Special cameras have therefore been designed to meet these particular requirements. The critical parameters defining the instrument quality are the signal-to-noise ratio and the resolution limit to small angles. The optics design originally proposed by Kratky (Kratky, 1954; Kratky and Stabinger, 1984) is certainly the most successful and widely used commercial instrument in the field. In addition to the optical quality, routine use in modern R&D or analytical laboratories calls for a high standard of user-friendliness: full turn-key system integration, ease of alignment and standardisation, direct measurement and control interface to a personal computer, and method-specific software support are features of today's standard in analytical instrumentation. A picture of the typical, commercial SWAX camera (HECUS-MBraun Graz X-Ray Systems, Graz, Austria) together with a representative data set is shown in Figure 11.9.

11.2.3.1 SAXS from dilute, monodisperse solutions

SAXS is perhaps the most widely applicable but least well-known technique of X-ray structure analysis. It has found its most popular applications in the field of biopolymers, mostly in studies on proteins and/or enzymes in solution (Trewhella, 1982; Pickover and Engelman, 1982; Lattman, 1994; Vachette, 1996; Svergun et al., 1997). Figure 11.10 shows schematically the basic concept behind this technique. It is evident that this could be an excellent tool for structural analysis in macromolecular chemistry because it is applicable to solutions, in which chemical or biochemical reactions can be performed. The investigation of conformational changes, of association–dissociation equilibria, and of other significant structural rearrangements becomes feasible with this approach.

Nothing comes free, however, and there are certain disadvantages, which have until recently hindered the broad diffusion of this technique into laboratories. First, the very fact that the method relies on the existence of non-interacting, i.e. very widely separated, particles in a solution, implies that the scattering signal is weak. Considering simply the number of scattering macromolecules in a dilute solution and in a crystal, given the same size of the irradiating X-ray beam, the crystal will contain far more molecules than the solution. Secondly, the particles in solution are present in random orientation; therefore, the total scattering pattern is the average of all individual particle scattering effects, and hence is continuously distributed over a scattering curve. This is in contrast to the

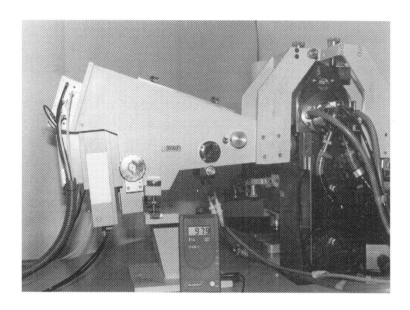

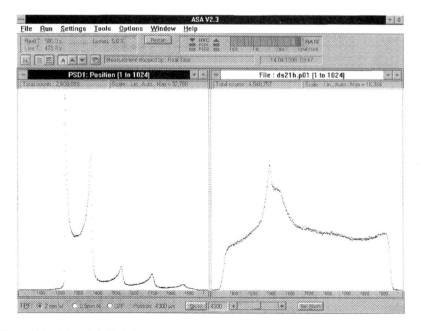

Figure 11.9 (Upper) A SWAX camera (Hecus M.Braun-Graz X-Ray Systems, with permission) (Lower) Typical data display format during exposure.

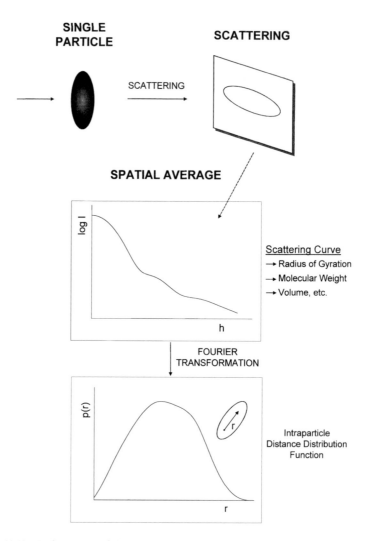

Figure 11.10 Basic concept of the scattering from a monodisperse, dilute solution of identical particles.

diffraction from crystals, where all particles have the same orientation and the diffraction effect is concentrated into discrete and separate peaks that have a comparatively high signal strength.

Solution scattering is unbeatable, however, in cases where crystals, for whatever reason, cannot be obtained. This is the case with several important classes of particles related to the field of lipids. For instance, plasma lipoproteins, the carrier particles of lipids (phospholipids, free cholesterol, cholesterylesters and triglycerides) have been notoriously resistant to crystallisation (Kostner and Laggner, 1989). This is partly due to the fact that the molecular components within these complexes are highly mobile and, hence, the ensemble structure is an average over many conformational substates, and partly to the fact that monodispersity is very difficult to achieve in the isolation procedures from blood plasma.

Figure 11.11 shows a representative solution scattering pattern of low-density lipoprotein (LDL), which is commonly known as the 'bad

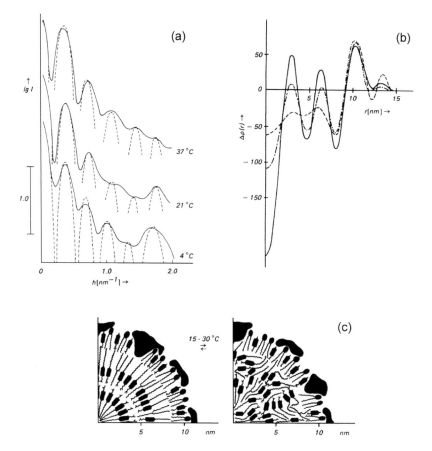

Figure 11.11 (a) Corrected X-ray small angle scattering curves of human LpB at 4, 21 and 37°C (full lines). Broken lines are the theoretical scattering curves calculated for spherical particles with the radial electron density distributions shown in (b). The vertical superposition of the curves is arbitrary. (b) Radial electron density distribution $\Delta\rho(r)$ as obtained from the experiments at 4°C (—), 21°C (·····) and 37°C (---) by Fourier transformation of the amplitudes. (c) Idealized cross-sectional view of LpB as derived from the radial electron density distributions below and above the transition. (From Laggner, 1995, with permission.)

cholesterol' in clinical practice. Here, the application of solution scattering, most actively pursued by several groups in the 1970s (Laggner and Müller, 1978; Atkinson and Small, 1986; Laggner, 1995) has led to a model that is still the generally accepted basis for structural discussions in this field. It has to be noted that the temperature dependence of the LDL structure, with its core lipid transition, is a feature that cannot be resolved by crystallographic studies, since the crystals would decay in the course of the transition. For a long time it appeared that the solution SAXS data would remain the only diffraction information about LDL structure. However, even there the progress in purification and crystallisation techniques has recently led to first reports about single crystal diffraction data, though still at low resolution (Prassl et al., 1996).

Another example where structural information on lipid systems is intrinsically non-crystallographic is phospholipid bilayer vesicles, which serve as model systems for studying biological membrane interactions (Lasic, 1993; Rosoff, 1996). A sketch of such vesicles and the typical small-angle scattering pattern of such systems is shown in Figure 11.12. To the eye of the SAXS experiment, the essential structural feature here is the bilayer cross-section electron density distribution. The samples need not be monodisperse in vesicle size; as long as they are composed of one type of bilayer structure, the scattering pattern will reflect this bilayer structure. Interesting features that can be studied here are, for example, variations of the bilayer structures upon changes in physical parameters (temperature, pressure) or chemical environment, for example as induced by the addition of low-molecular-mass additives (salts, drugs, peptides) (Laggner et al., 1979; Lewis and Engelman, 1983).

As a third example interesting to lipidologists, the wide area of micelles (Luzzati, 1968; Tadros, 1984; Friberg and Lindman, 1992) will be mentioned. These supramolecular assemblies of amphiphilic monomers are of central interest to colloid science, and industrially relevant in fields like detergents, cosmetics, drugs, food and feedstuffs. In general, these are rather tricky systems. The basic condition for single-particle scattering, as outlined above, is rarely met by these systems: they are frequently strongly interacting and their structure changes with dilution. In such systems it has become good practice to describe a solution or dispersion of finite concentration by a liquid structure feature $S(q)$ (analogous to the crystal lattice function) and a form factor $P(q)$ describing the single-particle scattering (equivalent to the solution scattering at infinite dilution). The situation is illustrated schematically in Figure 11.13. The total scattering effect $I(q)$ is the product of the former two components: $I(q) = P(q) \times S(q)$. In other words, if one wants to know the micellar structure one needs additional information to define the solution structure factor $S(q)$, or, if one knows that the particles are of one given

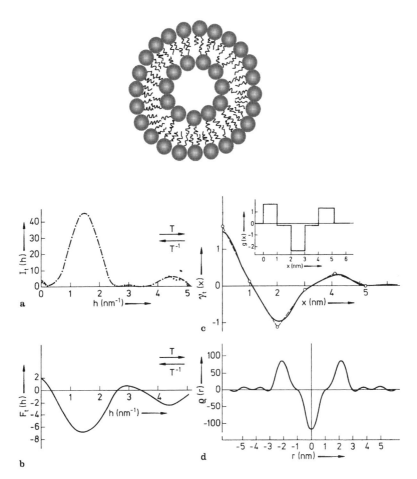

Figure 11.12 Small-angle scattering from a dilute, random dispersion of membranes (vesicles, top). (a) Corrected intensities or thickness factor obtained from the experimental intensity distribution $I(h)$ by multiplication with h^2. (b) Structure factor (amplitude function) with arbitrarily chosen signs $(+,-,+,-,)$. (c) Autocorrelation function of the electron density $\rho(x)$ profile across the membrane obtained by cosine transformation of $I_t(h)$; the insert shows the profile obtained by deconvolution. (d) Centrosymmetric electron density profile obtained by cosine transformation of $F_t(h)$. From a study on lipoprotein X, an assembly of unilamellar vesicles.

structure (e.g. a spherical core shell model), one can obtain the solution structure factor, which provides clues to the thermodynamics in particle interaction. Obviously, this approach lends itself very well to studies on physical and chemical effects on micellar systems. For illustrative examples, see Cotton (1991) and Tardieu (1994).

REAL SPACE

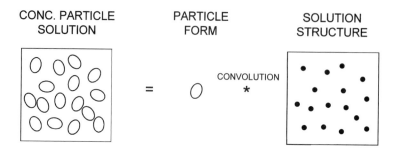

| CONC. PARTICLE SOLUTION | PARTICLE FORM | SOLUTION STRUCTURE |

SCATTERING SPACE

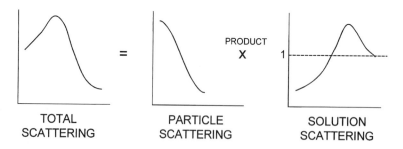

| TOTAL SCATTERING | PARTICLE SCATTERING | SOLUTION SCATTERING |

Lipids 11.13 Scattering from a concentrated solution of identical particles contains information on particle form and solution structure.

11.3   Some notes on instrumentation

The essential components of an X-ray diffraction instrument (sketched in Figure 11.14) are: a radiation source (X-ray tube, rotating anode, synchrotron); the optical system (collimator, focusing mirrors, mono-chromator); the sample stage (temperature stage, pressure cell, crystal rotation unit, etc.); and a detector (X-ray film, scintillation counter with goniometer, position sensitive detector).

The development of radiation sources is simply governed by the quest for ever-increasing photon flux at the sample. The traditional X-ray tubes (sealed and fixed anodes) essentially dominated the field from the times of

Source   Collimator   Sample   Detector

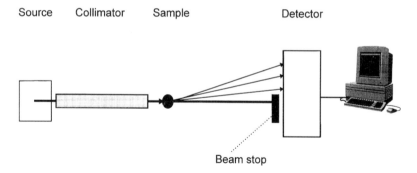

Beam stop

Lipids 11.14   Scheme of SAX instrumentation.

Wilhelm Roentgen until the late 1960s when the first rotating anodes came into use. These overcame the problem of heat dissipation from the target metal anode by distributing the heat load over the cylindrical envelope of a rotating anode. The power load and the ensuing photon flux could be enhanced by a factor of about 10, reducing the necessary exposure times from days to hours (in the normal case of a weakly scattering system). Synchrotron radiation sources have led to a tremendous increase in X-ray flux by about five or more orders of magnitude, reducing the required measuring times to seconds, milliseconds and, in favourable cases, even microseconds.

While the advance from the stable, fixed anode to the rotating anode was mainly a matter of increasing the throughput of samples in the laboratory, and one of convenience, the synchrotron sources added a new quality to X-ray structure analysis. For the first time this has made it possible to perform X-ray cinematography at the molecular scale by fast time-resolved X-ray diffraction. A separate paragraph is devoted to this methodology below.

The optical systems are largely determined by the nature of the experiment, i.e. whether one is dealing with crystallography, where a distribution of diffraction spots has to be collected over a large area, or, at the other extreme, whether one aims for a broad continuous halo around the direction of the primary beam, which is essentially the case with diffuse small-angle scattering. It would exceed the scope of this review to go into detail of the different designs developed for one or the other purpose. What is important, however, is to stress that there is no single system yet available, that would cover all different diffraction techniques in one instrument. It should also be noted that the design and dimensions of the optical system are mainly determined by the dimensions and brilliance of the source on the one hand and by the size and resolution of the detector system.

The sample holder in the case of lipids is in most cases a thin-walled Mark capillary made of glass or fused quartz, held in a thermostable cuvette. Special designs for pressurising or for controlled humidity environment have been reported (Katsaras, 1997; Pressl *et al.*, 1997b). The most important parameter in the choice of the sample environment is the signal-to-noise ratio, to be achieved by minimising window thickness and materials. Metals, most favourably beryllium (*note, however, its toxicity*), have to be kept to a minimum owing both to their high absorbance and to their background scattering, both of which reduce the signal-to-noise ratio. For small-angle scattering it is furthermore essential to evacuate the beam path as far as possible, because air scattering and absorption would both very efficiently reduce the signal-to-noise ratio. Such precautions are normally not necessary in the case of single-crystal diffraction or even in powder diffraction, where the signal intensity is so dominant that an elevated background matters only in very extreme cases of weak diffraction signals.

Next to the advances brought about by synchrotron radiation sources, the development of electronic position-sensitive detectors has revolutionised the methodology of X-ray diffraction (Walenta, 1991; Helliwell, 1992). While photographic film, as originally used, is rather insensitive (about 10^5 photons are needed to reach an optical density of 1 in the developed film) and therefore requires long exposure times, the modern position-sensitive detectors monitor the diffracted X-rays electronically at any position over the area of the detector, and the scattering or diffraction pattern can be stored in computer memory. The development of the diffraction pattern can be followed in real time on the computer screen. Obviously this has changed life for the X-ray experimentalist. In the laboratory, therefore, the new detectors together with rotating anodes have made exposure times with lipids or fats in the order of minutes the routine. This is a very important asset if one aims at the possibility of screening large numbers of samples, for instance in the exploration of a phase diagram or in the search for membrane-active substances (Latal *et al.*, 1997).

Future developments can be foreseen both at the high end of the technology, i.e. in the field of synchrotrons, and in the low end, where the transition from specialised instrumentation requiring separate X-ray laboratories to bench-top or field instruments is being pursued.

11.4   Industrial process control

So far X-ray diffraction has been practically exclusively a laboratory technique, either as a research tool in the analysis of natural or synthetic structures or as an analytical tool in the quality control laboratory. The latter is the closest to direct industrial application.

Actual in-line process control in the sense of permanent monitoring of a chemical product was for a long time restricted to very simple measurements, mostly based on gravimetric or optical properties. X-Ray techniques were considered too elaborate, because of their traditionally long measuring times, and unsuitable for this purpose; also, for many low-tech products, the traditional in-line or off-line control methods were entirely sufficient.

In the pursuit of producing ever more refined materials, however, where product quality is critically dependent on the nanostructure of the materials, the demand for suitable real-time monitoring methods with a resolution in the nanometre scale has recently gained strong momentum. This pertains especially to the field of soft organic materials such as gels, liquid crystals, dispersions in the product areas of food and feedstuffs, detergents, cosmetics and pharmaceutical products.

The reason why X-ray structure monitoring is in many cases considered superior to other physicochemical sensory techniques is as follows. Consider a dispersion of liposomes: there are two grossly distinct length scales that determine the product quality. One is the macroscopic length scale, determined by the size of the dispersed particles, which may be in the micrometre to millimetre domain. This determines mainly the rheological properties of the whole system. The more subtle and quality-relevant properties, such as the retention capacity for drugs or other additives, however, lie in the nanostructure, i.e. in the supramolecular packing of the constituent lipid molecules. In other words, the supramolecular structure of the active matter phase is the one that needs to be controlled to ensure the product quality specifications. Obviously, X-ray diffraction in general, or small-angle scattering in particular are superior to any other bulk techniques (optical, electrical, gravimetric), because the X-ray diffraction data are not influenced by the macro-heterogeneity of the system, which in a technical product often involves air bubbles or macro-inclusions of water. X-Ray diffraction will only detect the nanostructural properties, which in an increasing number of cases are the relevant ones.

So far no instruments have become commercially available that would fulfil the technical demands for serving as a monitoring tool to be implemented directly in a chemical plant. All laboratory instruments rely on a practically vibration-free, thermostated and climatized environment. For an in-line process control camera, these limitations have to be overcome in a stand-alone, self-calibrating system that is able to function reliably over long periods and is insensitive to mechanical and electronic noise as well as to severe temperature changes. The endeavour to develop such a product control camera has been taken up by a project group funded by the European Commission through the Standards, Measurements and Testing Programme (EUROSAX EU Project, 1996–1999). As an industrial test-area and the first important field for application, this in-line camera is aimed at monitoring the water content in highly concentrated surfactant products. The target specification is to measure continuously in intervals of a few minutes the lamellar repeat distance of the product flowing in the line to a precision of 0.01 nm at typical values between 2.5 and 6 nm. The actual readings of the instrument, taken continuously at intervals of 1 minute will serve as feedback control values for the process. By such techniques, the process can be made continuous, whereas so far it has been discontinuous: the uncontrolled product was collected as a batch, analysed, and adjusted in quality through mixing in separate tanks. The economic and ecological advantages of a feedback-controlled continuous system are straightforward: cutting out the extra production steps of adjusting and mixing means savings in energy and time (i.e. money); the ecological advantage lies both in the energy savings and in the avoidance of wastes incurred by irreversible overtitrations in the adjustment and mixing steps.

A prototype for an in-line production control SAXS camera has been built and is presently being tested both in the laboratory and in the industrial environment. Figure 11.15 shows the instrument and a set of representative results. The technological achievement in the design of this system lies in the maintenance of highest X-ray optical resolution, in line with the state of the art in laboratory type systems, while adding the properties of ruggedness and full automation. The price for its realisation is being paid by restricting the resolution window to the relatively narrow range of interest given by the actual product application, in this case between 2 and 6 nm, and sacrificing the optical variability of a research laboratory camera. On the other hand, the precision within this window of interest is kept at the maximum.

Cost-effectiveness and ecological constraints increasingly rule industrial production, particularly in the field of chemical products. As the degree of product refinement and quality enhancement pushes the emphasis in physical properties from the micrometre to the nanometre scale, industrial in-line X-ray diffraction techniques as real-time quality monitoring tools will attain great importance in the future.

11.5   Large installations—synchrotron radiation and neutron sources

During the past two decades two large-installation research tools have entered the scene of structural research: synchrotron radiation sources

(a)

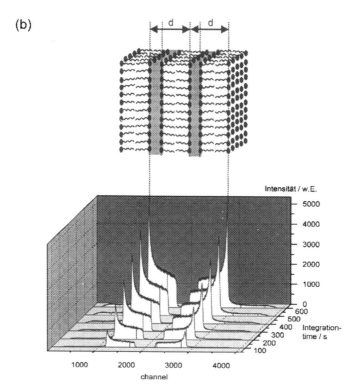

(b)

Figure 11.15 (a) Prototype of a SAX camera for in-line process control. (b) Scheme of diffraction pattern from a detergent–water system.

and high flux neutron reactors (Figure 11.16). Both of them have brought about considerable advances in the field of lipid and fat research, and therefore they shall be briefly presented here. Obviously, large installations have their own characteristics as far as the practical aspects of their use is concerned. Synchrotron radiation sources and neutron facilities are both multi-user systems and pose similar requirements on experimental planning. In general, this planning is at least a matter of 6 to 12 months: well argued proposals have to be submitted in order to obtain time slots at the respective instruments, which are normally in the range of one day to one week. The proposals have to be both exciting in their scientific objectives and well founded upon preliminary experiments. To meet both

Figure 11.16 View of two large installations on one site: foreground, the ring of ESRF and, middle right, the high-flux neutron reactor of ILL (Grenoble).

conditions is frequently difficult. Since the time for the experiments, once granted, is rigidly scheduled, every possible effort has to be taken to complete preparations for that date. If the material under study is synthetic and commercially available, this may not cause any problems. If difficult chemical, physicochemical or biochemical preparations precede the study, the whole operation has to be logistically planned. From the nature of the experimental environment, it is also clear that the actual time allowed for the experiments does not normally allow exploratory or altogether novel experiments. In most cases one has to know in great detail how the experiments are going to proceed, and to work hard so that high quality data are obtained.

This is the rule and one should try to benefit both from the disciplined mode of experimentation and from the exceptions to the rule, which always occur. It is advisable, therefore, in the use of large installations, to prepare a double strategy consisting of a set of safe and sound experiments that will work and provide data under any foreseeable conditions, and on the other hand to leave some space and time for exciting and conceptually new experiments. This strategy, if successfully pursued, is almost a guarantee for continued access to large installations. On the one hand, there are always enough data to fill the requested serious reports, and on the other hand there are new and exciting prospects from the high-risk exploratory experiments to form the basis for the next proposal.

One further consideration should help towards intelligent and productive use of large installations. The experiments one plans to do, be it with neutrons or synchrotron radiation, are relatively rare milestones in the course of the work-year, and certainly they can be the most exciting periods. However, they have to be complemented by closely related experiments at one's own laboratory. Only this provides a sound basis for the first part of the double strategy, i.e. the collection of high-quality data. Although there is also a tendency to use large installations as a substitute for the effort at the base laboratory, this is certainly not a good policy because it leads to an excessive demand of experimental time at large facilities for purposes they have not been designed or dedicated for.

In the following, the specific potentials and merits of synchrotron radiation sources and neutron facilities for structure research on fats and lipids will be illustrated briefly.

11.5.1 Synchrotron radiation

Synchrotrons are circular particle accelerators in which charged particles, i.e. electrons or positrons, are brought to very high energies and velocities close to the speed of light. With certain techniques the circulating bunches of charges can be stored in the rings over long times, i.e. with a half-life of 12 hours or more. These circulating charges are sources of electromagnetic radiation—i.e. light. For the observer at rest looking tangentially to the circulating particle bunch, the light will be strongly blue-shifted owing to the relativistic nature of the moving source (Margaritondo, 1988; Wiedemann, 1991; Raoux, 1993). Therefore, synchrotrons and storage rings are the most brilliant X-ray sources existing. Among other aspects (such as the continuous spectrum), this brilliance is the real *raison d'être* for synchrotron light sources, which are dedicated to the production of high-quality light.

Why should one want to have more powerful light sources when even laboratory sources are dangerous enough to require lead shielding and the like? Notwithstanding their danger to health or life as sources of potent ionizing radiation, normal laboratory X-ray tubes are relatively low in X-ray intensity. One typically quantifies the power of a light source in terms of the flux of photons that can be directed onto a unit area of sample. For laboratory X-ray generators this is in the order of 10^{7}–10^{8} photons s^{-1} mm^{-2}. On synchrotrons, this can be enhanced to 10^{12}–10^{13} presently, and further improvements are being made every year. This, in a comparison based on human sensory experience, corresponds to bright sunshine as opposed to weak moonlight.

The first and major argument for synchrotron radiation is, therefore, the potential for performing cine X-ray diffraction, i.e. to do experiments in a time-resolved manner during the process of a structural rearrangement. This cinematographic approach is particularly promising in the fields of powder diffraction and solution scattering, because their chemical or physical processes can be triggered in a controlled way, as will be shown below. For single-crystal studies, this approach is less attractive, although even there one can think of light-induced, simultaneous reactions throughout the crystal.

11.5.1.1 Time-resolved small-angle and wide-angle diffraction on lipids

The structural question underlying time-resolved studies on mesomorphic lipid phase transitions is the following: how does the transition process work in time and space between two topologically and geometrically different phase structures? This will be briefly illustrated by two studies on different phase transitions in phospholipid–water systems.

The pretransition of saturated phosphatidylcholine. This transition involves a structural change from a lamellar bilayer water phase with tilted quasi-hexagonally packed hydrocarbon chains to a rippled phase structure (Laggner, 1993; Rappolt and Rapp, 1996). Figure 11.17

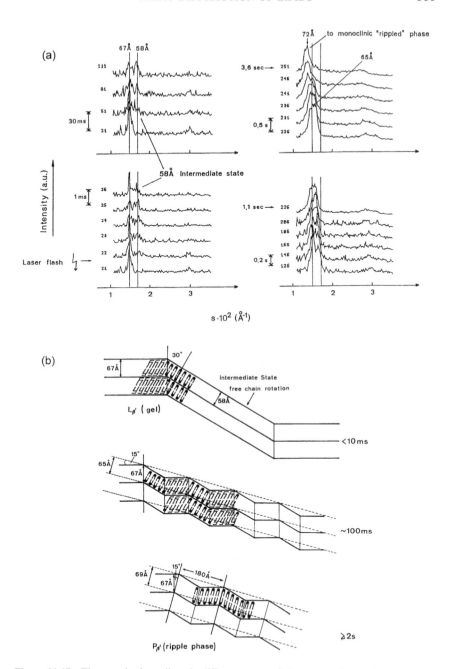

Figure 11.17 Time-resolved small-angle diffractograms of the pretransition of DPPC–water after a laser-induced temperature jump.

shows the time development of the first-order Bragg peak region in various sets of intervals. Immediately after the temperature jump, induced by a millisecond infrared laser pulse, the original first-order peak decreases, and a second peak appears, corresponding to about 4–6 Å smaller Bragg spacings. This coexists for variable periods of time, depending on the hydrocarbon chain length and the jump amplitude, but always for about 100 ms with the original lattice peak. This coexistence time is followed by a merger of the two peaks, and a shift towards larger d-spacings. Finally, after a few seconds, the pattern reaches that characteristic for the rippled bilayer phase. A structural model consistent with these observations is shown in Figure 11.17 b. The importance of this observation lies in the occurrence of an ordered structural intermediate involved in this transition process. Under equilibrium conditions such intermediates cannot be observed, indicating that the transition pathways between mesomorphic lipid phases can differ depending on the power level by which they are induced. Under low-power conditions close to equilibrium, the system moves randomly in a shallow valley of the potential landscape, whereas under high-power conditions, with a strong driving force towards the new equilibrium, the system follows an ordered pathway rapidly and cooperatively. The ordered intermediates appear to play the role of geometrical hinges between initial and final phases.

A second interesting case is the transition between a lamellar bilayer and a hexagonal tubular phase structure as given by the $L_\alpha \rightarrow H_{II}$ phase transition of phosphatidylethanolamine–water systems. Figure 11.18 shows the transition as it is seen by time-resolved synchrotron X-ray small-angle diffraction after an infrared laser-induced temperature jump. Here again the process involves at least two steps with different time scales. In the first few milliseconds after the temperature jump, the appearance of a Bragg peak (first and second orders are visible), with d-spacing about 5 Å smaller, indicates the formation of a lamellar packing with reduced water space between the bilayers. After a lag period of about 20–50 ms, the first signal of the characteristic hexagonal diffraction pattern is observed. This increases in intensity and shifts to its final position within about 1–2 s. Here again, the transition involves the formation of short-lived intermediates, which can be interpreted in a similar way to that discussed above for the case of planar bilayer to rippled phase transition.

The above-mentioned examples serve to illustrate the tremendous gain in information content of diffraction experiments achieved by adding the dimension of time. To benefit fully from this potential, however, not only the optical part of the experiment has to be optimized, but also the physical or chemical methods chosen to trigger the process. The above

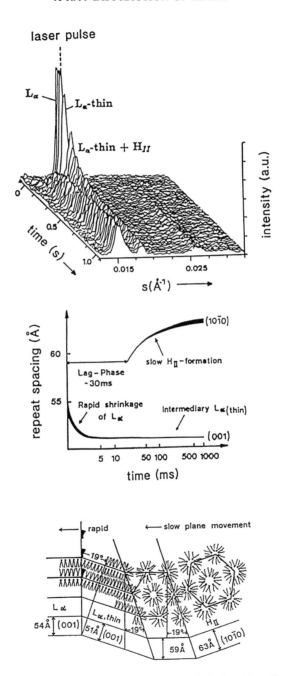

Figure 11.18 Time-resolved development of an IR laser pulse-induced L_α – H_{II} phase transition of SOPE. (a) X-ray diffraction patterns; (b) *d*-spacings as a function of time within 1 s after the pulse (time resolution 1 ms); (c) schematic model.

examples have illustrated that rapid high-power triggering leads to an ordered, 'regimented' response within the system, and the structural intermediates are highly interesting as they offer new aspects in discussing transition mechanisms and material properties, both technologically or biologically. Considerable efforts are therefore devoted to the design of intelligent and effective triggering methods used for such time-resolved X-ray diffraction studies.

The most frequently used jump-relaxation methods, i.e. temperature, pressure and concentration jumps, are illustrated in Figure 11.19. They all have their individual merits and drawbacks, and it is advisable to study a given process by the complementary use of several techniques. The temperature jump method, for instance, by infrared laser pulse (Kriechbaum et al., 1989), is certainly a very rapid method, heating a lipid–water system by about $10°C\,ms^{-1}$, but the method works intrinsically only in one direction (i.e. heating) while cooling jumps are obviously not possible. Secondly, and perhaps more limiting, the temperature jump is not homogeneous throughout the sample but follows the exponential absorption law over the optical thickness, which can be remedied in part by back-reflection of the radiation passing through the sample by a suitable mirror.

The pressure jump method does not suffer from these problems: it works in both directions and is homogeneous throughout the sample. The disadvantages apart from the more elaborate design of a pressure cell (Erbes et al., 1996; Steinhart et al., 1998) lie in the unavoidable temperature jump associated with adiabatic compression. The transition works against the pressure jump: in a compression experiment, the developed heat will work against the condensation of hydrocarbon chains, and in a dilatation experiment (pressure drop) the concomitant temperature drop will counteract the chain melting. These aspects have to be carefully considered to avoid potential misinterpretation of results.

The third possibility, stopped-flow mixing, is of course not applicable to highly viscous concentrated lipid systems and might be interesting only for studies on dilute, micellar systems, where interactions with additives, salts, drugs, etc. become feasible. This method has so far not been widely used in the fields of fats and lipids.

11.5.1.2 X-ray reflection and diffraction of Langmuir films

Amphipathic lipids, such as fatty acids, alcohols, amines or phospholipids, spread on air–water interfaces as monomolecular films, so-called Langmuir films, shown schematically in Figure 11.20. They are traditional standard objects for thermodynamic studies in colloid and interface science (Lvov et al., 1989; Ulman, 1991). Only with synchrotron radiation, owing to the high brightness of the X-ray beams, has it become possible to study such films by surface reflection and diffraction (Als-

T-Jump SWAX

p-Jump SWAX

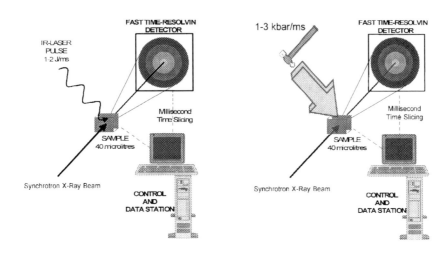

Stopped-Flow SWAX

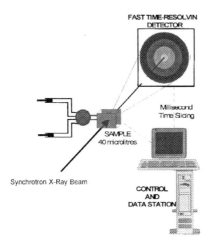

Figure 11.19 Scheme of some jump-relaxation studies possible with synchrotron radiation.

Nielsen and Möhwald, 1991). The principle, qualitatively, is as follows. At incident angles lower than the limiting angle of total reflection, the incident beam is reflected similarly as light is from a mirror plane

SOLID PHASE

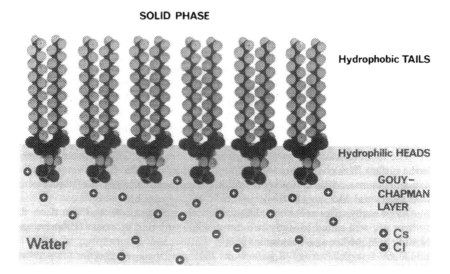

Figure 11.20   Scheme of monomolecular surface film. (From Als-Nielsen and Möhwald, 1991, with permission.)

('specular reflection'). If the surface is not an infinitely sharp step in electron density (refractive index), but has a defined electron density profile, then the reflected beam intensity measured as a function of angle, relative to the limiting angle of total reflection, contains the information about this electron density profile. The analysis is then similar to that of small-angle scattering from single-shelled bilayer vesicles, as indicated above. A second component of the deflection of the primary beam, the one in the plane of the surface, is caused by the in-plane structure of the monolayer, i.e. the side-by-side hydrocarbon chain packing, as shown schematically in Figure 11.21.

11.5.2   Neutron scattering and diffraction

Diffraction and scattering are both wave–matter interactions in their physical principle. X-rays as electromagnetic waves are scattered mainly by their interaction with electrons; consequently the results of a diffraction experiment reflect the electron density distribution within the matter investigated. Neutrons can also be ascribed wave nature, the wavelength of which is given by the de Broglie equation $mv = h/\lambda$. The velocity v of neutrons, emitted by a neutron source, e.g. by a reactor core, can be varied within wide limits by choosing suitable moderators, such as water or liquid hydrogen or deuterium, so that wavelengths in the order

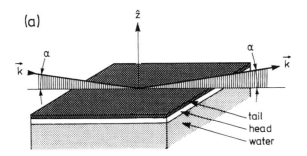

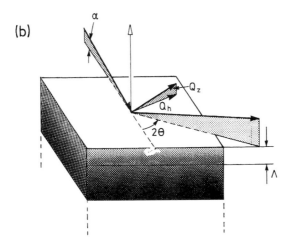

Figure 11.21 Schematic of surface X-ray diffraction: (a) specular reflection, (b) in-plane diffraction. (From Als-Nielsen and Möhwald, 1991, with permission.)

of atomic distances in molecules are obtained, which are suitable for probing molecular structure. In contrast to X-rays, which are scattered by electrons, neutrons are scattered by the atomic nuclei. For comprehensive treatises, see Dachs (1978), Kostorz (1979), Schoenborn (1984), Higgins and Benoît (1996).

Why should one want to use neutrons, which are certainly a rather exotic choice for 'light'? There is one compelling reason: neutrons interact most strongly with hydrogen which is, by numbers, the most frequent element in organic materials, but which is practically not seen by X-rays. Therefore, neutron diffraction is a highly complementary tool to X-ray diffraction, one method seeing what the other misses and vice versa.

The clue to the popularity of neutrons in organic materials research lies in the facile replacement of hydrogen by deuterium, and these two isotopes differ enormously in their scattering power for neutrons. This opens the possibility of judiciously varying the contrast of a given hydrogenous region by H/D replacement. One of the easiest ways to do this is by varying the H_2O/D_2O ratio in water. Two different sets of experiments are described in the following, which illustrate the potential of such approaches in lipid research.

Contrast variation/matching. If one considers micelles of amphipatic lipids as consisting essentially of two regions, one composed of the hydrated headgroups (in which heteroatoms, such as O, N, P or S can dominate) and the other containing the apolar hydrocarbon chains, one can conceive a simple experiment to study these two domains separately: by varying the H_2O/D_2O ratio in the solvent, one can practically match any desired region within such a micelle in terms of its scattering power, so that only the other, non-matched domain gives rise to the scattering effect. The mathematical formalism for determining the characteristic scattering functions from contrast variation was developed by Stuhrmann and Kirste (for review see Stuhrmann, 1982).

Contrast enhancement through selective deuteration. If, in a given composite supramolecular structure, the scattering power for X-rays or neutrons is rather similar for individual components, these components can be selectively enhanced in their scattering power by deuteration. As mentioned above, deuterium and hydrogen differ massively in their respective potential for neutron scattering. By this approach one can highlight a certain desired region within a composite structure and establish its structural arrangement within the total ensemble. This approach obviously depends on suitable chemical methods to reconstitute a composite structure from its individual components, and it has to be critically verified that the selectively deuterated structure is indeed isomorphic to the original non-deuterated one. Otherwise, the results from such an experiment consisting of the comparison of the neutron scattering pattern of the deuterated versus the non-deuterated system could be grossly misleading. A potent way to check on this isomorphous replacement is X-ray scattering, because there, hydrogen and deuterium are not distinctly visible and the scattering patterns of isomorphously deuterated structures should be identical to those of the non-deuterated system.

Both of these approaches have been used extensively in studies on lipoproteins and biological membranes. In lipoprotein they have, for instance, served to determine the mutual location of protein and phospholipid headgroup domains, which are not distinguishable by their scattering contrast with X-rays. Selective deuteration of the phospholipid headgroups has served to solve this problem (Laggner *et al.*, 1981). Another example from the same field concerns the distinguishing of various possible arrangements of cholesteryl fatty acyl esters in the apolar core of LDL (Laggner *et al.*, 1984). Many similar experiments have been performed to solve similar problems in the field of biological membranes.

11.6 Radiation damage

A general concern about the application of X-ray diffraction techniques and to a certain extent also of neutrons, is whether the samples are adversely influenced or degraded while they are under inspection. This is a particularly important consideration with modern synchrotron radiation sources with their X-ray fluxes that are many orders of magnitude higher than present in a conventional X-ray camera. One has to consider the physical and chemical reactions induced by ionizing X-rays. To a large extent this depends on the nature of the samples. Consider, for instance, a solid, predominantly crystalline fat: the highest probability of damage is given by the abundance of C—C and C—H bonds and, consequently, the initial damage products will be C-radicals. The fixation of this radical in a crystalline lattice, however, will rather effectively prevent propagation in the form of a chain reaction and, therefore, damage will be a slow process, despite the initially fast creation of radicals. Another extreme is given by hydrated liquid-crystalline phospholipid dispersions, where water is the major component. There the probability is predominant formation of OH radicals, which in turn can interact with phospholipid molecules by hydrogen abstraction, and the ensuing lipid radicals may combine by cross-linking. These local chemical and structural changes can lead to point or line defects and alter the supramolecular structure locally around the damage centres. This can have quite dramatic consequences in the sense of isothermally inducing an altogether new phase. If, for instance, short-chain hydrocarbons are formed as a consequence of radiation damage in a lamellar liquid crystalline phase closely below the lamellar to hexagonal transition, the hexagonal phase can be induced by these short-chain hydrocarbons, since they are known to stabilise the hexagonal structures. It is therefore highly recommended to control the integrity of the samples by independent chemical methods, such as thin-layer chromatography.

Another problem, specific for synchrotron radiation, is the simple heating of the samples by the absorption of the X-ray beam. At modern synchrotron X-ray beam lines, estimated radiation power densities on the order of $1\,mW\,mm^{-2}$ can be generated. Consequently, a heating effect of some tenths of a degree Celsius per second must be expected. This may not seem dramatic, since one generally aims in such experiments at the millisecond time domain or below, but one has to take all possible precautions to limit the irradiation time to the minimum by using beam shutters that cut off the primary beam whenever there is no active data accumulation. Both problems, radiation damage and sample heating, can be partly overcome by using shorter wavelengths than normally employed in the laboratory, which is technically possible for synchrotrons, where the wavelength spectrum is continuously tuneable within wide limits.

Finally, however, a positive note in favour of synchrotron sources is justified: radiation damage is primarily a dose-determined effect, which means that the same amount of primary damage is caused by a given dose, irrespective of whether this is obtained by low-flux or high-flux radiation. Thus, if no damage occurs during an exposure over hours at conventional sources, a corresponding exposure at a synchrotron source, which takes fractions of a second, will also be without damage. Indeed, the shorter exposure times may even be favourable, since all secondary reactions are diffusion controlled and therefore a brief exposure may produce valid data before the secondary destruction processes are complete.

References

Als-Nielsen, J. and Möhwald, H. (1991) Synchrotron X-ray scattering studies of Langmuir films, in *Handbook on Synchrotron Radiation*, vol. 4 (eds S. Ebashi, M. Koch and E. Rubenstein), North-Holland Elsevier Science, Amsterdam, pp 1-53.

Atkinson, D. and Small, D.M. (1986) Recombinant lipoproteins: implications for structure and assembly of native lipoproteins, in *Annual Review of Biophysics and Biophysical Chemistry*, vol. 15 (eds D.M. Engelman, C.R. Cantor and T.D. Pollard), Annual Reviews, Palo Alto, CA, pp 403-456.

Bóta, A., Kriechbaum, M. and Laggner, P. (1997) Transitional states in the pretransition range ($L_{\beta'} - P_{\beta'}$) of fully hydrated dipalmitoylphosphatidylcholine systems. *Models in Chemistry*, **134** (2–3) 299-315.

Baltá-Calleja, F.J. and Vonk, C.G. (eds) (1989) *X-Ray Scattering of Synthetic Polymers*, Elsevier, Amsterdam.

Cotton, J.P. (1991) Introduction to scattering experiments, in *Neutron, X-Ray and Light Scattering: Introduction to an Investigative Tool for Colloidal and Polymeric Systems* (eds P. Lindner and T. Zemb), North-Holland Delta Series, Elsevier Science, Amsterdam, pp 3-18.

Dachs, H. (ed.) (1978) *Neutron Diffraction*, Springer-Verlag, Berlin.

Erbes, J., Winter, R. and Rapp, G. (1996) Rate of phase transformations between mesophases of the 1:2 lecithin/fatty acid mixtures DMPC/MA and DPPC/PA—a time-resolved synchrotron X-ray diffraction study. *Ber. Bunsen-Ges. Phys. Chem.*, **100** (10) 1713-1722.

EUROSAX EU Project (1996–1997) *Novel Technique for In-Process Monitoring of Surfactant Production by Small-Angle X-Ray Scattering* (coordinator P. Laggner), Project No. SMT4-CT95-2009.

Feigin, L.A. and Svergun, D.I. (1987) *Structure Analysis by Small-Angle X-Ray and Neutron Scattering* (ed. G.W. Taylor) Plenum Press, New York.

Friberg, S.E. and Lindman, B. (eds) (1992) *Organized Solutions—Surfactants in Science and Technology*, Marcel Dekker, New York.

Glatter, O. and Kratky, O. (eds) (1982) *Small Angle X-ray Scattering*, Academic Press, London.

Guinier, A. (1963) *X-Ray Diffraction in Crystals, Imperfect Crystals, and Amorphous Bodies*, W. H. Freeman, San Francisco.

Helliwell, J.R. (ed.) (1992) *Macromolecular Crystallography with Synchrotron Radiation*, Cambridge University Press, Cambridge.

Higgins, J.S. and Benoît, H.C. (1996) *Polymers and Neutron Scattering*, Clarendon Press, Oxford.

Janiak, M.J., Small, D.M. and Shipley, G.G. (1976) Nature of the thermal pretransition of synthetic phospholipids: dimyristoyl- and dipalmitoyllecithin. *J. Biochem.*, **15** 4575-4580.

Jenkins, R. and Snyder, R.L. (eds) (1996) *Introduction to X-Ray Powder Diffractometry*, Wiley, New York.

Katsaras, J. (1997) Highly aligned lipid membrane systems in the physiologically relevant 'excess water' condition. *Biophys. J.*, **73** 2924-2929.

Komura, S. and Furukawa, H. (eds) (1988) *Dynamics of Ordering Processes in Condensed Matter*, Plenum Press, New York.

Kostner, G.M. and Laggner, P. (1989) Chemical and physical properties of lipoproteins, in *Human Plasma Lipoproteins (Clinical Biochemistry)* (eds J.C. Fruchart and J. Shepherd), Walter deGruyter, Berlin, pp 23-54.

Kostorz, G. (ed.) (1979) *Treatise on Materials Science and Technology*, Academic Press, New York.

Kratky, O. (1954) Neues Verfahren zur Herstellung von blendenstreuungsfreien Röntgen-Kleinwinkel-Aufnahmen. *Z. Elektrochem.*, **58** 49-53.

Kratky, O. and Laggner, P. (1992) X-ray small-angle scattering, in *Encyclopedia of Physical Science and Technology* (ed. R.A. Meyers) 2nd edn, Academic Press, New York, vol. 17, pp 727-781.

Kratky, O. and Stabinger, H. (1984) X-ray small-angle camera with block-collimation system, an instrument of colloid research. *Colloid Polymer Sci.*, **262** 345-360.

Kriechbaum, M., Rapp, G., Hendrix, J. and Laggner, P. (1989) Millisecond time-resolved X-ray diffraction on liquid-crystalline phase transitions using infrared laser t-jump technique and synchrotron radiation. *Rev. Sci. Instrum.*, **60** 2541-2544.

Laggner, P. (1993) Nonequilibrium phenomena in lipid membrane phase transitions. *J. de Phys.*, **IV C** 259-269.

Laggner, P. (1995) X-ray and neutron small-angle scattering on plasma lipoproteins, in *Modern Aspects of Small-Angle Scattering* (ed. H. Brumberger), Kluwer Academic, pp 371-386.

Laggner, P. and Müller, K. (1978) The structure of serum lipoproteins as analysed by X-ray small-angle scattering. *Q. Rev. Biophy.*, **11** (3) 371-425.

Laggner, P., Gotto Jr., A.M. and Morrisett, J.D. (1979) Structure of the dimyristoylphos-phatidylcholine vesicle and the complex formed by its interaction with apolipoprotein C-III: X-ray small-angle scattering studies. *Biochemistry*, **18** 164-171.

Laggner, P., Kostner, G.M., Degovics, G. and Worcester, D.L. (1984) Structure of the cholesteryl ester core of human plasma low density lipoproteins: selective deuterations and neutron small-angle scattering. *Proc. Natl. Acad. Sci. USA*, **81** 4389-4393.

Laggner, P., Kostner, G.M., Rakusch, U. and Worcester, D. (1981) Neutron small-angle scattering on selectively deuterated human plasma low density lipoproteins. The location of polar phospholipid headgroups. *J. Biol. Chem.*, **256** (22) 11832-11839.

Lasic, D.D. (1993) *Liposomes: From Physics to Applications*, Elsevier, Amsterdam.

Latal, A., Degovics, G., Epand, R.F., Epand, R.M. and Lohner, K. (1997) Structural aspects of the interaction of peptidyl-glycylleucine-carboxyamide, a highly potent antimicrobial peptide from frog skin, with lipids. *Eur. J. Biochem.*, **248** 938-946.

Lattman, E.E. (1994) Small angle scattering studies of protein folding. *Curr. Opin. Struct. Biol.*, **4** 87-92.

Lewis, B.A. and Engelman, D.M. (1983) Lipid bilayer thickness varies linearly with acyl chain length in fluid phosphatidylcholine vesicles. *J. Mol. Biol.*, **166** 211-217.

Lohner, K., Schuster, A., Degovics, G., Müller, K. and Laggner, P. (1987) Thermal phase behaviour and structure of hydrated mixtures between dipalmitoyl- and dihexadecylphosphatidylcholine. *Chem. Phys. Lipids*, **44** 61-70.

Luzzati, V. (1968) X-ray diffraction studies of lipid–water systems, in *Biological Membranes— Physical Fact and Function* (ed. D. Chapman) American Press, London, pp 71-123.

Lvov, Yu.M., Troitsky, V.I. and Feigin, L.A. (1989) Structure analysis of Langmuir–Blodgett films with alternating bilayers by means of small-angle X-ray scattering and electron diffraction, in *Molecular Crystals and Liquid Crystals*, vol. 172, Gordon and Breach Science Publishers, Philadelphia, pp 89-97.

Müller, A. (1923) The X-ray investigation of fatty acids. *J. Am. Chem. Soc.*, **123** 2043.

Margaritondo, G. (1988) *Introduction to Synchrotron Radiation*, Oxford University Press, New York.

Pascher, I., Lundmark, M., Nyholm, P.-G. and Sundell, S. (1992) Crystal Structures of membrane lipids. *Biochim. Biophys. Acta*, **1113** 339-373.

Pickover, C.A. and Engelman, D.M. (1982) On the interpretation and prediction of X-ray scattering profiles of biomolecules in solutions. *Biopolymers*, **21** 817-831.

Prassl, R., Chapman, J.M., Nigon, F., Sara, M., Eschenburg, S., Betzel, C., Saxena, A. and Laggner, P. (1996) Crystallization and preliminary X-ray analysis of a low density lipoprotein from human plasma. *J. Biol. Chem.*, **271** 28131-28133.

Pressl, K., Jørgensen, K. and Laggner, P. (1997a) Characterization of the sub–main-transition in distearoylphosphatidylcholine studied by simultaneous small- and wide-angle X-ray diffraction. *Biochim. Biophys. Acta*, **1325** 1-7.

Pressl, K., Kriechbaum, M., Steinhart, M. and Laggner, P. (1997b) High pressure cell for small- and wide-angle X-ray scattering. *Rev. Sci. Instrum.*, **68** (12) 4588-4592.

Raoux, D. (1993) Introduction to synchrotron radiation and to the physics of storage rings, in *Neutron and Synchrotron Radiation for Condensed Matter Studies—Theory, Instruments and Methods*, vol. I (eds J. Baruchel, J.L. Hodeau, M.S. Lehmann, J.R. Regnard and C. Schlenker), Les Editions de Physique, Springer-Verlag, Berlin, pp 37-78.

Rappolt, M. and Rapp, G. (1996) Structure of the stable and metastable ripple phase of dipalmitoylphosphatidylcholine. *Eur. Biophys. J.*, **24** 381-386.

Rosoff, M. (ed.) (1996) *Vesicles/Surfactant Science Series*, Marcel Dekker, New York.

Schoenborn, B.P. (ed.) (1984) *Neutrons in Biology*, Plenum Press, New York.

Sherwood, D. (1976) *Crystals, X-rays and Proteins*, Longman, London.

Small, D. (1986) *Handbook of Lipid Research: The Physical Chemistry of Lipids—From Alkanes to Phospholipids*, Plenum Press, New York.

Steinhart, M., Kriechbaum, M., Pressl, K., Amenitsch, H., Laggner, P. and Bernstorff, S. (1998) High-pressure instrument for small- and wide-angle X-ray scattering. II. Time-resolved experiments. *Rev. Sci. Instrum.* (submitted).

Stuhrmann, H.B. (1982) Contrast Variation, in *Small Angle X-Ray Scattering* (eds O. Glatter and O. Kratky), Academic Press, London, pp 197-213.

Svergun, D.I., Voklov, V.V., Kozin, M.B., Stuhrmann, H.B., Barberato, C. and Koch, M.H.J. (1997) Shape determination from solution scattering of biopolymers. *J. App. Crystallogr.*, **30** 798-802.

Tadros, Th.F. (1984) *Surfactants*, Academic Press, London.

Tardieu, A. (1994) Thermodynamics and structure—concentrated solutions—structured disorder in vision, in *Neutron and Synchrotron Radiation for Condensed Matter Studies—Applications to Soft Condensed Matter and Biology*, vol. III (eds J. Baruchel, J.L. Hodeau, M.S. Lehmann, J.R. Regnard and C. Schlenker), Les Editions de Physique, Springer-Verlag, Berlin, pp 145-160.

Trewhella, J. (1982) Insight into biomolecular function from small-angle scattering. *Curr. Opin. Struct. Biol.*, **7** 702-708.

Ulman, A. (1991) *An Introduction to Ultrathin Organic Films from Langmuir–Blodgett to Self-Assembly*, Academic Press, Boston.

Vachette, P. (1996) Small-angle X-ray scattering by solutions of biological macromolecules, in *Proceedings International School of Physics 'Enrico Fermi'* (eds E. Burattini and A. Balerna), IOS Press, Amsterdam, pp 269-292.

Walenta, A.H. (ed.) (1991) *Proceedings of the European Workshop on X-Ray Detectors for Synchrotron Radiation Sources*, ZESS, University of Siegen, Germany.

Warren, B. E. (1969) *X-Ray Diffraction*, Addison-Wesley, Reading, MA.

Wiedemann, H. (1991) Storage ring optimization, in *Handbook on Synchrotron Radiation*, vol. 3 (eds G.S. Brown and D.E. Moncton), North-Holland, Elsevier Science, Amsterdam, pp1-36.

12 Use of colorimetry

Alex A. Belbin

12.1 Why measure colour ?

Day-to-day routine colour measurement of edible oils and fats has been a standard procedure since the early 1900s: records show that cottonseed oil was being measured at least as early 1918 (Figure 12.1).

Colour in oils and fats is largely due to the presence of such pigments as chlorophylls, carotenoids, flavins and tocopherols, with the first two having the greater contribution. Some crude oils can have an unexpectantly high pigmentation, which is often attributed to adverse

Figure 12.1 Late nineteenth century tintometer.

growing conditions such as frost damage or too little or too much water (Meloy, 1953). The result of this is a tendency for the colour to darken in storage. This instability can cause problems when blending with prime oils as the darkening effect is carried over (Fash, 1934). Colour measurement before and after storage but before blending would help to avoid problems created by the blending of unsuitable oils.

These darkening effects or reversions can occur in tallows when they are bleached at too high a temperature (Zschau, 1990a). Tallow can also be affected by too much chlorophyll, giving it a distinctive dirty greenish appearance. This chlorophyll is the consequence of the grazing of animals immediately prior to slaughter when residual chlorophyll remains in the animal's stomach and is subsequently rendered down. A similar situation sometimes occurs with fish oil, resulting from algae eaten by the fish (Zschau, 1990b).

Much of the colour in oils and fats can be removed by the refining and bleaching operations and in the case of carotene by deodorising; it is not in the interest of the oil producer to have all the colour removed, as the 'correct' colour is a major part of its aesthetic quality, to say nothing of the saving in cost to be made by not over-refining or bleaching. Frequent colour measurement is the key to considerable savings in time and bleaching earths.

Colour is also a guide to the conditions of used edible oils and fats. Using the deep fat fryer in the home will change the colour of the oil after even the first cooking, but this is not normally a problem as the oil is discarded long before the quality begins to effect the taste of the food. However, this is not the case with many fast-food outlets or 'take-aways', where enormous quantities of oil are used and where, because of the cost, an oil change will often not take place until the appearance and flavour of the food begins to deteriorate. The deterioration is caused partly by oxidised fatty acids and carbonised food that darkens the oil. Fortunately, the colour change will almost always be from yellow to brown, making it relatively easy to place the oil colour on a suitable yellow-to-brown colour scale (see single number scales in Section 12.4). Oils from different producers or from different sources may have different shades of brown when they reach the end of their life. It will be up to each producer to determine these levels.

These are some of the reasons why it is necessary to measure the colour of oils, but ultimately the consideration is of the cost of processing, the quality of the product and what the product looks like to the end user. The end user may be a food producer who is concerned that oil with a colour that is not within his specification will have an adverse effect on the colour of his finished product. This is particularly so when the product itself has very little colour.

Many everyday consumer products are instantly recognisable by their colour and when that colour remains constant the consumer is reassured that the quality of the product remains constant also. Most consumers do not consciously notice that the colour of their favourite cooking oil or detergent is the same. It is only when the colour appears to be a different shade from normal that colour suddenly becomes all-important. As soon as a colour difference is perceived, an unconscious signal is received by the consumer implying that 'different' means 'not as good'. This is a situation that must be avoided if confidence in the product is to be maintained.

12.2   Instrumentation—visual

The colour of oils and fats is measured instrumentally; this is currently done using a well-established subtractive colorimeter called a Tintometer (Figure 12.2).

The Tintometer enables its operator to view an oil sample under controlled conditions. The lighting is standard. The angle of view is fixed. The optics are designed to help the operator view both the white reference field and the oil sample simultaneously using the most favourable part of the retina, the *fovea ocularis*. At a 2° subtention angle (Figure 12.3), the

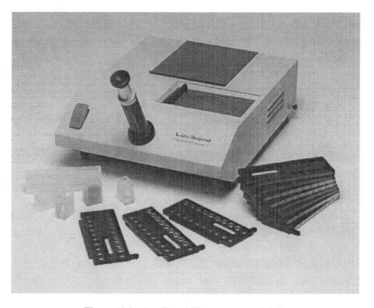

Figure 12.2   Lovibond Tintometer model F.

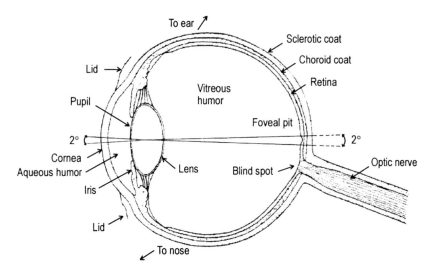

Figure 12.3 Diagram showing 2° subtention.

fovea ocularis is packed with colour-sensing cones without the influence of low luminosity- and shape-sensing rods. This phenomenon is common to all normal-vision observers, consequently making good agreement possible when seeing colour under these conditions.

The vital parts of the Tintometer are the series of red, yellow and blue permanently coloured glass standards. These standards vary from desaturated water-white colours to fully saturated deep reds, yellows and blues. Each standard is numbered and is subtly different from the one preceding and the one following it. These colour standards make up the arbitrary Lovibond colour scale.

The oil sample is placed in an optical glass cell or tube with a path length of 133.4 mm (5 $\frac{1}{4}$ inches) or less and then viewed by the operator before a colour match is made. This is done by superimposing a mixture of red and yellow standards over the reference field, which is adjacent to the field containing the oil sample. Blue is sometimes used if the oil has a tendency to be dull or greenish, and on occasions when the oil appears much brighter than the red and yellow combinations a neutral (grey) standard can be superimposed over the sample to reduce its brightness and bring it within the scope of the Tintometer, enabling a satisfactory colour match to be made.

The resulting colour match is reported as Lovibond units of red and yellow plus blue or neutral tint if used.

The Lovibond colour standards are accepted throughout much of the world as a proven means of assigning fairly precise colour values to edible oils and fats and consequently are used as a means of communication in the industry. Because they are visual, it is possible to appreciate rapidly the meaning of small colour differences in numerical terms. This small and often unappreciated benefit is not easily achieved when measurements are made objectively at selected wavelengths.

Several types of Tintometer units have been on the market and two in particular need to be explained here. The AF710 AOCS/Tintometer, based on the Wesson principle as described in AOCS Method Cc 13b-45, is used throughout the Americas and in many parts of the world where it is necessary to comply with the above method and where there is a strong American interest. With this instrument it is possible to achieve a colour match using only the red and yellow combination of standards. The field of view is rather restrictive, and the lack of blue and neutral tint standards make it necessary to ignore any difference in brightness and greenness. This instrument has been the recognised AOCS standard since 1962.

The second instrument is the Model E, (now superseded by the Model F Lovibond Tintometer) (Figure 12.4, see colour plate), which has become the accepted standard in most other countries.

The geometry and the colour scales in the two instruments are dissimilar and consequently the results obtained are not compatible. Steps are being taken to harmonise the instruments and colour scales, making for easier worldwide exchange of information regarding the colour status of oils and fats. The AOCS committee has already included the British Standard Method BS684/87 Section 1.14 for the colour measurement of edible oils and fats in the AOCS Standard Method Cc 13e-92.

The current visual instruments are those mentioned above: Model E or F complying with BS684 1.14 and AF710 complying with AOCS Method Cc 13b-45.

Over the years a number of different types of visual Tintometers have been used in the edible oils and fats industries. Many of these instruments are very old and are obsolete. Some of them should never have been used in the industry and many do not have the full complement of colour filters. The great majority are not used in accordance with the manufacturers' recommendations and, as a consequence of all these variables, problems of colour match agreements between instrument and/or users have arisen. This chapter is an attempt to clarify the situation and at the same time offer some suggestions towards normalising the methods of colour measurement within the industry.

Generally speaking, the edible oil and fats industry tends to overlook the fact that colour is a three-dimensional sensation and, for the sake of speed of operation and of convenience, only two dimensions are used

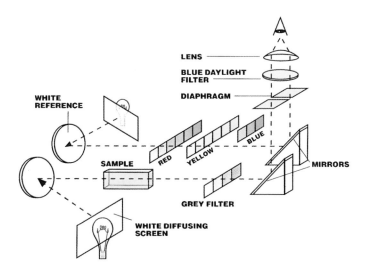

Figure 12.4 Schematic of the principle of the Model F Lovibond Tintometer.

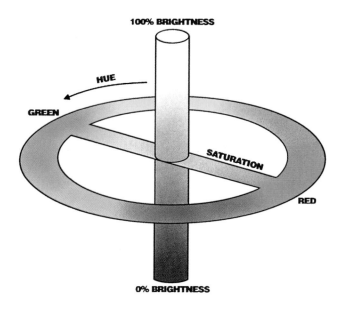

Figure 12.5 Saturation, hue and brightness polar diagram.

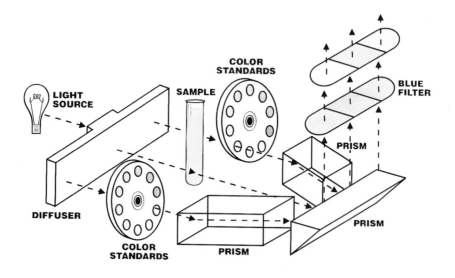

Figure 12.6 The principle of single-number colour scale measurement.

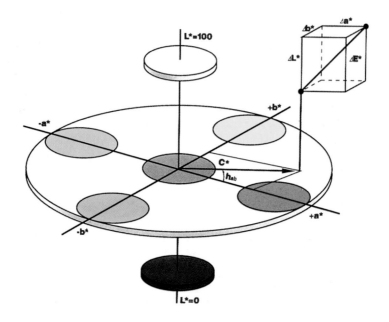

Figure 12.8 CIE L*a*b* colour space.

when making a colour match of a sample. This is not meant as a criticism of the industry, as until recent years this method of measuring and reporting the colour has worked very successfully and without too much discord. However, now that there is a slow but inexorable move toward non-visual electronic colour measurement, the disadvantages of using two-dimensional procedures for obtaining what is essentially a three-dimensional result are becoming increasingly apparent. The difficulties arise when attempting to correlate absolute measurements obtained from spectrophotometers or the like with arbitrary visual measurements that have been obtained using abridged methods.

Terminology for describing colour may need to be clarified at this point. The word 'colour' describes the sensation of visible light as seen by the eye of the observer and interpreted by the brain. As the quality and quantity of visible light changes and as the object illuminated by it has the property of absorbing or reflecting all or part of the light falling on it, the sensation of colour changes. The changes are broadly described by words denoting colour: red, yellow, green, and so on.

For most everyday needs the general word denoting colour needs to be further qualified: primrose yellow, grass green or burgundy red provide more specific descriptions, but even those are not enough for most manufacturers of coloured products. Colour is a three-dimensional sensation and when this is taken into account it becomes easier to describe a colour more precisely as a point in colour space (Figure 12.5, see colour plate).

The three dimensions of colour are hue, saturation and brightness. 'Hue' describes the type of colour: red, green, blue, etc. 'Saturation' describes the purity of the hue; a colour with a red hue can be a pale pink or a deep crimson. In some systems, the term 'chroma' is used to refer to saturation. 'Brightness' is a measure of the luminance of the colour, or how much light is being transmitted or reflected. In some systems, the term 'value' is used for brightness.

To obtain a complete picture of the colour, it is necessary to take all of the above into consideration. To make comparisons using just two of the three dimensions, i.e. hue and saturation, could mean that two colours having the same values for these will not necessarily look alike because the brightness may be different for each.

An example of this would be the comparison of the colours royal purple and lilac. They both have a similar hue and saturation, but the royal purple has a lower reflectance percentage, consequently making the colours appear totally dissimilar, which of course they are! For readers who are familiar with the Munsell Colour System, compare 7.5P 3/8 and 7.5P 7/8.

To carry the analogy to the use of the Tintometers, it will be necessary to visualise the effect of using the red, yellow, blue and neutral tint filters. The colours of edible oils tend to be a mixture of red and yellow Lovibond units so, for the purpose of this exercise, we can say that the combinations of the two will be a measure of the hue and saturation. Provided that all oils and fats always had proportionately the same brightness as the combinations of Lovibond red and yellow used to match them, there would be no problem.

Unfortunately, however, this is frequently not the case. Some oils are very bright and others appear quite dull; for example, compare olive oil with maize oil. When an oil is duller than the nearest combination of red and yellow Lovibond, it will not be possible to match it without adding a third factor. In this case, the third factor is Lovibond blue. The blue will have the effect of desaturating or dulling the red and yellow filters until their combined brightness is the same as that of the oil. Only when the brightness levels are the same will it be possible to decide whether the red and yellow values obtained are correct. If not, it is a simple task to make the appropriate adjustments.

A similar situation occurs when measuring an oil that is brighter than the nearest combination of Lovibond red and yellow values. In this case, the third factor will be the neutral tint filters. These can be used to reduce the brightness of the oil and bring it within the range of the Lovibond red and yellow filters.

When colour measurements are made ignoring the brightness factor, very often the results obtained are not colour matches but, instead, are approximations or intelligent guesswork. This may not cause problems if only one or two selected operators are used and if they do not have to communicate worldwide with operators with a different perception of intelligent guesswork.

Clear, colourless glass compensating slides
Some Tintometers are supplied with clear, colourless glass compensating slides fixed into the lower sections of the rack containing the colour standards. The reasons for this are explained as follows. When light passes through a glass standard, a small percentage is lost at each of the glass surfaces owing to scattering and refraction, as well as that which is lost through the internal transmission due to colour.

This loss of light is quite significant when making a colour match in the region of 1.0 red. The visual difference between 0.8 and 0.9 of red is relatively small compared with the difference between 1.0 and 1.1. The change in saturation is about the same, but the change in brightness is down by about 8%. Any brightness difference between 0.9 and 0.8 is due to a minute change in internal transmission, whereas the difference between 1.0 and 1.1 is caused by the internal transmission difference and the effect of the extra piece of glass used, i.e. 1.0 and 0.1. The use of clear glass compensating slides fitted into the lower half of the rack cancels out this brightness difference, as both the sample of oil and the reference field will have the same number of glass surfaces in the field of view no matter how many colour standards are used to complete the colour match. These compensating slides are specified in BS684 Section 1.14 Determination of Colour.

The use of compensating slides tends to reduce the amount of neutral tint that is often necessary to obtain a good colour match. In some instances, however, a small amount of blue may be required.

Metal sheath

Some Tintometers are supplied with a black shroud (metal sheath) to enclose the cell to shield it from all light with the exception of that which is being reflected from the standard white reference. The purpose of this is to obtain more theoretically correct viewing conditions. Making measurements under these condition tends to reduce the need for neutral tint and does increase the need to use blue in the colour match.

In comparing results obtained using compensating slides and/or metal sheaths with the generally adopted procedure of measuring the colour of the oils just using red and yellow, slight increases in the red reading may be noted when using compensating slides and slightly greater ones when using the metal sheath as well. The use of the metal sheath is optional but must be reported if used.

12.3 Colour matching considerations

12.3.1 Colour blindness

One of the most obvious, but often overlooked, reasons for disagreements between Tintometer operators is colour deficiency. As about 8% of the male population are colour deficient and 0.4% of the female, it is important that operators of the visual Tintometer instruments be tested for this deficiency. The Ishihara colour chart is a convenient method for carrying out this test.

12.3.2 Ambient lighting

Contrast or high ambient lighting causes unnecessary fatigue to the eye and also increases the adaptation time for the operator's eye to become accustomed to the conditions in the viewing tube of the Tintometer. It is often not possible to avoid overheard lighting in a busy laboratory, but one should at all costs avoid using the instrument facing a window.

12.3.3   Colour matching

The time taken to carry out a colour match must be kept to a minimum. Prolonged periods of viewing colour will cause eye fatigue and lead to erroneous results. It is recommended to view the sample for no more than 10 s and then take a few seconds' rest. The eye will soon recover and matching can then proceed. This procedure should be followed until a colour match is achieved.

12.3.4   Poor maintenance

The condition of the Tintometer also influences the appearance of the colour seen in the viewing tube. Some obvious faults are as follows:

  (i) white paintwork discoloured;
 (ii) lamps aged or blackened;
(iii) standard white halon or magnesium carbonate discoloured or dirty;
 (iv) glass colour standards affected by grease or bloom;
  (v) optical viewing tube mirrors marked or misaligned; lenses and blue filter dusty.

12.3.5   Visual perception

Care must be taken when considering differences obtained between instruments and between operators. While numerically the Lovibond units are additive, i.e. $1.0 + 2.0 = 3.0$ and $3.0 + 4.0 = 7.0$, the visual difference between 1.0 and 2.0 is not the same as between 3.0 and 4.0. As the Lovibond unit increases in value, difference between each preceding unit diminishes. This is especially the case with yellow. The following examples will help to clarify this:

  (i) 1.6 red 45.0 yellow;
 (ii) 1.6 red 70.0 yellow;
(iii) 1.6 red 70.0 yellow;
 (iv) 2.0 red 70.0 yellow.

There are approximately four just-perceptible differences between (i) and (ii) and also between (iii) and (iv). The numerical difference between (i) and (ii) is 25 yellow units and between (iii) and (iv) it is only 0.4 red units.
   The human eye is much more sensitive to changes in the red region of the Lovibond range than it is to changes in the yellow region.

Consequently, greater differences in yellow can be tolerated, while small changes in red are immediately apparent.

12.4 Single-number scales

Alternative methods for specifying colour have been on the market for some time. These include the one-dimensional or single-number colour scales. These scales consist of a series of predetermined colours that are representative of certain types of oil samples. The standards for these scales are sometimes in the form of coloured solutions placed in hermetically sealed glass tubes or, more often, in the form of permanently coloured glass.

The test involves placing an oil sample in an optical cell or test tube and comparing the sample directly with the colour standards under a standardised white diffuse light until a visual colour match is found. The standardised light is daylight or simulated daylight (Figure 12.6, see colour plate).

The Gardner Scale is one such single-number scale and often is used to categorise lecithins, natural and synthetic drying oils, fatty acids and some oil derivatives. The specifications are based on AOCS Methods Td 1a-64 and Ja 9-78.

Another well-established and well-used single-number scale is one that was introduced by Dr Hazen in 1892 (Judd, 1963) for grading the colour of natural waters and has now been taken up by a number of industries, including producers of oleochemicals, for near colourless liquids. Typical applications are for colour grading of glycerines, distilled fatty acids, myristic and oleic acids, stearic and palmitic acids. These oleochemicals are used in the production of cosmetics, toiletries, soaps and detergents, pharmaceuticals and food emulsions. The Hazen scale, or APHA scale as it is more popularly known in the United States, is made up from standard solutions of Platinum Cobalt which are quite colour stable if stored in a cool dark place. However, owing to the inconvenience of preparing these solutions and also to the market price of platinum, glass colour standards are frequently used instead. The specified APHA maximum levels are to be found in the AOCS Official Methods of Analysis 4th Edition 1990.

Two other scales often used in the oils and fats industry are the iodine scale and the FAC (Fatty Acid Committee) scale. The iodine colour scale is used in a number of European countries and is specified in DIN 6162. The FAC scale is used primarily for grading inedible tallows and dark oils and is specified as per AOCS Method Cc 13a-45. Manufacturers producing instruments using single-number scales include BYK Gardner of Silver Springs, Maryland, USA, Hellige GmbH of Germany, and Tintometer Limited in England.

Single-number scales serve a useful purpose if the hue and brightness of the sample do not differ too much from the standard. When this does happen observers may have difficulty in placing the sample on the colour scale. This can result in commercial disagreements.

12.5 Electronic colour measurement

The use of spectrophotometers or colorimeters would at first sight be an obvious solution to the uncertainties of using subjective means of quantifying colour, but before such instruments can be used successfully, a satisfactory correlation relating Lovibond red/yellow to any measurements obtained on the above instruments must be found. This requirement is mandatory as the 'Lovibond' colour is a necessary requisite for establishing that the oil or fat is within the required specification to satisfy both its quality and its market price.

During the mid 1970s, this problem was partly overcome with the introduction of a three-filter electronic instrument that was correlated directly to Lovibond units of red and yellow. This instrument worked very well with lightly coloured refined oils and, by agreement with Unilever, was manufactured and marketed by The Tintometer Limited. A similar instrument based on the same principles was marketed by McCloskey Scientific Industries Inc. of New Jersey, USA.

A much improved version of the Unilever instrument, the AF960, was introduced during the early 1980s by The Tintometer Limited, but was still limited in its capability to correlate with the visual Tintometer over the whole gamut of oil colours.

In recent years the problem of correlation has been eased considerably with the introduction of the CL500 computer-assisted instrument, and again with the 16-filter Lovibond PFX990 Tintometer (Figure 12.7). The latter, using sophisticated microprocessor technology, has a range of optional colour scales built into its program and the potential for adding more. This includes some of the single-numbered colour scales such as Gardner, FAC and APHA. It also encompasses the different categories of 'Lovibond' with particular regard to AF710 AOCS – Tintometer red and yellow scale and the Model E (now Model F) red, yellow, blue and neutral tint scale.

The use of 16 standard interference filters in the PFX990 extensively expands its capability for achieving much better correlation with the visual Tintometers when compared with the three filters used for colour

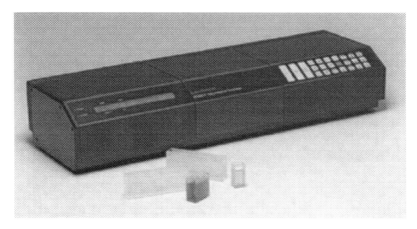

Figure 12.7 Lovibond Tintometer PFX990.

measurement in the AF960. This is confirmed by a comparative study carried out in 1995 by Wan and Parkarinen using the 16-filter Colourscan automated colorimeter. The subject of this study was refined and bleached cottonseed oils.

A further study took place in 1996 involving 30 laboratories and 14 countries (Wan *et al.*, 1997). In this study 18 samples of oil and 3 glass colour standards were measured in the PFX990 16-filter automated Tintometer, the AF710 AOCS-Tintometer and the Model E Tintometer. As a result of this study a new method for automated colour measurement of edible oils and fats has been approved (AOCS Official Method Cc 13j-97). A full report of this above study is available from the AOCS Technical Department (AOCS, n.d.).

12.6 Avoiding problems

When changing from visual colour measurement to electronic colour measurement, care must be taken not to carry over all of the working practices associated with visual Tintometers in a busy edible oils and fats laboratory. Visual Tintometers, while correctly described as scientific instruments, are often used in harsh environments where they are not treated as such. However, provided that they are maintained in the manner described earlier, they can absorb considerable abuse and misuse and still produce correct results.

This is not the case with electronic Tintometers or spectrophotometers, which are much more sensitive to such treatment. These instruments rely on a narrow beam of light passing through the cell containing the sample and reaching the photodetector without interference. The fact that the light may be obstructed in some way is not always easy to detect. It is therefore easier to take steps to ensure that this does not happen.

(1) Make sure that the instrument has had the desired warming up period after switching on.
(2) Calibrate and check calibration at frequent intervals: standard filters are often available for this.
(3) Make sure that any glass windows or other optics in the cell chamber are clean and free from grease and atmospheric bloom.
(4) Always clean up spillages as they occur to avoid ingress of oil between or behind glass surfaces.
(5) Make sure that the cell windows are free from smears, fingerprints and grease that would deflect the beam of light. The level of cleanliness must be greater than that needed for the visual Tintometers.
(6) Check that all oil samples are free from bubbles, stry, unmelted fat particles and heat thermals. Bubbles may be removed by warming the oil sample and gently stirring. As stry is normally caused by a residual of the previous sample, the solution is to dispose of the sample, wash the cell and remeasure with fresh oil. Unmelted fat particles cause light scatter and should be removed by heating the oil to 10°C above its melting point. Heat thermals are caused by the oil and the cell being at different temperatures creating cold spots. This problem can be overcome by heating both the cell and the oil to the same temperature.

A common mistake is to place the oil in the cell and heat it in a microwave oven. Microwaves will heat the oil but do not heat the glass cell, consequently thermals are formed where the oil is in contact with the glass surfaces.

12.7 Chlorophyll and carotene

As chlorophyll and carotene are among the main contributors to colour, it is sometimes necessary to determine actual levels present in parts per million. AOCS Official Method Cc 13d-55, revised 1991, is an accepted procedure for the determination of chlorophyll A, where absorbance at 670 and 710 nm is determined using a spectrophotometer. The PFX990 Tintometer also uses this method modified to accommodate cell sizes up to 133.4 mm ($5\frac{1}{4}$ inches).

Chlorophyll B may also be determined by using an additional absorbance at 595 nm. However, when chlorophyll A is detected at levels higher than 0.03, chlorophyll B is reported as zero.

In cases where more accurate measurements are required, and in particular where the chlorophyll derivatives pheophytins, pheophorbides and pyropheophytins need to be isolated, high-performance liquid chromatography (HPLC) may be used (Fraser and Frankl, n.d.).

Carotene determinations can be made by using the British Standard Method BS684 Section 2.20 Determination of Carotene in Vegetable Oils. This is an uncomplicated spectrophotometric method using a dilution of the oil in a 10 mm cell and absorbance at 445 nm. The PFX990 and AF960 use a modification of this method.

12.8   Spectrophotometric correlations

The possibility of measuring the colour of oils and fats without referring to Lovibond units is attractive to some Tintometer users, especially those involved in research. The electronic Tintometers have to some extent satisfied this need but, even so, the end result is still based on arbitrary values.

The AOCS Color-Spectrophotometric Method Cc 13c-50, reapproved 1989, does achieve a degree of success in this direction by measuring some oils at four selected wavelengths and arriving at a photometric colour index. The oils measured were cottonseed, soybean and peanut. The collaborative study showed a correlation coefficient between Lovibond colour and spectrophotometric colour of 0.993.

12.9   CIELAB absolute colour measurement

The Lovibond colour system is unique. It is deeply embedded in the colour methodology of the edible oils and fats industry and has served it well for nearly one hundred years. However, there are many workers, particularly those involved with research and development, who would much prefer to switch to a non-arbitrary, non-subjective means of measuring colour, where the results will be totally objective and absolute.

In 1931, an international conference of the Commision Internationale de l'Éclairage (CIE) agreed upon a standard for the numerical specification of a colour as it appears to a 'standard observer'. This system was designed as a common basis for all colour measuring instruments independent of the colour vision of any particular individual.

Since 1931 a number of satellite systems have sprung up allowing the user to calculate and visualise uniform colour differences within a particular colour space. One such system is the CIELAB L*a*b* (Figure 12.8, see colour plate). Colour measurements made using this system enable the user to measure small differences between a standard and sample and present the data as a single number. This number, Delta E (ΔE) will indicate the number of visible differences between the standard and sample no matter where they are placed within the three-dimensional colour space and irrespective of hue, saturation and brightness differences. This can be a very useful tool when all that is required is to establish whether the sample is within the tolerance provided.

More information about the colour can be obtained from the L*a*b* data. L* is a measure of lightness or brightness, a* is a measure of greenness to redness and b* is a measure of yellowness to blueness (Figure 12.9).

This system would be amenable to research laboratories interested in the effects of heat, contaminants and unwanted pigmentation. As the system is already used for match prediction in the dyeing and paint industries, it seems reasonable to believe that it could prove useful in predicting the colour-removing power of bleaching earth.

Making the transition from Lovibond to CIELAB in the production laboratory will be tedious and time consuming but not impossible. With access to instruments such as the PFX990, which included both Lovibond colour and L*a*b*, in its program, measurements in both scales could be made in less than one minute. Enough data could be gathered over a period of months to obtain a firm correlation.

The present technology of red, yellow and blue colour measurement in its various modes is affordable and comprehensive to both developed and developing nations in all parts of the world. Any changes that are made must be considered carefully if colour data communication worldwide is to be maintained.

Appendix 12.A: Glossary of colour terms

Absorbance   See Optical density
Absorptance   The ratio of absorbed radiant flux to the incident flux
Absorption   The transformation of radiant energy to a different form of energy by interaction
Brightness   The quantity of light reflected (a bright colour reflects more light whereas a dim colour reflects less)
Chroma   See Saturation
Chromatic   Perceived as having a hue; not white, grey or black
Chromaticity   Chromaticity is that part of a colour specification that does not involve the amount of light energy; it is specified by pairs of chromaticity coordinates or dominant wavelength and purity

Chromaticity coordinates   The ratio of each of the tristimulus values of a colour to the sum of the tristimulus values (designated x, y, z in the CIE system)

CIE, Commission International de l'Éclairage   The international organisation concerned with light, vision and colour

Colour difference   The magnitude of the difference between two object colours under specified conditions

Colour grading   Identifying a sample by a colour grade or score that is specific to the colour or the material graded

Colour matching   Procedure for providing a trial colour that is indistinguishable, within a specified tolerance, from a reference colour

Colour measurement   The process of deriving, by visual or electronic means, a set of three numbers that describe the attributes of a colour

Gardner colour scale   A colour scale for clear, light-yellow liquids, defined by the chromaticities of glass standards numbered from 1 for the lightest to 18 for the darkest

Hazen colour scale   See Platinum-Cobalt colour scale

Hue or colour   The type of colour, whether it is red, blue, green, purple, etc

Lovibond colour system   A system of colour specification based on a combination of yellow, red and blue filter glasses

Optical density   Logarithm to the base 10 of the reciprocal of the transmittance

Photometer   An instrument for measuring light

Platinum-Cobalt colour scale   A colour scale for clear, light yellow liquids defined by specified dilutions of a platinum-cobalt stock solution, ranging from 5 for the lightest colour to 500 for the darkest

Saturation or chroma   Refers to the strength or amount of the hue: how much the full hue is desaturated with white

Saybolt colour   The colour of a clear petroleum liquid based on a scale of -16 (darkest) to $+30$ (lightest)

Spectrophotometry   Measurement of the relative amounts of radiant flux at each wavelength of the spectrum

Standard illuminants   Relative spectral power distributions defining illuminants for use in colorimetric computations

Transmission   Passage of radiation through a medium without change of frequency

Transmittance   The ratio of transmitted flux to incident flux under specified conditions

Tristimulus values   The amounts of three specified stimuli required to match a colour

Turbidity   Reduction of transparency of a specimen

Appendix 12.B: Standard references

Standard	Scope	Colour Scale
AOCS Cc 13a-43	Dark fats, oils and tallows	FAC
AOCS Ea 9-65	Refined glycerine	Pt-Co/Hazen/APHA
AOCS Ja 9-87	Non-granular lecithin products	Gardner
AOCS Td 1a-64	Drying oils, fatty acids, oil derivatives	Gardner
AOCS Td 3a-64	Fatty acids after heating	Gardner
ASTM D 29	Lac resins	Gardner
ASTM D 234	Raw linseed oil	Gardner
ASTM D 960	Castor oil	Gardner
ASTM D 1045	Water	Pt-Co/Hazen/APHA
ASTM D 1209	Clear liquids	Pt-Co/Hazen/APHA
ASTM D 1462	Soybean oil	Gardner
ASTM D 1544	Transparent liquids	Gardner
ASTM D 1967	Drying oils after heating	Gardner
ASTM D 1981	Fatty acids after heating	Gardner
ASTM D 3169	Sunflower oil	Gardner
ASTM D 5237	Fabric softeners	Gardner
BS 2690	Water	Pt-Co/Hazen/APHA
BS 4835	Plasticiser esters	Pt-Co/Hazen/APHA
BS 5339	Liquid chemicals	Pt-Co/Hazen/APHA
DIN 53409	Clear liquids	Pt-Co/Hazen/APHA
DIN 6162	Solvents, plasticisers, resins, oils and fatty acids	Iodine
ISO 6271	Clear liquids	Pt-Co/Hazen/APHA

Appendix 12.C: Colour scales used in the analysis of non-mineral oils and fats

Colour scale	Example references	Scope	Lovibond instruments
Pt-Co/Hazen/APHA	AOCS Method Ea 9-65	Refined glycerine	PFX190
	AOCS Method Td 1b-64	Industrial oils and derivatives with light colours, e.g. certain fatty nitrogen compounds	PFX190 2000 Comparator + Daylight 2000 Nessleriser
Gardner	AOCS Method Ja 9-87	Non-granular lecithin products	PFX990
	AOCS Method Td 1a-64	Natural and synthetic drying oils, fatty acids and oil derivatives	PFX190 Gardner 3000 Comparator 2000 Comparator
Gardner (heated samples)	AOCS Method Td 3a-64	Fatty acids heated in Gardner tubes	PFX190 Gardner 3000 Comparator

Appendix 12.C: (Continued)

Colour scale	Example references	Scope	Lovibond instruments
FAC	AOCS Method Cc 13a-43	Animal fats and all fats and oils too dark to be graded by the Wesson method	PFX990 PFX190 FAC 3000 Comparator
Lovibond	BS 684 Section 1.14 BS 7207 AOCS Method Cc 13e-92	Fats and fatty oils Crude vegetable fats All normal animal and vegetable fats and oils	Tintometer Model F (BS684)
	—	Animal and vegetable fats and oils	PFX990
	AOCS Method Cc 13j-97	Refined and refined–bleached–deodorised	Tintometer Model F PFX990
AOCS-Tintometer	AOCS Method Cc 13b-45 (Wesson Method)	Special red and yellow version of the Lovibond scale for all normal fats and oils	PFX990
	AOCS Method Cc 8d-55	Determination of colour after treatment with alkali and bleaching earth.	Tintometer AF 710-2
	AOCS Method Cc 13j-97	Applicable to tallows and greases intended for soap production Refined and refined–bleached–deodorised	PFX990
AF960 AOCS		Similar to AOCS-Tinto-meter scale but modified for use on early electronic instruments (AF960)	PFX990
AF960-Lovibond		Abridged red and yellow scale used in early electronic instruments	PFX990
Chlorophyll A & B		Direct measurement of chlorophyll content (parts per million)	PFX990
beta Carotene		Direct measurement of β-carotene content (parts per million)	PFX990

Appendix 12.D: the Commission Internationale de l'Éclairage System

The CIE system was introduced to provide an objective and inter-nationally accepted method for defining any given colour in terms of certain mathematical coordinates. The specification for the system was first established at a conference held in Cambridge in 1931, and revised at further meetings, most notably in 1964 and 1976. It recognises the variables that affect our identification of a colour—illuminant, viewing angle and the observer—and sets out recommendations for them.

Three illuminants, A, B, and C, were initially selected to represent typical sources of illumination. However, because of some dissatisfaction with illuminant C, a fourth illuminant was specified (D65).

CIE illuminant	Description
A	Representative of a tungsten lamp with a colour temperature of 2856 K.
B	Representative of midday sunlight at a colour temperature of 4874 K, produced by a tungsten lamp with a colour temperature of 2856 K and a blue daylight correction filter.
C	Representative of average daylight from an overcast sky at a colour temperature of 6774 K, produced by a tungsten lamp with a colour temperature of 2856 K and a blue daylight correction filter.
D65	Theoretical average daylight with a colour temperature of approximately 6500 K.

To improve reproducibility of results between observers, a set of colour matching functions was derived using a number of observers matching spectral colours at a viewing angle of 2°. The average of this data set is known as the CIE 1931 colour matching functions x, y, z. Later work in 1964 also included a 10° field of observation.

The CIE system is based on a modified Maxwell triangle in which each of the three primary colours—red, green and blue—occupies a corner of the triangle. Each colour is at maximum saturation at the apex, diminishing in intensity to zero as it reaches the centre. As each primary colour mixes with the others, they create a continuous spectrum of colour within the triangle. The introduction of coordinates makes it possible to pinpoint and define any colour.

In practice, it was found that these coordinates did not all lie within the triangle and a modified shape was developed to take this into account. The wavelengths for the three primaries were set at 435.8 nm (blue), 546.1 nm (green) and 700 nm (red). When delivered at the right intensities, these produce a true white at the centre of the triangle. Subsequent revisions to the CIE system have resulted in the introduction of further series of coordinates designed to take account of various distortions in the chromaticity diagram. A number of these scales are briefly described below.

X Y Z Tristimulus values

These values are the theoretical reference stimuli. The *Y* value is the most commonly referenced value as it is equivalent to the percentage luminance factor.

Chromaticity coordinates x, y, z

The co-ordinates *x*, *y* and *z* are derived from *X*, *Y*, *Z* by the following calculation, such that $x + y + z = 1$,

$$x = \frac{X}{X+Y+Z} \qquad y = \frac{Y}{X+Y+Z} \qquad z = \frac{Z}{X+Y+Z}$$

The values of *x* and *y* can then be used to pinpoint a colour in the *x*, *y* coordinate system. The *x* and *y* chromaticity coordinates are generally reported along with the value of luminance factor *Y*.

The uniform chromaticity diagram; u', v'

The *u'*, *v'* chromaticity coordinates were derived to aid the prediction of the magnitude of the perceived colour difference between two objects that are found to mismatch in colour. These modify the *x* and *y* chromaticity coordinates so that the colour difference anywhere in the diagram will have the same appearance of difference.

The CIE 1976 (L*a*b*) space and colour difference formula

The a* axis runs from green to red, the b* axis runs from yellow to blue and L* (lightness) runs from black to white. As the L*a*b* Diagram is a three dimensional diagram, the colour difference between two points can be obtained in all directions. This colour difference is expressed as Delta E, where a value of 1.0 approximates to a just-perceptible colour difference.

References

AOCS (n.d.) *International Collaborative Study, Oil Color—Automated Method Versus Visual Methods*, AOCS Technical Department, Champaign, IL.

Fash, R.H. (1934) *Oil and Soap*, **11** 106.

Fraser, M.S. and Frankl, G. (n.d.) *Detection of Chlorophyll Derivatives in Soybean Oil by HPLC*, Hunt Wesson Foods, Fullerton, CA 92634.

Judd, D.B. (1963) *Color in Business, Science and Industry*, 2nd edn, p 258.

Meloy, G.W. (1953) *Cotton and Cotton Oil Press*, **44** (23) 14.

Wan, P.J. and Parkarinen, D.R. (1995) Comparison of visual and automated colorimeter for refined and bleached cottonseed oils. *J. Am. Oil Chem. Soc.*, **72** 455-458.

Wan, P.J., Horley, T.W., Gay, J.D. and Berner, D.L. (1997) Comparison of visual and automated colorimeters—an international study. *J. Am. Oil Chem. Soc.*, **74** (6) 731-738.

Zschau, W. (1990a) *Inform*, **1** 638.

Zschau, W. (1990b) *Inform*, **1** 643.

Index